Introductory Statistics for the Life and Biomedical Sciences

First Edition

Julie Vu

Preceptor in Statistics
Harvard University

David Harrington

Professor of Biostatistics (Emeritus)
Harvard T.H. Chan School of Public Health
Dana-Farber Cancer Institute

Table of Contents

Foreword

The past year has been challenging for the health sciences in ways that we could not have imagined when we started writing 5 years ago. The rapid spread of the SARS coronavirus (SARS-CoV-2) worldwide has upended the scientific research process and highlighted the need for maintaining a balance between speed and reliability. Major medical journals have dramatically increased the pace of publication; the urgency of the situation necessitates that data and research findings be made available as quickly as possible to inform public policy and clinical practice. Yet it remains essential that studies undergo rigorous review; the retraction of two high-profile coronavirus studies[1,2] sparked widespread concerns about data integrity, reproducibility, and the editorial process.

In parallel, deepening public awareness of structural racism has caused a re-examination of the role of race in published studies in health and medicine. A recent review of algorithms used to direct treatment in areas such as cardiology, obstetrics and oncology uncovered examples of race used in ways that may lead to substandard care for people of color.[3] The SARS-CoV-2 pandemic has reminded us once again that marginalized populations are disproportionately at risk for bad health outcomes. Data on 17 million patients in England[4] suggest that Blacks and South Asians have a death rate that is approximately 50% higher than white members of the population.

Understanding the SARS coronavirus and tackling racial disparities in health outcomes are but two of the many areas in which Biostatistics will play an important role in the coming decades. Much of that work will be done by those now beginning their study of Biostatistics. We hope this book provides an accessible point of entry for students planning to begin work in biology, medicine, or public health. While the material presented in this book is essential for understanding the foundations of the discipline, we advise readers to remember that a mastery of technical details is secondary to choosing important scientific questions, examining data without bias, and reporting results that transparently display the strengths and weaknesses of a study.

---

[1]Mandeep R. Mehra et al. "Retraction: Cardiovascular Disease, Drug Therapy, and Mortality in Covid-19. N Engl J Med. DOI: 10.1056/NEJMoa2007621." In: *New England Journal of Medicine* 382.26 (2020), pp. 2582–2582. DOI: 10.1056/NEJMc2021225.

[2]Mandeep R Mehra et al. "RETRACTED:Hydroxychloroquine or chloroquine with or without a macrolide for treatment of COVID-19: a multinational registry analysis". In: *The Lancet* (2020). DOI: https://doi.org/10.1016/S0140-6736(20)31180-6.

[3]Darshali A. Vyas et al. "Hidden in Plain Sight — Reconsidering the Use of Race Correction in Clinical Algorithms". In: *New England Journal of Medicine* (2020). DOI: 10.1056/NEJMms2004740.

[4]Elizabeth J. Williamson et al. "OpenSAFELY: factors associated with COVID-19 death in 17 million patients". In: *Nature* (2020). ISSN: 1476-4687.

Preface

This text introduces statistics and its applications in the life sciences and biomedical research. It is based on the freely available *OpenIntro Statistics*, and, like *OpenIntro*, it may be downloaded at no cost.[5] In writing *Introduction to Statistics for the Life and Biomedical Sciences*, we have added substantial new material, but also retained some examples and exercises from *OpenIntro* that illustrate important ideas even if they do not relate directly to medicine or the life sciences. Because of its link to the original *OpenIntro* project, this text is often referred to as *OpenIntro Biostatistics* in the supplementary materials.

This text is intended for undergraduate and graduate students interested in careers in biology or medicine, and may also be profitably read by students of public health or medicine. It covers many of the traditional introductory topics in statistics, in addition to discussing some newer methods being used in molecular biology.

Statistics has become an integral part of research in medicine and biology, and the tools for summarizing data and drawing inferences from data are essential both for understanding the outcomes of studies and for incorporating measures of uncertainty into that understanding. An introductory text in statistics for students who will work in medicine, public health, or the life sciences should be more than simply the usual introduction, supplemented with an occasional example from biology or medical science. By drawing the majority of examples and exercises in this text from published data, we hope to convey the value of statistics in medical and biological research. In cases where examples draw on important material in biology or medicine, the problem statement contains the necessary background information.

Computing is an essential part of the practice of statistics. Nearly everyone entering the biomedical sciences will need to interpret the results of analyses conducted in software; many will also need to be capable of conducting such analyses. The text and associated materials separate those two activities to allow students and instructors to emphasize either or both skills. The text discusses the important features of figures and tables used to support an interpretation, rather than the process of generating such material from data. This allows students whose main focus is understanding statistical concepts not to be distracted by the details of a particular software package. In our experience, however, we have found that many students enter a research setting after only a single course in statistics. These students benefit from a practical introduction to data analysis that incorporates the use of a statistical computing language. The' self-paced learning labs associated with the text provide such an introduction; these are described in more detail later in this preface. The datasets used in this book are available via the R openintro package available on CRAN[6] and the R oibiostat package available via GitHub.

---

[5]PDF available at `https://www.openintro.org/book/biostat/` and source available at `https://github.com/OI-Biostat/oi_biostat_text`.

[6]Diez DM, Barr CD, Çetinkaya-Rundel M. 2012. openintro: OpenIntro data sets and supplement functions. `http://cran.r-project.org/web/packages/openintro`.

Textbook overview

The chapters of this book are as follows:

1. **Introduction to data.** Data structures, basic data collection principles, numerical and graphical summaries, and exploratory data analysis.

2. **Probability.** The basic principles of probability.

3. **Distributions of random variables.** Introduction to random variables, distributions of discrete and continuous random variables, and distributions for pairs of random variables.

4. **Foundations for inference.** General ideas for statistical inference in the context of estimating a population mean.

5. **Inference for numerical data.** Inference for one-sample and two-sample means with the t-distribution, power calculations for a difference of means, and ANOVA.

6. **Simple linear regression.** An introduction to linear regression with a single explanatory variable, evaluating model assumptions, and inference in a regression context.

7. **Multiple linear regression.** General multiple regression model, categorical predictors with more than two values, interaction, and model selection.

8. **Inference for categorical data.** Inference for single proportions, inference for two or more groups, and outcome-based sampling.

Examples, exercises, and appendices

Examples in the text help with an understanding of how to apply methods:

EXAMPLE 0.1

This is an example. When a question is asked here, where can the answer be found?

The answer can be found here, in the solution section of the example.

When we think the reader would benefit from working out the solution to an example, we frame it as Guided Practice.

GUIDED PRACTICE 0.2

The reader may check or learn the answer to any Guided Practice problem by reviewing the full solution in a footnote.[7]

There are exercises at the end of each chapter that are useful for practice or homework assignments. Solutions to odd numbered problems can be found in Appendix A. Readers will notice that there are fewer end of chapter exercises in the last three chapters. The more complicated methods, such as multiple regression, do not always lend themselves to hand calculation, and computing is increasingly important both to gain practical experience with these methods and to explore complex datasets. For students more interested in concepts than computing, however, we have included useful end of chapter exercises that emphasize the interpretation of output from statistical software.

Probability tables for the normal, t, and chi-square distributions are in Appendix B, and PDF copies of these tables are also available from **openintro.org** for anyone to download, print, share, or modify. The labs and the text also illustrate the use of simple R commands to calculate probabilities from common distributions.

---

[7] Guided Practice problems are intended to stretch your thinking, and you can check yourself by reviewing the footnote solution for any Guided Practice.

Self-paced learning labs

The labs associated with the text can be downloaded from github.com/OI-Biostat/oi_biostat_labs. They provide guidance on conducting data analysis and visualization with the R statistical language and the computing environment RStudio, while building understanding of statistical concepts. The labs begin from first principles and require no previous experience with statistical software. Both R and RStudio are freely available for all major computing operating systems, and the Unit 0 labs (00_getting_started) provide information on downloading and installing them. Information on downloading and installing the packages may also be found at **openintro.org**.

The labs for each chapter all have the same structure. Each lab consists of a set of three documents: a handout with the problem statements, a template to be used for working through the lab, and a solution set with the problem solutions. The handout and solution set are most easily read in PDF format (although Rmd files are also provided), while the template is an Rmd file that can be loaded into RStudio. Each chapter of labs is accompanied by a set of "Lab Notes", which provides a reference guide of all new R functions discussed in the labs.

Learning is best done, of course, if a student attempts the lab exercises before reading the solutions. The "Lab Notes" may be a useful resource to refer to while working through problems.

OpenIntro, online resources, and getting involved

OpenIntro is an organization focused on developing free and affordable education materials. The first project, *OpenIntro Statistics*, is intended for introductory statistics courses at the high school through university levels. Other projects examine the use of randomization methods for learning about statistics and conducting analyses (*Introductory Statistics with Randomization and Simulation*) and advanced statistics that may be taught at the high school level (*Advanced High School Statistics*).

We encourage anyone learning or teaching statistics to visit **openintro.org** and get involved by using the many online resources, which are all free, or by creating new material. Students can test their knowledge with practice quizzes, or try an application of concepts learned in each chapter using real data and the free statistical software R. Teachers can download the source for course materials, labs, slides, datasets, R figures, or create their own custom quizzes and problem sets for students to take on the website. Everyone is also welcome to download the book's source files to create a custom version of this textbook or to simply share a PDF copy with a friend or on a website. All of these products are free, and anyone is welcome to use these online tools and resources with or without this textbook as a companion.

Acknowledgements

The *OpenIntro* project would not have been possible without the dedication of many people, including the authors of *OpenIntro Statistics*, the OpenIntro team and the many faculty, students, and readers who commented on all the editions of *OpenIntro Statistics*.

This text has benefited from feedback from Andrea Foulkes, Raji Balasubramanian, Curry Hilton, Michael Parzen, Kevin Rader, and the many excellent teaching fellows at Harvard College who assisted in courses using the book. The cover design was provided by Pierre Baduel.

Chapter 1

Introduction to data

Making observations and recording **data** form the backbone of empirical research, and represent the beginning of a systematic approach to investigating scientific questions. As a discipline, statistics focuses on addressing the following three questions in a rigorous and efficient manner: How can data best be collected? How should data be analyzed? What can be inferred from data?

This chapter provides a brief discussion on the principles of data collection, and introduces basic methods for summarizing and exploring data.

For labs, slides, and other resources, please visit
www.openintro.org/book/biostat

1.1   Case study: preventing peanut allergies

The proportion of young children in Western countries with peanut allergies has doubled in the last 10 years. Previous research suggests that exposing infants to peanut-based foods, rather than excluding such foods from their diets, may be an effective strategy for preventing the development of peanut allergies. The "Learning Early about Peanut Allergy" (LEAP) study was conducted to investigate whether early exposure to peanut products reduces the probability that a child will develop peanut allergies.[1]

The study team enrolled children in the United Kingdom between 2006 and 2009, selecting 640 infants with eczema, egg allergy, or both. Each child was randomly assigned to either the peanut consumption (treatment) group or the peanut avoidance (control) group. Children in the treatment group were fed at least 6 grams of peanut protein daily until 5 years of age, while children in the control group avoided consuming peanut protein until 5 years of age.

At 5 years of age, each child was tested for peanut allergy using an oral food challenge (OFC): 5 grams of peanut protein in a single dose. A child was recorded as passing the oral food challenge if no allergic reaction was detected, and failing the oral food challenge if an allergic reaction occurred. These children had previously been tested for peanut allergy through a skin test, conducted at the time of study entry; the main analysis presented in the paper was based on data from 530 children with an earlier negative skin test.[2]

Individual-level data from the study are shown in Figure 1.1 for 5 of the 530 children—each row represents a participant and shows the participant's study ID number, treatment group assignment, and OFC outcome.[3]

participant.ID	treatment.group	overall.V60.outcome
LEAP_100522	Peanut Consumption	PASS OFC
LEAP_103358	Peanut Consumption	PASS OFC
LEAP_105069	Peanut Avoidance	PASS OFC
LEAP_994047	Peanut Avoidance	PASS OFC
LEAP_997608	Peanut Consumption	PASS OFC

Figure 1.1: Individual-level LEAP results, for five children.

The data can be organized in the form of a two-way summary table; Figure 1.2 shows the results categorized by treatment group and OFC outcome.

	FAIL OFC	PASS OFC	Sum
Peanut Avoidance	36	227	263
Peanut Consumption	5	262	267
Sum	41	489	530

Figure 1.2: Summary of LEAP results, organized by treatment group (either peanut avoidance or consumption) and result of the oral food challenge at 5 years of age (either pass or fail).

---

[1] Du Toit, George, et al. Randomized trial of peanut consumption in infants at risk for peanut allergy. New England Journal of Medicine 372.9 (2015): 803-813.

[2] Although a total of 542 children had an earlier negative skin test, data collection did not occur for 12 children.

[3] The data are available as LEAP in the R package oibiostat.

The summary table makes it easier to identify patterns in the data. Recall that the question of interest is whether children in the peanut consumption group are more or less likely to develop peanut allergies than those in the peanut avoidance group. In the avoidance group, the proportion of children failing the OFC is 36/263 = 0.137 (13.7%); in the consumption group, the proportion of children failing the OFC is 5/267 = 0.019 (1.9%). Figure 1.3 shows a graphical method of displaying the study results, using either the number of individuals per category from Figure 1.2 or the proportion of individuals with a specific OFC outcome in a group.

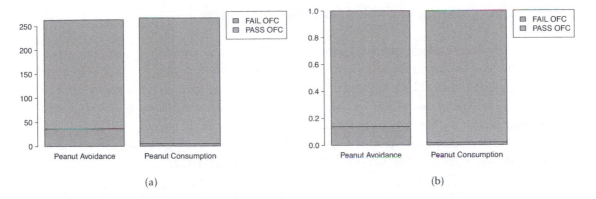

Figure 1.3: (a) A bar plot displaying the number of individuals who failed or passed the OFC in each treatment group. (b) A bar plot displaying the proportions of individuals in each group that failed or passed the OFC.

The proportion of participants failing the OFC is 11.8% higher in the peanut avoidance group than the peanut consumption group. Another way to summarize the data is to compute the ratio of the two proportions (0.137/0.019 = 7.31), and conclude that the proportion of participants failing the OFC in the avoidance group is more than 7 times as large as in the consumption group; i.e., the risk of failing the OFC was more than 7 times as great for participants in the avoidance group relative to the consumption group.

Based on the results of the study, it seems that early exposure to peanut products may be an effective strategy for reducing the chances of developing peanut allergies later in life. It is important to note that this study was conducted in the United Kingdom at a single site of pediatric care; it is not clear that these results can be generalized to other countries or cultures.

The results also raise an important statistical issue: does the study provide definitive evidence that peanut consumption is beneficial? In other words, is the 11.8% difference between the two groups larger than one would expect by chance variation alone? The material on inference in later chapters will provide the statistical tools to evaluate this question.

1.2   Data basics

Effective organization and description of data is a first step in most analyses. This section introduces a structure for organizing data and basic terminology used to describe data.

1.2.1   Observations, variables, and data matrices

In evolutionary biology, parental investment refers to the amount of time, energy, or other resources devoted towards raising offspring. This section introduces the frog dataset, which originates from a 2013 study about maternal investment in a frog species.[4] Reproduction is a costly process for female frogs, necessitating a trade-off between individual egg size and total number of eggs produced. Researchers were interested in investigating how maternal investment varies with altitude and collected measurements on egg clutches found at breeding ponds across 11 study sites; for 5 sites, the body size of individual female frogs was also recorded.

	altitude	latitude	egg.size	clutch.size	clutch.volume	body.size
1	3,462.00	34.82	1.95	181.97	177.83	3.63
2	3,462.00	34.82	1.95	269.15	257.04	3.63
3	3,462.00	34.82	1.95	158.49	151.36	3.72
150	2,597.00	34.05	2.24	537.03	776.25	NA

Figure 1.4: Data matrix for the frog dataset.

Figure 1.4 displays rows 1, 2, 3, and 150 of the data from the 431 clutches observed as part of the study.[5] Each row in the table corresponds to a single clutch, indicating where the clutch was collected (altitude and latitude), egg.size, clutch.size, clutch.volume, and body.size of the mother when available. "NA" corresponds to a missing value, indicating that information on an individual female was not collected for that particular clutch. The recorded characteristics are referred to as **variables**; in this table, each column represents a variable.

variable	description
altitude	Altitude of the study site in meters above sea level
latitude	Latitude of the study site measured in degrees
egg.size	Average diameter of an individual egg to the 0.01 mm
clutch.size	Estimated number of eggs in clutch
clutch.volume	Volume of egg clutch in mm^3
body.size	Length of mother frog in cm

Figure 1.5: Variables and their descriptions for the frog dataset.

It is important to check the definitions of variables, as they are not always obvious. For example, why has clutch.size not been recorded as whole numbers? For a given clutch, researchers counted approximately 5 grams' worth of eggs and then estimated the total number of eggs based on the mass of the entire clutch. Definitions of the variables are given in Figure 1.5.[6]

---

[4]Chen, W., et al. Maternal investment increases with altitude in a frog on the Tibetan Plateau. Journal of evolutionary biology 26.12 (2013): 2710-2715.

[5]The frog dataset is available in the R package oibiostat.

[6]The data discussed here are in the original scale; in the published paper, some values have undergone a natural log transformation.

The data in Figure 1.4 are organized as a **data matrix**. Each row of a data matrix corresponds to an observational unit, and each column corresponds to a variable. A piece of the data matrix for the LEAP study introduced in Section 1.1 is shown in Figure 1.1; the rows are study participants and three variables are shown for each participant. Data matrices are a convenient way to record and store data. If the data are collected for another individual, another row can easily be added; similarly, another column can be added for a new variable.

1.2.2 Types of variables

The Functional polymorphisms Associated with human Muscle Size and Strength study (FAMuSS) measured a variety of demographic, phenotypic, and genetic characteristics for about 1,300 participants.[7] Data from the study have been used in a number of subsequent studies,[8] such as one examining the relationship between muscle strength and genotype at a location on the ACTN3 gene.[9]

The famuss dataset is a subset of the data for 595 participants.[10] Four rows of the famuss dataset are shown in Figure 1.6, and the variables are described in Figure 1.7.

	sex	age	race	height	weight	actn3.r577x	ndrm.ch
1	Female	27	Caucasian	65.0	199.0	CC	40.0
2	Male	36	Caucasian	71.7	189.0	CT	25.0
3	Female	24	Caucasian	65.0	134.0	CT	40.0
595	Female	30	Caucasian	64.0	134.0	CC	43.8

Figure 1.6: Four rows from the famuss data matrix.

variable	description
sex	Sex of the participant
age	Age in years
race	Race, recorded as African Am (African American), Caucasian, Asian, Hispanic or Other
height	Height in inches
weight	Weight in pounds
actn3.r577x	Genotype at the location r577x in the ACTN3 gene.
ndrm.ch	Percent change in strength in the non-dominant arm, comparing strength after to before training

Figure 1.7: Variables and their descriptions for the famuss dataset.

The variables age, height, weight, and ndrm.ch are **numerical variables**. They take on numerical values, and it is reasonable to add, subtract, or take averages with these values. In contrast, a variable reporting telephone numbers would not be classified as numerical, since sums, differences, and averages in this context have no meaning. Age measured in years is said to be **discrete**, since it can only take on numerical values with jumps; i.e., positive integer values. Percent change in strength in the non-dominant arm (ndrm.ch) is **continuous**, and can take on any value within a specified range.

---

[7]Thompson PD, Moyna M, Seip, R, et al., 2004. Functional Polymorphisms Associated with Human Muscle Size and Strength. Medicine and Science in Sports and Exercise 36:1132 - 1139.

[8]Pescatello L, et al. Highlights from the functional single nucleotide polymorphisms associated with human muscle size and strength or FAMuSS study, BioMed Research International 2013.

[9]Clarkson P, et al., Journal of Applied Physiology 99: 154-163, 2005.

[10]The subset is from Foulkes, Andrea S. Applied statistical genetics with R: for population-based association studies. Springer Science & Business Media, 2009. The full version of the data is available at http://people.umass.edu/foulkes/asg/data.html.

Figure 1.8: Breakdown of variables into their respective types.

The variables sex, race, and actn3.r577x are **categorical variables**, which take on values that are names or labels. The possible values of a categorical variable are called the variable's **levels**.[11] For example, the levels of actn3.r577x are the three possible genotypes at this particular locus: CC, CT, or TT. Categorical variables without a natural ordering are called **nominal categorical variables**; sex, race, and actn3.r577x are all nominal categorical variables. Categorical variables with levels that have a natural ordering are referred to as **ordinal categorical variables**. For example, age of the participants grouped into 5-year intervals (15-20, 21-25, 26-30, etc.) is an ordinal categorical variable.

EXAMPLE 1.1

Classify the variables in the frog dataset: altitude, latitude, egg.size, clutch.size, clutch.volume, and body.size.

The variables egg.size, clutch.size, clutch.volume, and body.size are continuous numerical variables, and can take on all positive values.

In the context of this study, the variables altitude and latitude are best described as categorical variables, since the numerical values of the variables correspond to the 11 specific study sites where data were collected. Researchers were interested in exploring the relationship between altitude and maternal investment; it would be reasonable to consider altitude an ordinal categorical variable.

GUIDED PRACTICE 1.2

Characterize the variables treatment.group and overall.V60.outcome from the LEAP study (discussed in Section 1.1).[12]

GUIDED PRACTICE 1.3

Suppose that on a given day, a research assistant collected data on the first 20 individuals visiting a walk-in clinic: age (measured as less than 21, 21 - 65, and greater than 65 years of age), sex, height, weight, and reason for the visit. Classify each of the variables.[13]

---

[11]Categorical variables are sometimes called **factor variables**.

[12]These variables measure non-numerical quantities, and thus are categorical variables with two levels.

[13]Height and weight are continuous numerical variables. Age as measured by the research assistant is ordinal categorical. Sex and the reason for the visit are nominal categorical variables.

1.2.3 Relationships between variables

Many studies are motivated by a researcher examining how two or more variables are related. For example, do the values of one variable increase as the values of another decrease? Do the values of one variable tend to differ by the levels of another variable?

One study used the famuss data to investigate whether ACTN3 genotype at a particular location (residue 577) is associated with change in muscle strength. The ACTN3 gene codes for a protein involved in muscle function. A common mutation in the gene at a specific location changes the cytosine (C) nucleotide to a thymine (T) nucleotide; individuals with the TT genotype are unable to produce any ACTN3 protein.

Researchers hypothesized that genotype at this location might influence muscle function. As a measure of muscle function, they recorded the percent change in non-dominant arm strength after strength training; this variable, ndrm.ch, is the **response variable** in the study. A response variable is defined by the particular research question a study seeks to address, and measures the outcome of interest in the study. A study will typically examine whether the values of a response variable differ as values of an **explanatory variable** change, and if so, how the two variables are related. A given study may examine several explanatory variables for a single response variable.[14] The explanatory variable examined in relation to ndrm.ch in the study is actn3.r557x, ACTN3 genotype at location 577.

EXAMPLE 1.4

In the maternal investment study conducted on frogs, researchers collected measurements on egg clutches and female frogs at 11 study sites, located at differing altitudes, in order to investigate how maternal investment varies with altitude. Identify the response and explanatory variables in the study.

E

The variables egg.size, clutch.size, and clutch.volume are response variables indicative of maternal investment.

The explanatory variable examined in the study is altitude.

While latitude is an environmental factor that might potentially influence features of the egg clutches, it is not a variable of interest in this particular study.

Female body size (body.size) is neither an explanatory nor response variable.

GUIDED PRACTICE 1.5

G

Refer to the variables from the famuss dataset described in Figure 1.7 to formulate a question about the relationships between these variables, and identify the response and explanatory variables in the context of the question.[15]

---

[14]Response variables are sometimes called dependent variables and explanatory variables are often called independent variables or predictors.

[15]Two sample questions: (1) Does change in participant arm strength after training seem associated with race? The response variable is ndrm.ch and the explanatory variable is race. (2) Do male participants appear to respond differently to strength training than females? The response variable is ndrm.ch and the explanatory variable is sex.

1.3 Data collection principles

The first step in research is to identify questions to investigate. A clearly articulated research question is essential for selecting subjects to be studied, identifying relevant variables, and determining how data should be collected.

1.3.1 Populations and samples

Consider the following research questions:

1. Do bluefin tuna from the Atlantic Ocean have particularly high levels of mercury, such that they are unsafe for human consumption?

2. For infants predisposed to developing a peanut allergy, is there evidence that introducing peanut products early in life is an effective strategy for reducing the risk of developing a peanut allergy?

3. Does a recently developed drug designed to treat glioblastoma, a form of brain cancer, appear more effective at inducing tumor shrinkage than the drug currently on the market?

Each of these questions refers to a specific target **population**. For example, in the first question, the target population consists of all bluefin tuna from the Atlantic Ocean; each individual bluefin tuna represents a case. It is almost always either too expensive or logistically impossible to collect data for every case in a population. As a result, nearly all research is based on information obtained about a sample from the population. A **sample** represents a small fraction of the population. Researchers interested in evaluating the mercury content of bluefin tuna from the Atlantic Ocean could collect a sample of 500 bluefin tuna (or some other quantity), measure the mercury content, and use the observed information to formulate an answer to the research question.

(G)

GUIDED PRACTICE 1.6

Identify the target populations for the remaining two research questions.[16]

---

[16]In Question 2, the target population consists of infants predisposed to developing a peanut allergy. In Question 3, the target population consists of patients with glioblastoma.

1.3.2 Anecdotal evidence

Anecdotal evidence typically refers to unusual observations that are easily recalled because of their striking characteristics. Physicians may be more likely to remember the characteristics of a single patient with an unusually good response to a drug instead of the many patients who did not respond. The dangers of drawing general conclusions from anecdotal information are obvious; no single observation should be used to draw conclusions about a population.

While it is incorrect to generalize from individual observations, unusual observations can sometimes be valuable. E.C. Heyde was a general practitioner from Vancouver who noticed that a few of his elderly patients with aortic-valve stenosis (an abnormal narrowing) caused by an accumulation of calcium had also suffered massive gastrointestinal bleeding. In 1958, he published his observation.[17] Further research led to the identification of the underlying cause of the association, now called Heyde's Syndrome.[18]

An anecdotal observation can never be the basis for a conclusion, but may well inspire the design of a more systematic study that could be definitive.

---

[17] Heyde EC. Gastrointestinal bleeding in aortic stenosis. N Engl J Med 1958;259:196.

[18] Greenstein RJ, McElhinney AJ, Reuben D, Greenstein AJ. Co-lonic vascular ectasias and aortic stenosis: coincidence or causal relationship? Am J Surg 1986;151:347-51.

1.3.3   Sampling from a population

Sampling from a population, when done correctly, provides reliable information about the characteristics of a large population. The US Centers for Disease Control (US CDC) conducts several surveys to obtain information about the US population, including the Behavior Risk Factor Surveillance System (BRFSS).[19] The BRFSS was established in 1984 to collect data about health-related risk behaviors, and now collects data from more than 400,000 telephone interviews conducted each year. Data from a recent BRFSS survey are used in Chapter 4. The CDC conducts similar surveys for diabetes, health care access, and immunization. Likewise, the World Health Organization (WHO) conducts the World Health Survey in partnership with approximately 70 countries to learn about the health of adult populations and the health systems in those countries.[20]

The general principle of sampling is straightforward: a sample from a population is useful for learning about a population only when the sample is **representative** of the population. In other words, the characteristics of the sample should correspond to the characteristics of the population.

Suppose that the quality improvement team at an integrated health care system, such as Harvard Pilgrim Health Care, is interested in learning about how members of the health plan perceive the quality of the services offered under the plan. A common pitfall in conducting a survey is to use a **convenience sample**, in which individuals who are easily accessible are more likely to be included in the sample than other individuals. If a sample were collected by approaching plan members visiting an outpatient clinic during a particular week, the sample would fail to enroll generally healthy members who typically do not use outpatient services or schedule routine physical examinations; this method would produce an unrepresentative sample (Figure 1.9).

Figure 1.9: Instead of sampling from all members equally, approaching members visiting a clinic during a particular week disproportionately selects members who frequently use outpatient services.

Random sampling is the best way to ensure that a sample reflects a population. In a **simple random sample**, each member of a population has the same chance of being sampled. One way to achieve a simple random sample of the health plan members is to randomly select a certain number of names from the complete membership roster, and contact those individuals for an interview (Figure 1.10).

---

[19]https://www.cdc.gov/brfss/index.html
[20]http://www.who.int/healthinfo/survey/en/

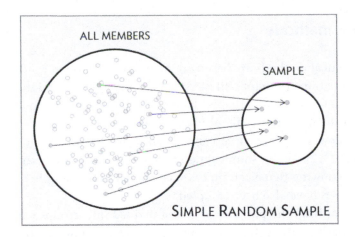

Figure 1.10: Five members are randomly selected from the population to be interviewed.

Even when a simple random sample is taken, it is not guaranteed that the sample is representative of the population. If the **non-response** rate for a survey is high, that may be indicative of a biased sample. Perhaps a majority of participants did not respond to the survey because only a certain group within the population is being reached; for example, if questions assume that participants are fluent in English, then a high non-response rate would be expected if the population largely consists of individuals who are not fluent in English (Figure 1.11). Such **non-response bias** can skew results; generalizing from an unrepresentative sample may likely lead to incorrect conclusions about a population.

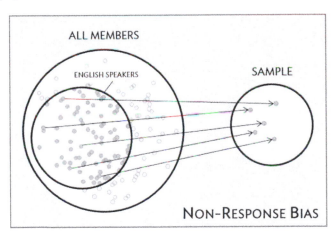

Figure 1.11: Surveys may only reach a certain group within the population, which leads to non-response bias. For example, a survey written in English may only result in responses from health plan members fluent in English.

GUIDED PRACTICE 1.7

It is increasingly common for health care facilities to follow-up a patient visit with an email providing a link to a website where patients can rate their experience. Typically, less than 50% of patients visit the website. If half of those who respond indicate a negative experience, do you think that this implies that at least 25% of patient visits are unsatisfactory?[21]

---

[21] It is unlikely that the patients who respond constitute a representative sample from the larger population of patients. This is not a random sample, because individuals are selecting themselves into a group, and it is unclear that each person has an equal chance of answering the survey. If our experience is any guide, dissatisfied people are more likely to respond to these informal surveys than satisfied patients.

1.3.4 Sampling methods

Almost all statistical methods are based on the notion of implied randomness. If data are not sampled from a population at random, these statistical methods – calculating estimates and errors associated with estimates – are not reliable. Four random sampling methods are discussed in this section: simple, stratified, cluster, and multistage sampling.

In a **simple random sample**, each case in the population has an equal chance of being included in the sample (Figure 1.12). Under simple random sampling, each case is sampled independently of the other cases; i.e., knowing that a certain case is included in the sample provides no information about which other cases have also been sampled.

In **stratified sampling**, the population is first divided into groups called **strata** before cases are selected within each stratum (typically through simple random sampling) (Figure 1.12). The strata are chosen such that similar cases are grouped together. Stratified sampling is especially useful when the cases in each stratum are very similar with respect to the outcome of interest, but cases between strata might be quite different.

Suppose that the health care provider has facilities in different cities. If the range of services offered differ by city, but all locations in a given city will offer similar services, it would be effective for the quality improvement team to use stratified sampling to identify participants for their study, where each city represents a stratum and plan members are randomly sampled from each city.

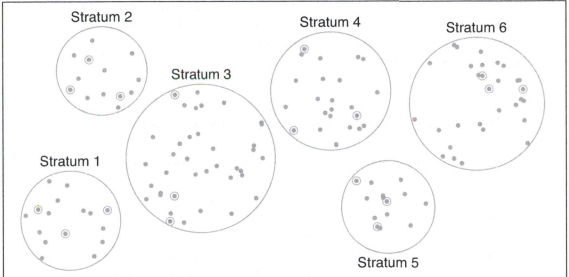

Figure 1.12: Examples of simple random and stratified sampling. In the top panel, simple random sampling is used to randomly select 18 cases (circled orange dots) out of the total population (all dots). The bottom panel illustrates stratified sampling: cases are grouped into six strata, then simple random sampling is employed within each stratum.

In a **cluster sample**, the population is first divided into many groups, called **clusters**. Then, a fixed number of clusters is sampled and all observations from each of those clusters are included in the sample (Figure 1.13). A **multistage sample** is similar to a cluster sample, but rather than keeping all observations in each cluster, a random sample is collected within each selected cluster (Figure 1.13).

Unlike with stratified sampling, cluster and multistage sampling are most helpful when there is high case-to-case variability within a cluster, but the clusters themselves are similar to one another. For example, if neighborhoods in a city represent clusters, cluster and multistage sampling work best when the population within each neighborhood is very diverse, but neighborhoods are relatively similar.

Applying stratified, cluster, or multistage sampling can often be more economical than only drawing random samples. However, analysis of data collected using such methods is more complicated than when using data from a simple random sample; this text will only discuss analysis methods for simple random samples.

EXAMPLE 1.8

Suppose researchers are interested in estimating the malaria rate in a densely tropical portion of rural Indonesia. There are 30 villages in the area, each more or less similar to the others. The goal is to test 150 individuals for malaria. Evaluate which sampling method should be employed.

A simple random sample would likely draw individuals from all 30 villages, which could make data collection extremely expensive. Stratified sampling is not advisable, since there is not enough information to determine how strata of similar individuals could be built. However, cluster sampling or multistage sampling are both reasonable options. For example, with multistage sampling, half of the villages could be randomly selected, and then 10 people selected from each village. This strategy is more efficient than a simple random sample, and can still provide a sample representative of the population of interest.

1.3.5 Introducing experiments and observational studies

The two primary types of study designs used to collect data are experiments and observational studies.

In an **experiment**, researchers directly influence how data arise, such as by assigning groups of individuals to different treatments and assessing how the outcome varies across treatment groups. The LEAP study is an example of an experiment with two groups, an experimental group that received the intervention (peanut consumption) and a control group that received a standard approach (peanut avoidance). In studies assessing effectiveness of a new drug, individuals in the control group typically receive a **placebo**, an inert substance with the appearance of the experimental intervention. The study is designed such that on average, the only difference between the individuals in the treatment groups is whether or not they consumed peanut protein. This allows for observed differences in experimental outcome to be directly attributed to the intervention and constitute evidence of a causal relationship between intervention and outcome.

In an **observational study**, researchers merely observe and record data, without interfering with how the data arise. For example, to investigate why certain diseases develop, researchers might collect data by conducting surveys, reviewing medical records, or following a **cohort** of many similar individuals. Observational studies can provide evidence of an association between variables, but cannot by themselves show a causal connection. However, there are many instances where randomized experiments are unethical, such as to explore whether lead exposure in young children is associated with cognitive impairment.

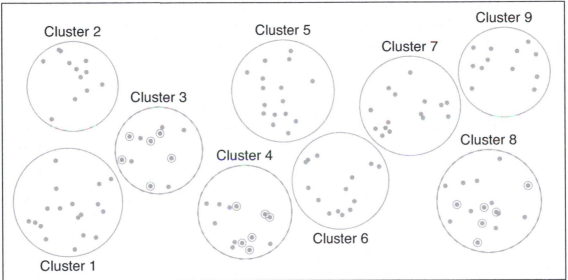

Figure 1.13: Examples of cluster and multistage sampling. The top panel illustrates cluster sampling: data are binned into nine clusters, three of which are sampled, and all observations within these clusters are sampled. The bottom panel illustrates multistage sampling, which differs from cluster sampling in that only a subset from each of the three selected clusters are sampled.

1.3.6 Experiments

Experimental design is based on three principles: control, randomization, and replication.

Control. When selecting participants for a study, researchers work to **control** for extraneous variables and choose a sample of participants that is representative of the population of interest. For example, participation in a study might be restricted to individuals who have a condition that suggests they may benefit from the intervention being tested. Infants enrolled in the LEAP study were required to be between 4 and 11 months of age, with severe eczema and/or allergies to eggs.

Randomization. Randomly assigning patients to treatment groups ensures that groups are balanced with respect to both variables that can and cannot be controlled. For example, randomization in the LEAP study ensures that the proportion of males to females is approximately the same in both groups. Additionally, perhaps some infants were more susceptible to peanut allergy because of an undetected genetic condition; under randomization, it is reasonable to assume that such infants were present in equal numbers in both groups. Randomization allows differences in outcome between the groups to be reasonably attributed to the treatment rather than inherent variability in patient characteristics, since the treatment represents the only systematic difference between the two groups.

In situations where researchers suspect that variables other than the intervention may influence the response, individuals can be first grouped into **blocks** according to a certain attribute and then randomized to treatment group within each block; this technique is referred to as **blocking** or **stratification**. The team behind the LEAP study stratified infants into two cohorts based on whether or not the child developed a red, swollen mark (a wheal) after a skin test at the time of enrollment; afterwards, infants were randomized between peanut consumption and avoidance groups. Figure 1.14 illustrates the blocking scheme used in the study.

Replication. The results of a study conducted on a larger number of cases are generally more reliable than smaller studies; observations made from a large sample are more likely to be representative of the population of interest. In a single study, **replication** is accomplished by collecting a sufficiently large sample. The LEAP study randomized a total of 640 infants.

Randomized experiments are an essential tool in research. The US Food and Drug Administration typically requires that a new drug can only be marketed after two independently conducted randomized trials confirm its safety and efficacy; the European Medicines Agency has a similar policy. Large randomized experiments in medicine have provided the basis for major public health initiatives. In 1954, approximately 750,000 children participated in a randomized study comparing polio vaccine with a placebo.[22] In the United States, the results of the study quickly led to the widespread and successful use of the vaccine for polio prevention.

---

[22]Meier, Paul. "The biggest public health experiment ever: the 1954 field trial of the Salk poliomyelitis vaccine." *Statistics: a guide to the unknown*. San Francisco: Holden-Day (1972): 2-13.

Figure 1.14: A simplified schematic of the blocking scheme used in the LEAP study, depicting 640 patients that underwent randomization. Patients are first divided into blocks based on response to the initial skin test, then each block is randomized between the avoidance and consumption groups. This strategy ensures an even representation of patients in each group who had positive and negative skin tests.

1.3.7 Observational studies

In observational studies, researchers simply observe selected potential explanatory and response variables. Participants who differ in important explanatory variables may also differ in other ways that influence response; as a result, it is not advisable to make causal conclusions about the relationship between explanatory and response variables based on observational data. For example, while observational studies of obesity have shown that obese individuals tend to die sooner than individuals with normal weight, it would be misleading to conclude that obesity causes shorter life expectancy. Instead, underlying factors are probably involved; obese individuals typically exhibit other health behaviors that influence life expectancy, such as reduced exercise or unhealthy diet.

Suppose that an observational study tracked sunscreen use and incidence of skin cancer, and found that the more sunscreen a person uses, the more likely they are to have skin cancer. These results do not mean that sunscreen causes skin cancer. One important piece of missing information is sun exposure – if someone is often exposed to sun, they are both more likely to use sunscreen and to contract skin cancer. Sun exposure is a **confounding variable**: a variable associated with both the explanatory and response variables.[23] There is no guarantee that all confounding variables can be examined or measured; as a result, it is not advisable to draw causal conclusions from observational studies.

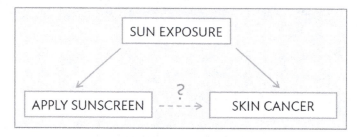

Confounding is not limited to observational studies. For example, consider a randomized study comparing two treatments (varenicline and buproprion) against a placebo as therapies for aiding smoking cessation.[24] At the beginning of the study, participants were randomized into groups: 352 to varenicline, 329 to buproprion, and 344 to placebo. Not all participants successfully completed the assigned therapy: 259, 225, and 215 patients in each group did so, respectively. If an analysis were based only on the participants who completed therapy, this could introduce confounding; it is possible that there are underlying differences between individuals who complete the therapy and those who do not. Including all randomized participants in the final analysis maintains the original randomization scheme and controls for differences between the groups.[25]

GUIDED PRACTICE 1.9

As stated in Example 1.4, female body size (body.size) in the parental investment study is neither an explanatory nor a response variable. Previous research has shown that larger females tend to produce larger eggs and egg clutches; however, large body size can be costly at high altitudes. Discuss a possible reason for why the study team chose to measure female body size when it is not directly related to their main research question.[26]

---

[23]Also called a **lurking variable**, **confounding factor**, or a **confounder**.

[24]Jorenby, Douglas E., et al. "Efficacy of varenicline, an $\alpha 4\beta 2$ nicotinic acetylcholine receptor partial agonist, vs placebo or sustained-release bupropion for smoking cessation: a randomized controlled trial." JAMA 296.1 (2006): 56-63.

[25]This strategy, commonly used for analyzing clinical trial data, is referred to as an intention-to-treat analysis.

[26]Female body size is a potential confounding variable, since it may be associated with both the explanatory variable (altitude) and response variables (measures of maternal investment). If the study team observes, for example, that clutch size tends to decrease at higher altitudes, they should check whether the apparent association is not simply due to frogs at higher altitudes having smaller body size and thus, laying smaller clutches.

Observational studies may reveal interesting patterns or associations that can be further investigated with follow-up experiments. Several observational studies based on dietary data from different countries showed a strong association between dietary fat and breast cancer in women. These observations led to the launch of the Women's Health Initiative (WHI), a large randomized trial sponsored by the US National Institutes of Health (NIH). In the WHI, women were randomized to standard versus low fat diets, and the previously observed association was not confirmed.

Observational studies can be either prospective or retrospective. A **prospective study** identifies participants and collects information at scheduled times or as events unfold. For example, in the Nurses' Health Study, researchers recruited registered nurses beginning in 1976 and collected data through administering biennial surveys; data from the study have been used to investigate risk factors for major chronic diseases in women.[27] **Retrospective studies** collect data after events have taken place, such as from medical records. Some datasets may contain both retrospectively- and prospectively-collected variables. The Cancer Care Outcomes Research and Surveillance Consortium (CanCORS) enrolled participants with lung or colorectal cancer, collected information about diagnosis, treatment, and previous health behavior, but also maintained contact with participants to gather data about long-term outcomes.[28]

---

[27]www.channing.harvard.edu/nhs

[28]Ayanian, John Z., et al. "Understanding cancer treatment and outcomes: the cancer care outcomes research and surveillance consortium." Journal of Clinical Oncology 22.15 (2004): 2992-2996

1.4 Numerical data

This section discusses techniques for exploring and summarizing numerical variables, using the frog data from the parental investment study introduced in Section 1.2.

1.4.1 Measures of center: mean and median

The **mean**, sometimes called the average, is a measure of center for a **distribution** of data. To find the average clutch volume for the observed egg clutches, add all the clutch volumes and divide by the total number of clutches.[29]

$$\bar{x} = \frac{177.8 + 257.0 + \cdots + 933.3}{431} = 882.5 \text{ mm}^3.$$

$\bar{x}$
sample mean
μ
population mean

The sample mean is often labeled $\bar{x}$, to distinguish it from μ, the mean of the entire population from which the sample is drawn. The letter x is being used as a generic placeholder for the variable of interest, clutch.volume.

MEAN

The sample mean of a numerical variable is the sum of the values of all observations divided by the number of observations:

$$\bar{x} = \frac{x_1 + x_2 + \cdots + x_n}{n}, \qquad (1.10)$$

where $x_1, x_2, \ldots, x_n$ represent the n observed values.

The **median** is another measure of center; it is the middle number in a distribution after the values have been ordered from smallest to largest. If the distribution contains an even number of observations, the median is the average of the middle two observations. There are 431 clutches in the dataset, so the median is the clutch volume of the 216th observation in the sorted values of clutch.volume: 831.8 mm^3.

---

[29]For computational convenience, the volumes are rounded to the first decimal.

1.4.2 Measures of spread: standard deviation and interquartile range

The spread of a distribution refers to how similar or varied the values in the distribution are to each other; i.e., whether the values are tightly clustered or spread over a wide range.

The standard deviation for a set of data describes the typical distance between an observation and the mean. The distance of a single observation from the mean is its **deviation**. Below are the deviations for the 1st, 2nd, 3rd, and 431st observations in the clutch.volume variable.

$$x_1 - \overline{x} = 177.8 - 882.5 = -704.7$$
$$x_2 - \overline{x} = 257.0 - 882.5 = -625.5$$
$$x_3 - \overline{x} = 151.4 - 882.5 = -731.1$$
$$\vdots$$
$$x_{431} - \overline{x} = 933.2 - 882.5 = 50.7$$

The sample **variance**, the average of the squares of these deviations, is denoted by s^2:

s^2
sample variance

$$s^2 = \frac{(-704.7)^2 + (-625.5)^2 + (-731.1)^2 + \cdots + (50.7)^2}{431 - 1}$$
$$= \frac{496,602.09 + 391,250.25 + 534,507.21 + \cdots + 2570.49}{430}$$
$$= 143,680.9.$$

The denominator is $n-1$ rather than n; this mathematical nuance accounts for the fact that sample mean has been used to estimate the population mean in the calculation. Details on the statistical theory can be found in more advanced texts.

The sample **standard deviation** s is the square root of the variance:

$$s = \sqrt{143,680.9} = 379.05\text{mm}^3.$$

s
sample standard deviation

Like the mean, the population values for variance and standard deviation are denoted by Greek letters: σ^2 for the variance and σ for the standard deviation.

σ^2
population variance

σ
population standard deviation

STANDARD DEVIATION

The sample standard deviation of a numerical variable is computed as the square root of the variance, which is the sum of squared deviations divided by the number of observations minus 1.

$$s = \sqrt{\frac{(x_1 - \overline{x})^2 + (x_2 - \overline{x})^2 + \cdots + (x_n - \overline{x})^2}{n - 1}}, \tag{1.11}$$

where $x_1, x_2, \ldots, x_n$ represent the n observed values.

Variability can also be measured using the **interquartile range** (IQR). The IQR for a distribution is the difference between the first and third quartiles: $Q_3 - Q_1$. The first quartile (Q_1) is equivalent to the 25th percentile; i.e., 25% of the data fall below this value. The third quartile (Q_3) is equivalent to the 75th percentile. By definition, the median represents the second quartile, with half the values falling below it and half falling above. The IQR for clutch.volume is $1096.0 - 609.6 = 486.4$ mm^3.

Measures of center and spread are ways to summarize a distribution numerically. Using numerical summaries allows for a distribution to be efficiently described with only a few numbers.[30] For example, the calculations for clutch.volume indicate that the typical egg clutch has volume of about 880 mm^3, while the middle 50% of egg clutches have volumes between approximately 600 mm^3 and 1100.0 mm^3.

1.4.3   Robust estimates

Figure 1.15 shows the values of clutch.volume as points on a single axis. There are a few values that seem extreme relative to the other observations: the four largest values, which appear distinct from the rest of the distribution. How do these extreme values affect the value of the numerical summaries?

Figure 1.15: **Dot plot** of clutch volumes from the frog data.

Figure 1.16 shows the summary statistics calculated under two scenarios, one with and one without the four largest observations. For these data, the median does not change, while the IQR differs by only about 6 mm^3. In contrast, the mean and standard deviation are much more affected, particularly the standard deviation.

scenario	robust		not robust	
	median	IQR	$\bar{x}$	s
original data (with extreme observations)	831.8	486.9	882.5	379.1
data without four largest observations	831.8	493.9	867.9	349.2

Figure 1.16: A comparison of how the median, IQR, mean ($\bar{x}$), and standard deviation (s) change when extreme observations are present.

The median and IQR are referred to as **robust estimates** because extreme observations have little effect on their values. For distributions that contain extreme values, the median and IQR will provide a more accurate sense of the center and spread than the mean and standard deviation.

---

[30]Numerical summaries are also known as summary statistics.

1.4.4 Visualizing distributions of data: histograms and boxplots

Graphs show important features of a distribution that are not evident from numerical summaries, such as asymmetry or extreme values. While dot plots show the exact value of each observation, histograms and boxplots graphically summarize distributions.

In a **histogram**, observations are grouped into bins and plotted as bars. Figure 1.17 shows the number of clutches with volume between 0 and 200 mm^3, 200 and 400 mm^3, etc. up until 2,600 and 2,800 mm^3.[31] These binned counts are plotted in Figure 1.18.

Clutch volumes	0-200	200-400	400-600	600-800	$\cdots$	2400-2600	2600-2800
Count	4	29	69	99	$\cdots$	2	1

Figure 1.17: The counts for the binned clutch.volume data.

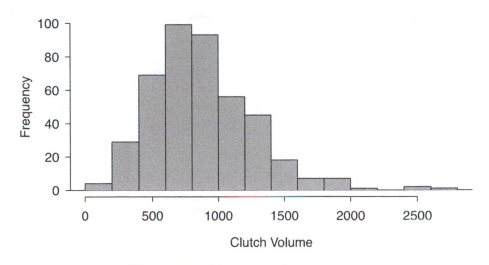

Figure 1.18: A histogram of clutch.volume.

Histograms provide a view of the **data density**. Higher bars indicate more frequent observations, while lower bars represent relatively rare observations. Figure 1.18 shows that most of the egg clutches have volumes between 500-1,000 mm^3, and there are many more clutches with volumes smaller than 1,000 mm^3 than clutches with larger volumes.

Histograms show the **shape** of a distribution. The tails of a **symmetric** distribution are roughly equal, with data trailing off from the center roughly equally in both directions. Asymmetry arises when one tail of the distribution is longer than the other. A distribution is said to be **right skewed** when data trail off to the right, and **left skewed** when data trail off to the left.[32] Figure 1.18 shows that the distribution of clutch volume is right skewed; most clutches have relatively small volumes, and only a few clutches have high volumes.

---

[31] By default in R, the bins are left-open and right-closed; i.e., the intervals are of the form (a, b]. Thus, an observation with value 200 would fall into the 0-200 bin instead of the 200-400 bin.

[32] Other ways to describe data that are skewed to the right/left: **skewed to the right/left** or **skewed to the positive/negative end**.

A **mode** is represented by a prominent peak in the distribution.[33] Figure 1.19 shows histograms that have one, two, or three major peaks. Such distributions are called **unimodal**, **bimodal**, and **multimodal**, respectively. Any distribution with more than two prominent peaks is called multimodal. Note that the less prominent peak in the unimodal distribution was not counted since it only differs from its neighboring bins by a few observations. Prominent is a subjective term, but it is usually clear in a histogram where the major peaks are.

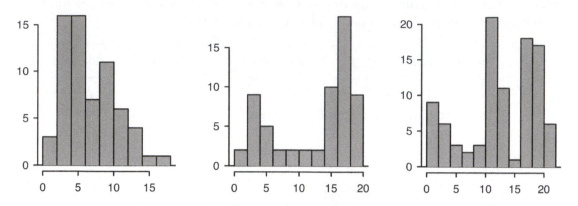

Figure 1.19: From left to right: unimodal, bimodal, and multimodal distributions.

A **boxplot** indicates the positions of the first, second, and third quartiles of a distribution in addition to extreme observations.[34] Figure 1.20 shows a boxplot of clutch.volume alongside a vertical dot plot.

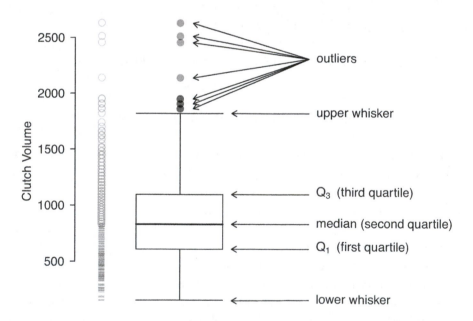

Figure 1.20: A boxplot and dot plot of clutch.volume. The horizontal dashes indicate the bottom 50% of the data and the open circles represent the top 50%.

---

[33]Another definition of mode, which is not typically used in statistics, is the value with the most occurrences. It is common that a dataset contains *no* observations with the same value, which makes this other definition impractical for many datasets.

[34]Boxplots are also known as box-and-whisker plots.

In a boxplot, the interquartile range is represented by a rectangle extending from the first quartile to the third quartile, and the rectangle is split by the median (second quartile). Extending outwards from the box, the **whiskers** capture the data that fall between $Q_1 - 1.5 \times IQR$ and $Q_3 + 1.5 \times IQR$. The whiskers must end at data points; the values given by adding or subtracting $1.5 \times IQR$ define the maximum reach of the whiskers. For example, with the clutch.volume variable, $Q_3 + 1.5 \times IQR = 1,096.5 + 1.5 \times 486.4 = 1,826.1$ mm^3. However, there was no clutch with volume 1,826.1 mm^3; thus, the upper whisker extends to 1,819.7 mm^3, the largest observation that is smaller than $Q_3 + 1.5 \times IQR$.

Any observation that lies beyond the whiskers is shown with a dot; these observations are called outliers. An **outlier** is a value that appears extreme relative to the rest of the data. For the clutch.volume variable, there are several large outliers and no small outliers, indicating the presence of some unusually large egg clutches.

The high outliers in Figure 1.20 reflect the right-skewed nature of the data. The right skew is also observable from the position of the median relative to the first and third quartiles; the median is slightly closer to the first quartile. In a symmetric distribution, the median will be halfway between the first and third quartiles.

GUIDED PRACTICE 1.12

(G) Use the histogram and boxplot in Figure 1.21 to describe the distribution of height in the famuss data, where height is measured in inches.[35]

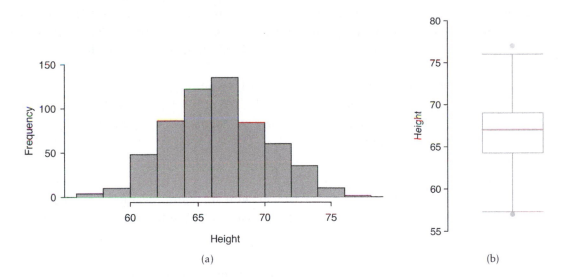

(a)   (b)

Figure 1.21: A histogram and boxplot of height in the famuss data.

---

[35]The data are roughly symmetric (the left tail is slightly longer than the right tail), and the distribution is unimodal with one prominent peak at about 67 inches. The middle 50% of individuals are between 5.5 feet and just under 6 feet tall. There is one low outlier and one high outlier, representing individuals that are unusually short/tall relative to the other individuals.

1.4.5 Transforming data

When working with strongly skewed data, it can be useful to apply a **transformation**, and rescale the data using a function. A natural log transformation is commonly used to clarify the features of a variable when there are many values clustered near zero and all observations are positive.

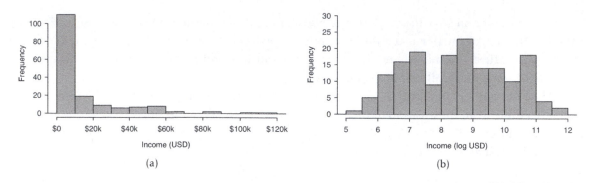

Figure 1.22: (a) Histogram of per capita income. (b) Histogram of the log-transformed per capita income.

For example, income data are often skewed right; there are typically large clusters of low to moderate income, with a few large incomes that are outliers. Figure 1.22(a) shows a histogram of average yearly per capita income measured in US dollars for 165 countries in 2011.[36] The data are heavily right skewed, with the majority of countries having average yearly per capita income lower than $10,000. Once the data are log-transformed, the distribution becomes roughly symmetric (Figure 1.22(b)).[37]

For symmetric distributions, the mean and standard deviation are particularly informative summaries. If a distribution is symmetric, approximately 70% of the data are within one standard deviation of the mean and 95% of the data are within two standard deviations of the mean; this guideline is known as the **empirical rule**.

EXAMPLE 1.13

On the log-transformed scale, mean log income is 8.50, with standard deviation 1.54. Apply the empirical rule to describe the distribution of average yearly per capita income among the 165 countries.

According to the empirical rule, the middle 70% of the data are within one standard deviation of the mean, in the range (8.50 - 1.54, 8.50 + 1.54) = (6.96, 10.04) log(USD). 95% of the data are within two standard deviations of the mean, in the range (8.50 - 2(1.54), 8.50 + 2(1.54)) = (5.42, 11.58) log(USD).

Undo the log transformation. The middle 70% of the data are within the range $(e^{6.96}, e^{10.04})$ = ($1,054, $22,925). The middle 95% of the data are within the range $(e^{5.42}, e^{11.58})$ = ($226, $106,937).

Functions other than the natural log can also be used to transform data, such as the square root and inverse.

---

[36] The data are available as wdi.2011 in the R package oibiostat.
[37] In statistics, the natural logarithm is usually written log. In other settings it is sometimes written as ln.

1.5 Categorical data

This section introduces tables and plots for summarizing categorical data, using the `famuss` dataset introduced in Section 1.2.2.

A table for a single variable is called a **frequency table**. Figure 1.23 is a frequency table for the `actn3.r577x` variable, showing the distribution of genotype at location r577x on the ACTN3 gene for the FAMuSS study participants.

In a **relative frequency table** like Figure 1.24, the proportions per each category are shown instead of the counts.

	CC	CT	TT	Sum
Counts	173	261	161	595

Figure 1.23: A frequency table for the `actn3.r577x` variable.

	CC	CT	TT	Sum
Proportions	0.291	0.439	0.271	1.000

Figure 1.24: A relative frequency table for the `actn3.r577x` variable.

A bar plot is a common way to display a single categorical variable. The left panel of Figure 1.25 shows a **bar plot** of the counts per genotype for the `actn3.r577x` variable. The plot in the right panel shows the proportion of observations that are in each level (i.e. in each genotype).

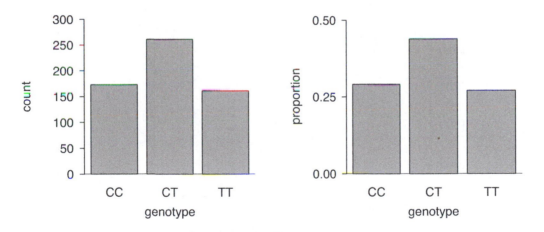

Figure 1.25: Two bar plots of `actn3.r577x`. The left panel shows the counts, and the right panel shows the proportions for each genotype.

1.6   Relationships between two variables

This section introduces numerical and graphical methods for exploring and summarizing relationships between two variables. Approaches vary depending on whether the two variables are both numerical, both categorical, or whether one is numerical and one is categorical.

1.6.1   Two numerical variables

Scatterplots

In the frog parental investment study, researchers used clutch volume as a primary variable of interest rather than egg size because clutch volume represents both the eggs and the protective gelatinous matrix surrounding the eggs. The larger the clutch volume, the higher the energy required to produce it; thus, higher clutch volume is indicative of increased maternal investment. Previous research has reported that larger body size allows females to produce larger clutches; is this idea supported by the frog data?

A **scatterplot** provides a case-by-case view of the relationship between two numerical variables. Figure 1.26 shows clutch volume plotted against body size, with clutch volume on the y-axis and body size on the x-axis. Each point represents a single case. For this example, each case is one egg clutch for which both volume and body size (of the female that produced the clutch) have been recorded.

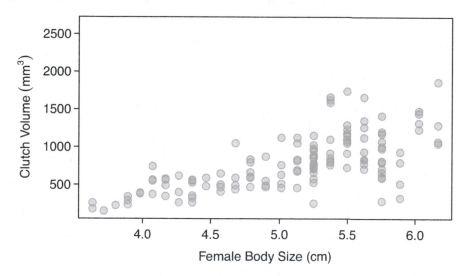

Figure 1.26: A scatterplot showing clutch.volume (vertical axis) vs. body.size (horizontal axis).

The plot shows a discernible pattern, which suggests an **association**, or relationship, between clutch volume and body size; the points tend to lie in a straight line, which is indicative of a **linear association**. Two variables are **positively associated** if increasing values of one tend to occur with increasing values of the other; two variables are **negatively associated** if increasing values of one variable occurs with decreasing values of the other. If there is no evident relationship between two variables, they are said to be **uncorrelated** or **independent**.

As expected, clutch volume and body size are positively associated; larger frogs tend to produce egg clutches with larger volumes. These observations suggest that larger females are capable of investing more energy into offspring production relative to smaller females.

The National Health and Nutrition Examination Survey (NHANES) consists of a set of surveys and measurements conducted by the US CDC to assess the health and nutritional status of adults and children in the United States. The following example uses data from a sample of 500 adults (individuals ages 21 and older) from the NHANES dataset.[38]

EXAMPLE 1.14

Body mass index (BMI) is a measure of weight commonly used by health agencies to assess whether someone is overweight, and is calculated from height and weight.[39] Describe the relationships shown in Figure 1.27. Why is it helpful to use BMI as a measure of obesity, rather than weight?

Figure 1.27(a) shows a positive association between height and weight; taller individuals tend to be heavier. Figure 1.27(b) shows that height and BMI do not seem to be associated; the range of BMI values observed is roughly consistent across height.

Weight itself is not a good measure of whether someone is overweight; instead, it is more reasonable to consider whether someone's weight is unusual relative to other individuals of a comparable height. An individual weighing 200 pounds who is 6 ft tall is not necessarily an unhealthy weight; however, someone who weighs 200 pounds and is 5 ft tall is likely overweight. It is not reasonable to classify individuals as overweight or obese based only on weight.

BMI acts as a relative measure of weight that accounts for height. Specifically, BMI is used as an estimate of body fat. According to US National Institutes of Health (US NIH) and the World Health Organization (WHO), a BMI between 25.0 - 29.9 is considered overweight and a BMI over 30 is considered obese.[40]

(a)   (b)

Figure 1.27: (a) A scatterplot showing height versus weight from the 500 individuals in the sample from NHANES. One participant 163.9 cm tall (about 5 ft, 4 in) and weighing 144.6 kg (about 319 lb) is highlighted. (b) A scatterplot showing height versus BMI from the 500 individuals in the sample from NHANES. The same individual highlighted in (a) is marked here, with BMI 53.83.

---

[38]The sample is available as nhanes.samp.adult.500 in the R oibiostat package.

[39]$BMI = \dfrac{weight_{kg}}{height_m^2} = \dfrac{weight_{lb}}{height_{in}^2} \times 703.$

[40]https://www.nhlbi.nih.gov/health/educational/lose_wt/risk.htm

EXAMPLE 1.15

Figure 1.28 is a scatterplot of life expectancy versus annual per capita income for 165 countries in 2011. Life expectancy is measured as the expected lifespan for children born in 2011 and income is adjusted for purchasing power in a country. Describe the relationship between life expectancy and annual per capita income; do they seem to be linearly associated?

Life expectancy and annual per capita income are positively associated; higher per capita income is associated with longer life expectancy. However, the two variables are not linearly associated. When income is low, small increases in per capita income are associated with relatively large increases in life expectancy. However, once per capita income exceeds approximately $20,000 per year, increases in income are associated with smaller gains in life expectancy.

In a linear association, change in the y-variable for every unit of the x-variable is consistent across the range of the x-variable; for example, a linear association would be present if an increase in income of $10,000 corresponded to an increase in life expectancy of 5 years, across the range of income.

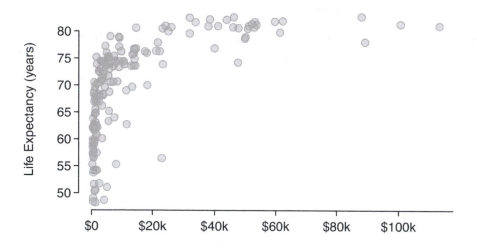

Figure 1.28: A scatterplot of life expectancy (years) versus annual per capita income (US dollars) in the wdi.2011 dataset.

Correlation

Correlation is a numerical summary statistic that measures the strength of a linear relationship between two variables. It is denoted by r, the **correlation coefficient**, which takes on values between -1 and 1.

r
correlation
coefficient

If the paired values of two variables lie exactly on a line, $r = \pm 1$; the closer the correlation coefficient is to ± 1, the stronger the linear association. When two variables are positively associated, with paired values that tend to lie on a line with positive slope, $r > 0$. If two variables are negatively associated, $r < 0$. A value of r that is 0 or approximately 0 indicates no apparent association between two variables.[41]

---

[41] If paired values lie perfectly on either a horizontal or vertical line, there is no association and r is mathematically undefined.

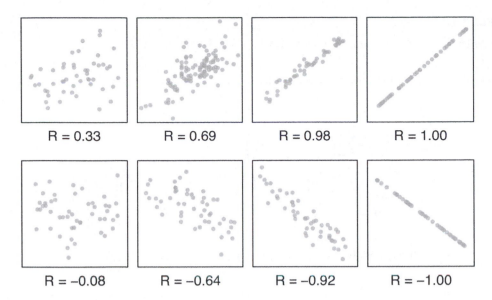

Figure 1.29: Scatterplots and their correlation coefficients. The first row shows positive associations and the second row shows negative associations. From left to right, strength of the linear association between x and y increases.

The correlation coefficient quantifies the strength of a linear trend. Prior to calculating a correlation, it is advisable to confirm that the data exhibit a linear relationship. Although it is mathematically possible to calculate correlation for any set of paired observations, such as the life expectancy versus income data in Figure 1.28, correlation cannot be used to assess the strength of a nonlinear relationship.

CORRELATION

The correlation between two variables x and y is given by:

$$r = \frac{1}{n-1} \sum_{i=1}^{n} \left(\frac{x_i - \overline{x}}{s_x} \right) \left(\frac{y_i - \overline{y}}{s_y} \right), \tag{1.16}$$

where $(x_1, y_1), (x_2, y_2), \ldots, (x_n, y_n)$ are the n paired values of x and y, and s_x and s_y are the sample standard deviations of the x and y variables, respectively.

EXAMPLE 1.17

Calculate the correlation coefficient of x and y, plotted in Figure 1.30.

Calculate the mean and standard deviation for x and y: $\bar{x} = 2, \bar{y} = 3, s_x = 1$, and $s_y = 2.65$.

$$
\begin{aligned}
r &= \frac{1}{n-1} \sum_{i=1}^{n} \left(\frac{x_i - \bar{x}}{s_x} \right) \left(\frac{y_i - \bar{y}}{s_y} \right) \\
&= \frac{1}{3-1} \left[\left(\frac{1-2}{1} \right) \left(\frac{5-3}{2.65} \right) + \left(\frac{2-2}{1} \right) \left(\frac{4-3}{2.65} \right) + \left(\frac{3-2}{1} \right) \left(\frac{0-3}{2.65} \right) \right] \\
&= -0.94.
\end{aligned}
$$

The correlation is -0.94, which reflects the negative association visible from the scatterplot in Figure 1.30.

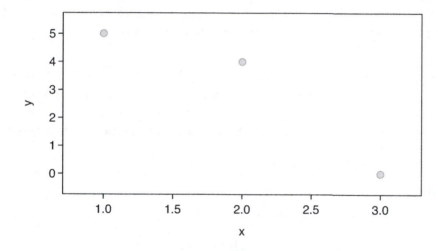

Figure 1.30: A scatterplot showing three points: (1, 5), (2, 4), and (3, 0).

EXAMPLE 1.18

Is it appropriate to use correlation as a numerical summary for the relationship between life expectancy and income after a log transformation is applied to both variables? Refer to Figure 1.31.

Figure 1.31 shows an approximately linear relationship; a correlation coefficient is a reasonable numerical summary of the relationship. As calculated from statistical software, $r = 0.79$, which is indicative of a strong linear relationship.

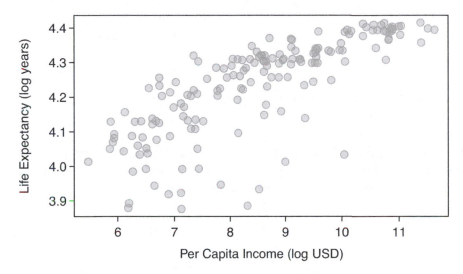

Figure 1.31: A scatterplot showing `log(income)` (horizontal axis) vs. `log(life.expectancy)` (vertical axis).

1.6.2 Two categorical variables

Contingency tables

A **contingency table** summarizes data for two categorical variables, with each value in the table representing the number of times a particular combination of outcomes occurs.[42] Figure 1.32 summarizes the relationship between race and genotype in the famuss data.

The **row totals** provide the total counts across each row and the **column totals** are the total counts for each column; collectively, these are the **marginal totals**.

	CC	CT	TT	Sum
African Am	16	6	5	27
Asian	21	18	16	55
Caucasian	125	216	126	467
Hispanic	4	10	9	23
Other	7	11	5	23
Sum	173	261	161	595

Figure 1.32: A contingency table for race and `actn3.r577x`.

---

[42]Contingency tables are also known as **two-way tables**.

Like relative frequency tables for the distribution of one categorical variable, contingency tables can also be converted to show proportions. Since there are two variables, it is necessary to specify whether the proportions are calculated according to the row variable or the column variable.

Figure 1.33 shows the row proportions for Figure 1.32; these proportions indicate how genotypes are distributed within each race. For example, the value of 0.593 in the upper left corner indicates that of the African Americans in the study, 59.3% have the CC genotype.

	CC	CT	TT	Sum
African Am	0.593	0.222	0.185	1.000
Asian	0.382	0.327	0.291	1.000
Caucasian	0.268	0.463	0.270	1.000
Hispanic	0.174	0.435	0.391	1.000
Other	0.304	0.478	0.217	1.000

Figure 1.33: A contingency table with row proportions for the race and actn3.r577x variables.

Figure 1.34 shows the column proportions for Figure 1.32; these proportions indicate the distribution of races within each genotype category. For example, the value of 0.092 indicates that of the CC individuals in the study, 9.2% are African American.

	CC	CT	TT
African Am	0.092	0.023	0.031
Asian	0.121	0.069	0.099
Caucasian	0.723	0.828	0.783
Hispanic	0.023	0.038	0.056
Other	0.040	0.042	0.031
Sum	1.000	1.000	1.000

Figure 1.34: A contingency table with column proportions for the race and actn3.r577x variables.

EXAMPLE 1.19

For African Americans in the study, CC is the most common genotype and TT is the least common genotype. Does this pattern hold for the other races in the study? Do the observations from the study suggest that distribution of genotypes at r577x vary between populations?

The pattern holds for Asians, but not for other races. For the Caucasian individuals sampled in the study, CT is the most common genotype at 46.3%. CC is the most common genotype for Asians, but in this population, genotypes are more evenly distributed: 38.2% of Asians sampled are CC, 32.7% are CT, and 29.1% are TT. The distribution of genotypes at r577x seems to vary by population.

GUIDED PRACTICE 1.20

As shown in Figure 1.34, 72.3% of CC individuals in the study are Caucasian. Do these data suggest that in the general population, people of CC genotype are highly likely to be Caucasian?[43]

---

[43] No, this is not a reasonable conclusion to draw from the data. The high proportion of Caucasians among CC individuals primarily reflects the large number of Caucasians sampled in the study – 78.5% of the people sampled are Caucasian. The uneven representation of different races is one limitation of the famuss data.

Segmented bar plots

A **segmented bar plot** is a way of visualizing the information from a contingency table. Figure 1.35 graphically displays the data from Figure 1.32; each bar represents a level of actn3.r577x and is divided by the levels of race. Figure 1.35(b) uses the row proportions to create a standardized segmented bar plot.

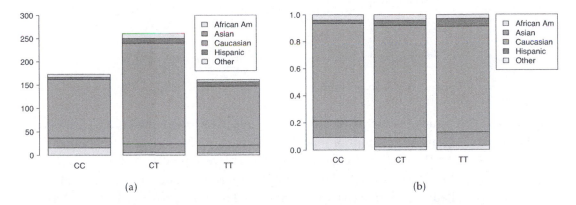

Figure 1.35: (a) Segmented bar plot for individuals by genotype, with bars divided by race. (b) Standardized version of Figure (a).

Alternatively, the data can be organized as shown in Figure 1.36, with each bar representing a level of race. The standardized plot is particularly useful in this case, presenting the distribution of genotypes within each race more clearly than in Figure 1.36(a).

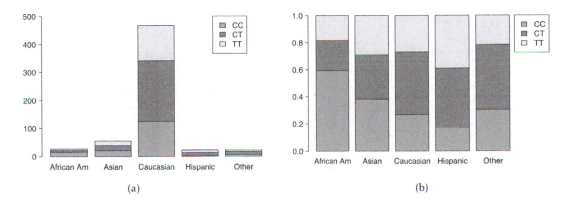

Figure 1.36: (a) Segmented bar plot for individuals by race, with bars divided by genotype. (b) Standardized version of Figure (a).

Two-by-two tables: relative risk

The results from medical studies are often presented in **two-by-two tables** (2×2 tables), contingency tables for categorical variables that have two levels. One of the variables defines two groups of participants, while the other represents the two possible outcomes. Figure 1.37 shows a hypothetical two-by-two table of outcome by group.

	Outcome A	Outcome B	Sum
Group 1	a	b	$a + b$
Group 2	c	d	$c + d$
Sum	$a + c$	$b + d$	$a + b + c + d = n$

Figure 1.37: A hypothetical two-by-two table of outcome by group.

In the LEAP study, participants are divided into two groups based on treatment (peanut avoidance versus peanut consumption), while the outcome variable records whether an individual passed or failed the oral food challenge (OFC). The results of the LEAP study as shown in Figure 1.2 are in the form of a 2×2 table; the table is reproduced below as Figure 1.38.

A statistic called the **relative risk** (RR) can be used to summarize the data in a 2×2 table; the relative risk is a measure of the risk of a certain event occurring in one group relative to the risk of the event occurring in another group.[44]

	FAIL OFC	PASS OFC	Sum
Peanut Avoidance	36	227	263
Peanut Consumption	5	262	267
Sum	41	489	530

Figure 1.38: Results of the LEAP study, described in Section 1.1.

The question of interest in the LEAP study is whether the risk of developing peanut allergy (i.e., failing the OFC) differs between the peanut avoidance and consumption groups. The relative risk of failing the OFC equals the ratio of the proportion of individuals in the avoidance group who failed the OFC to the proportion of individuals in the consumption group who failed the OFC.

EXAMPLE 1.21

Using the results from the LEAP study, calculate and interpret the relative risk of failing the oral food challenge, comparing individuals in the avoidance group to individuals in the consumption group.

$$RR_{\text{failing OFC}} = \frac{\text{proportion in avoidance group who failed OFC}}{\text{proportion in consumption group who failed OFC}} = \frac{36/263}{5/267} = 7.31.$$

The relative risk is 7.31. The risk of failing the oral food challenge was more than 7 times greater for participants in the peanut avoidance group than for those in the peanut consumption group.

---

[44]Chapter 8 discusses another numerical summary for 2×2 tables, the **odds ratio**.

EXAMPLE 1.22

An observational study is conducted to assess the association between smoking and cardiovascular disease (CVD), in which researchers identified a cohort of individuals and categorized them according to smoking and disease status. If the relative risk of CVD is calculated as the ratio of the proportion of smokers with CVD to the proportion of non-smokers with CVD, interpret the results of the study if the relative risk equals 1, is less than 1, or greater than 1.

A relative risk of 1 indicates that the risk of CVD is equal for smokers and non-smokers.

A relative risk less than 1 indicates that smokers are at a lower risk of CVD than non-smokers; i.e., the proportion of individuals with CVD among smokers is lower than the proportion among non-smokers.

A relative risk greater than 1 indicates that smokers are at a higher risk of CVD than non-smokers; i.e., the proportion of individuals with CVD among smokers is higher than the proportion among non-smokers.

GUIDED PRACTICE 1.23

For the study described in Example 1.22, suppose that of the 231 individuals, 111 are smokers. 40 smokers and 32 non-smokers have cardiovascular disease. Calculate and interpret the relative risk of CVD.[45]

Relative risk relies on the assumption that the observed proportions of an event occurring in each group are representative of the risk, or incidence, of the event occurring within the populations from which the groups are sampled. For example, in the LEAP data, the relative risk assumes that the proportions 33/263 and 5/267 are estimates of the proportion of individuals who would fail the OFC among the larger population of infants who avoid or consume peanut products.

EXAMPLE 1.24

Suppose another study to examine the association between smoking and cardiovascular disease is conducted, but researchers use a different study design than described in Example 1.22. For the new study, 90 individuals with CVD and 110 individuals without CVD are recruited. 40 of the individuals with CVD are smokers, and 80 of the individuals without CVD are non-smokers. Should relative risk be used to summarize the observations from the new study?

Relative risk should not be calculated for these observations. Since the number of individuals with and without CVD is fixed by the study design, the proportion of individuals with CVD within a certain group (smokers or non-smokers) as calculated from the data is not a measure of CVD risk for that population.

---

[45]The relative risk of CVD, comparing smokers to non-smokers, is $(40/111)/(32/120) = 1.35$. Smoking is associated with a 35% increase in the probability of CVD; in other words, the risk of CVD is 35% greater in smokers compared to non-smokers.

GUIDED PRACTICE 1.25

For a study examining the association between tea consumption and esophageal carcinoma, researchers recruited 300 patients with carcinoma and 571 without carcinoma and administered a questionnaire about tea drinking habits.[46] Of the 47 individuals who reported that they regularly drink green tea, 17 had carcinoma. Of the 824 individuals who reported that they never, or very rarely, drink green tea, 283 had carcinoma. Evaluate whether the proportions 17/47 and 283/824 are representative of the incidence rate of carcinoma among individuals who drink green tea regularly and those who do not.[47]

RELATIVE RISK

The relative risk of Outcome A in the hypothetical two-by-two table (Figure 1.37) can be calculated using either Group 1 or Group 2 as the reference group:

$$RR_{A, \text{ comparing Group 1 to Group 2}} = \frac{a/(a+b)}{c/(c+d)}$$

$$RR_{A, \text{ comparing Group 2 to Group 1}} = \frac{c/(c+d)}{a/(a+b)}$$

The relative risk should only be calculated for data where the proportions $a/(a+b)$ and $c/(c+d)$ represent the incidence of Outcome A within the populations from which Groups 1 and 2 are sampled.

1.6.3   A numerical variable and a categorical variable

Methods for comparing numerical data across groups are based on the approaches introduced in Section 1.4. **Side-by-side boxplots** and **hollow histograms** are useful for directly comparing how the distribution of a numerical variable differs by category.

Recall the question introduced in Section 1.2.3: is ACTN3 genotype associated with variation in muscle function? Figure 1.39 visually shows the relationship between muscle function (measured as percent change in non-dominant arm strength) and ACTN3 genotype in the famuss data with side-by-side boxplots and hollow histograms. The hollow histograms highlight how the shapes of the distributions of ndrm.ch for each genotype are essentially similar, although the distribution for the CC genotype has less right skewing. The side-by-side boxplots are especially useful for comparing center and spread, and reveal that the T allele appears to be associated with greater muscle function; median percent change in non-dominant arm strength increases across the levels from CC to TT.

GUIDED PRACTICE 1.26

Using Figure 1.40, assess how maternal investment varies with altitude.[48]

---

[46]Tea drinking habits and oesophageal cancer in a high risk area in northern Iran: population based casecontrol study, Islami F, et al., BMJ (2009), doi 10.1136/bmj.b929

[47]The proportions calculated from the study data should not be used as estimates of the incidence rate of esophageal carcinoma among individuals who drink green tea regularly and those who do not, since the study selected participants based on carcinoma status.

[48]As a general rule, clutches found at higher altitudes have greater volume; median clutch volume tends to increase as altitude increases. This suggests that increased altitude is associated with a higher level of maternal investment.

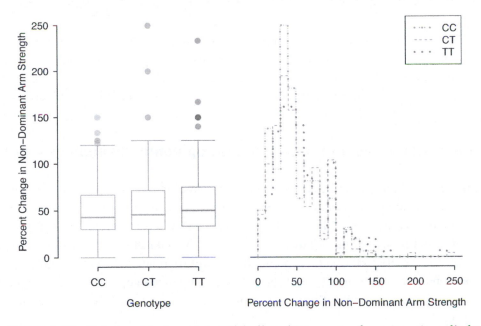

Figure 1.39: Side-by-side boxplot and hollow histograms for ndrm.ch, split by levels of actn3.r577x.

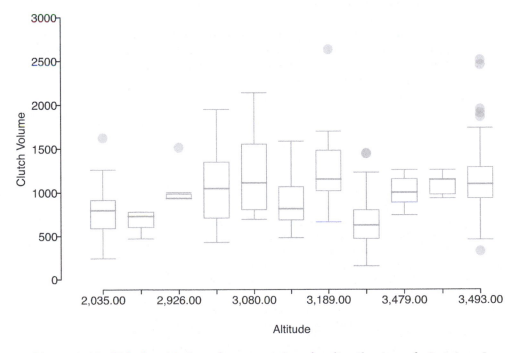

Figure 1.40: Side-by-side boxplot comparing the distribution of clutch.volume for different altitudes.

1.7   Exploratory data analysis

The simple techniques for summarizing and visualizing data that have been introduced in this chapter may not seem especially powerful, but when applied in practice, they can be instrumental for gaining insight into the interesting features of a dataset. This section provides three examples of data-driven research questions that can be investigated through exploratory data analysis.

1.7.1   Case study: discrimination in developmental disability support

In the United States, individuals with developmental disabilities typically receive services and support from state governments. The State of California allocates funds to developmentally-disabled residents through the California Department of Developmental Services (DDS); individuals receiving DDS funds are referred to as 'consumers'. The dataset dds.discr represents a sample of 1,000 DDS consumers (out of a total population of approximately 250,000), and includes information about age, gender, ethnicity, and the amount of financial support per consumer provided by the DDS.[49] Figure 1.41 shows the first five rows of the dataset, and the variables are described in Figure 1.42.

A team of researchers examined the mean annual expenditures on consumers by ethnicity, and found that the mean annual expenditures on Hispanic consumers was approximately one-third of the mean expenditures on White non-Hispanic consumers. As a result, an allegation of ethnic discrimination was brought against the California DDS.

Does this finding represent sufficient evidence of ethnic discrimination, or might there be more to the story? This section will illustrate the process behind conducting an exploratory analysis that not only investigates the relationship between two variables of interest, but also considers whether other variables might be influencing that relationship.

	id	age.cohort	age	gender	expenditures	ethnicity
1	10210	13-17	17	Female	2113	White not Hispanic
2	10409	22-50	37	Male	41924	White not Hispanic
3	10486	0-5	3	Male	1454	Hispanic
4	10538	18-21	19	Female	6400	Hispanic
5	10568	13-17	13	Male	4412	White not Hispanic

Figure 1.41: Five rows from the dds.discr data matrix.

variable	description
id	Unique identification code for each resident
age.cohort	Age as sorted into six groups, 0-5 years, 6-12 years, 13-17 years, 18-21 years, 22-50 years, and 51+ years
age	Age, measured in years
gender	Gender, either Female or Male
expenditures	Amount of expenditures spent by the State on an individual annually, measured in USD
ethnicity	Ethnic group, recorded as either American Indian, Asian, Black, Hispanic, Multi Race, Native Hawaiian, Other, or White Not Hispanic

Figure 1.42: Variables and their descriptions for the dds.discr dataset.

---

[49]The dataset is based on actual attributes of consumers, but has been altered to maintain consumer privacy.

Distributions of single variables

To begin understanding a dataset, start by examining the distributions of single variables using numerical and graphical summaries. This process is essential for developing a sense of context; in this case, examining variables individually addresses questions such as "What is the range of annual expenditures?", "Do consumers tend to be older or younger?", and "Are there more consumers from one ethnic group versus another?".

Figure 1.43 illustrates the right skew of expenditures, indicating that for the majority of consumers, expenditures are relatively low; most are within the $0 - $5,000 range. There are some consumers for which expenditures are much higher, such as within the $60,000 - $80,000 range. Precise numerical summaries can be calculated using statistical software: the quartiles for expenditures are $2,899, $7,026, and $37,710.

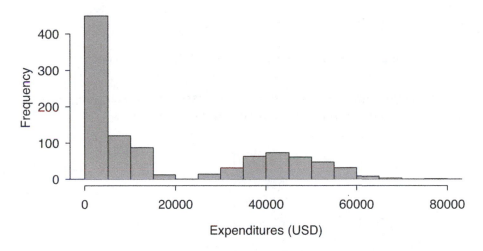

Figure 1.43: A histogram of expenditures.

A consumer's age is directly recorded as the variable age; in the age.cohort variable, consumers are assigned to one of six age cohorts. The cohorts are indicative of particular life phases. In the first three cohorts, consumers are still living with their parents as they move through preschool age, elementary/middle school age, and high school age. In the 18-21 cohort, consumers are transitioning from their parents' homes to living on their own or in supportive group homes. From ages 22-50, individuals are mostly no longer living with their parents but may still receive some support from family. In the 51+ cohort, consumers often have no living parents and typically require the most amount of support.

Figure 1.44 reveals the right-skewing of age. Most consumers are younger than 30. The plot in Figure 1.44(b) graphically shows the number of individuals in each age cohort. There are approximately 200 individuals in each of the middle four cohorts, while there are about 100 individuals in the other two cohorts.

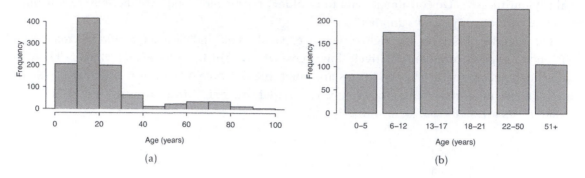

(a)   (b)

Figure 1.44: (a) Histogram of age. (b) Plot of age.cohort.

There are eight ethnic groups represented in dds.discr. The two largest groups, Hispanic and White non-Hispanic, together represent about 80% of the consumers.

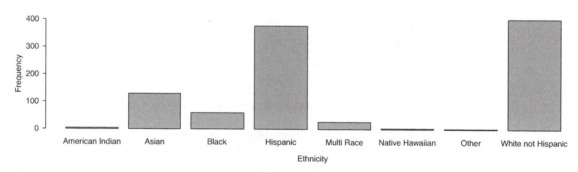

Figure 1.45: A plot of ethnicity.

GUIDED PRACTICE 1.27

Using Figure 1.46, does gender appear to be balanced in the dds.discr dataset?[50]

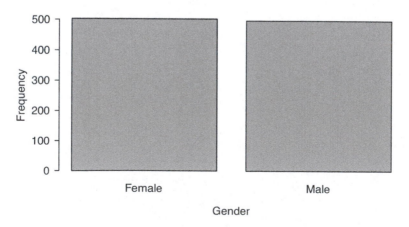

Figure 1.46: A plot of gender.

---

[50]Yes, approximately half of the individuals are female and half are male.

Relationships between two variables

After examining variables individually, explore how variables are related to each other. While there exist methods for summarizing more than two variables simultaneously, focusing on two variables at a time can be surprisingly effective for making sense of a dataset. It is useful to begin by investigating the relationships between the primary response variable of interest and the exploratory variables. In this case study, the response variable is expenditures, the amount of funds the DDS allocates annually to each consumer. How does expenditures vary by age, ethnicity, and gender?

Figure 1.47 shows a side-by-side boxplot of expenditures by age cohort. There is a clear upward trend, in which older individuals tend to receive more DDS funds. This reflects the underlying context of the data. The purpose of providing funds to developmentally disabled individuals is to help them maintain a quality of life similar to those without disabilities; as individuals age, it is expected their financial needs will increase. Some of the observed variation in expenditures can be attributed to the fact that the dataset includes a wide range of ages. If the data included only individuals in one cohort, such as the 22-50 cohort, the distribution of expenditures would be less variable, and range between $30,000 and $60,000 instead of from $0 and $80,000.

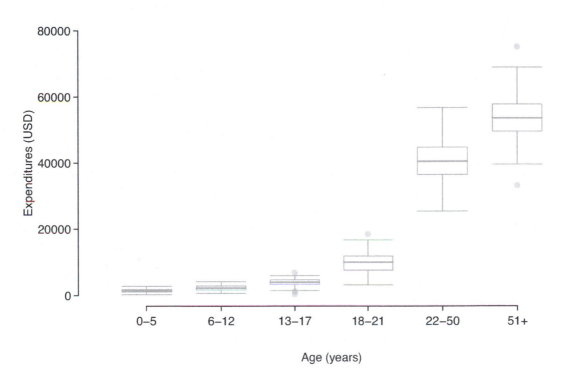

Figure 1.47: A plot of expenditures by age.cohort.

How does the distribution of expenditures vary by ethnic group? Does there seem to be a difference in the amount of funding that a person receives, on average, between different ethnicities? A side-by-side boxplot of expenditures by ethnicity (Figure 1.48) reveals that the distribution of expenditures is quite different between ethnic groups. For example, there is very little variation in expenditures for the Multi Race, Native Hawaiian, and Other groups. Additionally, the median expenditures are not the same between groups; the medians for American Indian and Native Hawaiian individuals are about $40,000, as compared to medians of approximately $10,000 for Asian and Black consumers.

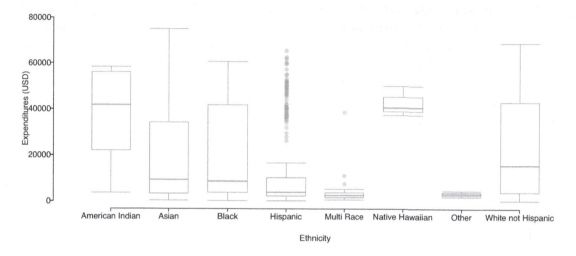

Figure 1.48: A plot of expenditures by ethnicity.

The trend visible in Figure 1.48 seems potentially indicative of ethnic discrimination. Before proceeding with the analysis, however, it is important to take into account the fact that two of the groups, Hispanic and White non-Hispanic, comprise the majority of the data; some ethnic groups represent less than 10% of the observations (Figure 1.45). For ethnic groups with relatively small sample sizes, it is possible that the observed samples are not representative of the larger populations. The rest of this analysis will focus on comparing how expenditures varies between the two largest groups, White non-Hispanic and Hispanic.

GUIDED PRACTICE 1.28

Using Figure 1.49, do annual expenditures seem to vary by gender?[51]

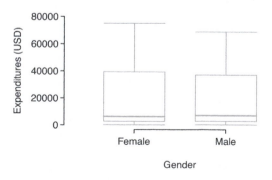

Figure 1.49: A plot of expenditures by gender.

---

[51]No, the distribution of expenditures within males and females is very similar; both are right skewed, with approximately equal median and interquartile range.

Figure 1.50 compares the distribution of expenditures between Hispanic and White non-Hispanic consumers. Most Hispanic consumers receive between about $0 to $20,000 from the California DDS; individuals receiving amounts higher than this are upper outliers. However, for White non-Hispanic consumers, median expenditures is at $20,000, and the middle 50% of consumers receive between $5,000 and $40,000. The precise summary statistics can be calculated from computing software, as shown in the corresponding R lab. The mean expenditures for Hispanic consumers is $11,066, while the mean expenditures for White non-Hispanic consumers is over twice as large at $24,698. On average, a Hispanic consumer receives less financial support from the California DDS than a White non-Hispanic consumer. Does this represent evidence of discrimination?

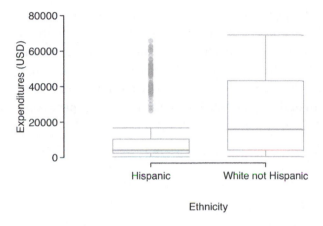

Figure 1.50: A plot of expenditures by ethnicity, showing only Hispanics and White Non-Hispanics.

Recall that expenditures is strongly associated with age—older individuals tend to receive more financial support. Is there also an association between age and ethnicity, for these two ethnic groups? When using data to investigate a question, it is important to explore not only how explanatory variables are related to the response variable(s), but also how explanatory variables influence each other.

Figures 1.51 and 1.52 show the distribution of age within Hispanics and White non-Hispanics. Hispanics tend to be younger, with most Hispanic consumers falling into the 6-12, 13-17, and 18-21 age cohorts. In contrast, White non-Hispanics tend to be older; most consumers in this group are in the 22-50 age cohort, and relatively more White non-Hispanic consumers are in the 51+ age cohort as compared to Hispanics.

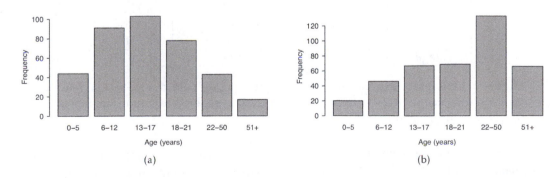

Figure 1.51: (a) Plot of age.cohort within Hispanics. (b) Plot of age.cohort within White non-Hispanics.

Age Cohort	Hispanic	White Non-Hispanic
0-5	44/376 = 12%	20/401 = 5%
6-12	91/376 = 24%	46/401 = 11%
13-17	103/376 = 27%	67/401 = 17%
18-21	78/376 = 21%	69/401 = 17%
22-50	43/376 = 11%	133/401 = 33%
51+	17/376 = 5%	66/401 = 16%
Sum	376/376 = 100%	401/401 = 100%

Figure 1.52: Consumers by ethnicity and age cohort, shown both as counts and proportions.

Recall that a confounding variable is a variable that is associated with the response variable and the explanatory variable under consideration; confounding was initially introduced in the context of sunscreen use and incidence of skin cancer, where sun exposure is a confounder. In this setting, age is a confounder for the relationship between expenditures and ethnicity. Just as it would be incorrect to claim that sunscreen causes skin cancer, it is essential here to recognize that there is more to the story than the apparent association between expenditures and ethnicity.

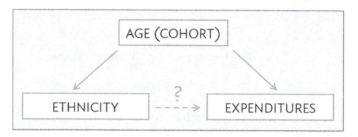

For a closer look at the relationship between age, ethnicity, and expenditures, subset the data further to compare how expenditures differs by ethnicity within each age cohort. If age is indeed the primary source of the observed variation in expenditures, then there should be little difference in average expenditures between individuals in different ethnic groups but the same age cohort.

Figure 1.53 shows the average expenditures within each age cohort, for Hispanics versus White non-Hispanics. The last column contains the difference between the two averages (calculated as White Non-Hispanics average - Hispanics average).

Age Cohort	Hispanics	White non-Hispanics	Difference
0-5	1,393	1,367	-26
6-12	2,312	2,052	-260
13-17	3,955	3,904	-51
18-21	9,960	10,133	173
22-50	40,924	40,188	-736
51+	55,585	52,670	-2915
Average	11,066	24,698	13,632

Figure 1.53: Average expenditures by ethnicity and age cohort, in USD ($). For all age cohorts except 18-21 years, average expenditures for White non-Hispanics is lower than for Hispanics.

When expenditures is compared within age cohorts, there are not large differences between mean expenditures for White non-Hispanics versus Hispanics. Comparing individuals of similar ages reveals that the association between ethnicity and expenditures is not nearly as strong as it seemed from the initial comparison of overall averages.

Instead, it is the difference in age distributions of the two populations that is driving the observed discrepancy in expenditures. The overall average of expenditures for the Hispanic consumers is lower because the population of Hispanic consumers is relatively young compared to the population of White non-Hispanic consumers, and the amount of expenditures for younger consumers tends to be lower than for older consumers. Based on an exploratory analysis that accounts for age as a confounding variable, there does not seem to be evidence of ethnic discrimination.

Identifying confounding variables is essential for understanding data. Confounders are often context-specific; for example, age is not necessarily a confounder for the relationship between ethnicity and expenditures in a different population. Additionally, it is rarely immediately obvious which variables in a dataset are confounders; looking for confounding variables is an integral part of exploring a dataset.

Chapter 7 introduces multiple linear regression, a method that can directly summarize the relationship between ethnicity, expenditures, and age, in addition to the tools for evaluating whether the observed discrepancies within age cohorts are greater than would be expected by chance variation alone.

Simpson's paradox

These data represent an extreme example of confounding known as **Simpson's paradox**, in which an association observed in several groups may disappear or reverse direction once the groups are combined. In other words, an association between two variables X and Y may disappear or reverse direction once data are partitioned into subpopulations based on a third variable Z (i.e., a confounding variable).

Figure 1.53 shows how mean expenditures is higher for Hispanics than White non-Hispanics in all age cohorts except one. Yet, once all the data are aggregated, the average expenditures for White non-Hispanics is over twice as large as the average for Hispanics. The paradox can be explored from a mathematical perspective by using weighted averages, where the average expenditure for each cohort is weighted by the proportion of the population in that cohort.

EXAMPLE 1.29

Using the proportions in Figure 1.52 and the average expenditures for each cohort in Figure 1.53, calculate the overall weighted average expenditures for Hispanics and for White non-Hispanics.[52]

For Hispanics:

$$1,393(.12) + 2,312(.24) + 3,955(.27) + 9,960(.21) + 40,924(.11) + 55,585(.05) = \$11,162.$$

For White non-Hispanics:

$$1,367(0.05) + 2,052(.11) + 3,904(.17) + 10,133(.17) + 40,188(.33) + 52,760(.16) = \$24,384.$$

The weights for the youngest four cohorts, which have lower expenditures, are higher for the Hispanic population than the White non-Hispanic population; additionally, the weights for the oldest two cohorts, which have higher expenditures, are higher for the White non-Hispanic population. This leads to overall average expenditures for the White non-Hispanics being higher than for Hispanics.

---

[52]Due to rounding, the overall averages calculated via this method will not exactly equal \$11,066 and \$24,698.

1.7.2   Case study: molecular cancer classification

The genetic code stored in DNA contains the necessary information for producing the proteins that ultimately determine an organism's observable traits (phenotype). Although nearly every cell in an organism contains the same genes, cells may exhibit different patterns of gene expression. Not only can genes be switched on or off in certain tissues, but they can also be expressed at varying levels. These variations in gene expression underlie the wide range of physical, biochemical, and developmental differences that characterize specific cells and tissues.

Originally, scientists were limited to monitoring the expression of only a single gene at a time. The development of microarray technology in the 1990's made it possible to examine the expression of thousands of genes simultaneously. While newer genomic technologies have started to replace microarrays for gene expression studies, microarrays continue to remain clinically relevant as a tool for genetic diagnosis. For example, a 2002 study examined the effectiveness of gene expression profiling as a tool for predicting disease outcome in breast cancer patients, reporting that the expression data from 70 genes constituted a more powerful predictor of survival than standard systems based on clinical criteria.[53]

This section introduces the principles behind DNA microarrays and discusses the 1999 Golub leukemia study, which represents one of the earliest applications of microarray technology for diagnostic purposes.

DNA microarrays

Microarray technology is based on hybridization, a basic property of nucleic acids in which complementary nucleotide sequences specifically bind together. Each microarray consists of a glass or silicon slide dotted with a grid of short (25-40 base pairs long), single-stranded DNA fragments, known as probes. The probes in a single spot are present in millions of copies, and optimized to uniquely correspond to a gene.

To measure the gene expression profile of a sample, mRNA is extracted from the sample and converted into complementary-DNA (cDNA). The cDNA is then labeled with a fluorescent dye and added to a microarray. When cDNA from the sample encounters complementary DNA probes, the two strands will hybridize, allowing the cDNA to adhere to specific spots on the slide. Once the chip is illuminated (to activate the fluorescence) and scanned, the intensity of fluorescence detected at each spot corresponds to the amount of bound cDNA.

Microarrays are commonly used to compare gene expression between an experimental sample and a reference sample. Suppose that the reference sample is taken from healthy cells and the experimental sample from cancer cells. First, the cDNA from the samples are differentially labeled, such as green dye for the healthy cells and red dye for the cancer cells. The samples are then mixed together and allowed to bind to the slide. If the expression of a particular gene is higher in the experimental sample than in the reference sample, then the corresponding spot on the microarray will appear red. In contrast, the spot will appear green if expression in the experimental sample is lower than in the reference sample. Equal expression levels result in a yellow spot, while no expression in either sample shows as a black dot. The fluorescence intensity data provide a relative measure of gene expression, showing which genes on the chip seem to be more or less active in relation to each other.

---

[53]van de Vijver MJ, He YD, van't Veer LJ, et al. A gene-expression sign as a predictor of survival in breast cancer. *New England Journal of Medicine* 2002;347:1999-2009.

The raw data produced by a microarray is messy, due to factors such as imperfections during chip manufacturing or unpredictable probe behavior. It is also possible for inaccuracies to be introduced from cDNA binding to probes that are not precise sequence matches; this nonspecific binding will contribute to observed intensity, but not reflect the expression level of a gene. Methods to improve microarray accuracy by reducing the frequency of nonspecific binding include using longer probes or multiple probes per gene that correspond to different regions of the gene sequence.[54] The Affymetrix company developed a different strategy involving the use of probe pairs; one set of probes are a perfect match to the gene sequence (PM probes), while the mismatch probes contain a single base difference in the middle of the sequence (MM probes). The MM probes act as a control for any cDNA that exhibit nonspecific binding; subtracting the MM probe intensity from the PM intensity (PM - MM) provides a more accurate measure of fluorescence produced by specific hybridization.

Considerable research has been done to develop methods for pre-processing microarray data to adjust for various errors and produce data that can be analyzed. When analyzing "cleaned" data from any experiment, it is important to be aware that the reliability of any conclusions drawn from the data depends, to a large extent, on the care that has been taken in collecting and processing the data.

Golub leukemia study

Accurate cancer classification is critical for determining an appropriate course of therapy. The chemotherapy regimens for acute leukemias differs based on whether the leukemia affects blood-forming cells (acute myeloid leukemia, AML) or white blood cells (acute lymphoblastic leukemia, ALL). At the time of the Golub study, no single diagnostic test was sufficient for distinguishing between AML and ALL. To investigate whether gene expression profiling could be a tool for classifying acute leukemia type, Golub and co-authors used Affymetrix DNA microarrays to measure the expression level of 7,129 genes from children known to have either AML or ALL.[55]

The original data (after some initial pre-processing) are available from the Broad Institute.[56] The version of the data presented in this text have undergone further processing; the expression levels have been normalized to adjust for the variability between the separate arrays used for each sampled individual.[57] Figure 1.54 describes the variables in the first six columns of the Golub data. The last 7,129 columns of the dataset contain the expression data for the genes examined in the study; each column is named after the probe corresponding to a specific gene.

variable	description
Samples	Sample number; unique to each patient.
BM.PB	Type of patient material. BM for bone marrow; PB for peripheral blood.
Gender	F for female, M for male.
Source	Hospital where the patient was treated.
tissue.mf	Combination of BM.PB and Gender
cancer	Leukemia type; aml is acute myeloid leukemia, allB is acute lymphoblastic leukemia with B-cell origin, and allT is acute lymphoblastic leukemia with T-cell origin.

Figure 1.54: Variables and their descriptions for the patient descriptors in Golub dataset.

[54]Chou, C.C. et al. Optimization of probe length and the number of probes per gene for optimal microarray analysis of gene expression. *Nucleic Acids Research* 2004; **32**: e99.

[55]Golub, Todd R., et al. Molecular classification of cancer: class discovery and class prediction by gene expression monitoring. Science 286 (1999): 531-537.

[56]http://www-genome.wi.mit.edu/mpr/data_set_ALL_AML.html

[57]John Maindonald, W. John Braun. *Data Analysis and Graphics using R: An Example-Based Approach.*

Figure 1.55 shows five rows and seven columns from the dataset. Each row corresponds to a patient. These five patients were all treated at the Dana Farber Cancer Institute (DFCI) (Source) for ALL with B-cell origin (cancer), and samples were taken from bone marrow (BM.PB). Four of the patients were female and one was male (Gender). The last row in the table shows the normalized gene expression level for the gene corresponding to the probe AFFX.BioB.5.at.

Samples	BM.PB	Gender	Source	tissue.mf	cancer	AFFX-BioB-5_at
39	BM	F	DFCI	BM:f	allB	-1363.28
40	BM	F	DFCI	BM:f	allB	-796.29
42	BM	F	DFCI	BM:f	allB	-679.14
47	BM	M	DFCI	BM:m	allB	-1164.40
48	BM	F	DFCI	BM:f	allB	-1299.65

Figure 1.55: Five rows and seven columns from the Golub data.

The goal of the Golub study was to develop a procedure for distinguishing between AML and ALL based only on the gene expression levels of a patient. There are two major issues to be addressed:

1. *Which genes are the most informative for making a prediction?* If a gene is differentially expressed between individuals with AML versus ALL, then measuring the expression level of that gene may be informative for diagnosing leukemia type. For example, if a gene tends to be highly expressed in AML individuals, but only expressed at low levels in ALL individuals, it is more likely to be a good predictor of leukemia type than a gene that is expressed at similar levels in both AML and ALL patients.

2. *How can leukemia type be predicted from expression data?* Suppose that a patient's expression profile is measured for a group of genes. In an ideal scenario, all the genes measured would exhibit AML-like expression, or ALL-like expression, making a prediction obvious. In reality, however, a patient's expression profile will not follow an idealized pattern. Some of the genes may have expression levels more typical of AML, while others may suggest ALL. It is necessary to clearly define a strategy for translating raw expression data into a prediction of leukemia type.

Even though the golub dataset is relatively small by modern standards, it is already too large to feasibly analyze without the use of statistical computing software. In this section, conceptual details will be demonstrated with a small version of the dataset (golub.small) that contains only the data for 10 patients and 10 genes. Figure 1.56 shows the cancer type and expression data in golub.small; the expression values have been rounded to the nearest whole number, and the gene probes are labeled A-J for convenience.

cancer	A	B	C	D	E	F	G	H	I	J
allB	39308	35232	41171	35793	-593	-1053	-513	-537	1702	1120
allT	32282	41432	59329	49608	-123	-511	265	-272	3567	-489
allB	47430	35569	56075	42858	-208	-712	32	-313	433	400
allB	25534	16984	28057	32694	89	-534	-24	195	3355	990
allB	35961	24192	27638	22241	-274	-632	-488	20	2259	348
aml	46178	6189	12557	34485	-331	-776	-551	-48	4074	-578
aml	43791	33662	38380	29758	-47	124	1118	3425	7018	1133
aml	53420	26109	31427	23810	396	108	1040	1915	4095	-709
aml	41242	37590	47326	30099	15	-429	784	-532	1085	-1912
aml	41301	49198	66026	56249	-418	-948	-340	-905	877	745

Figure 1.56: Leukemia type and expression data from golub.small.

To start understanding how gene expression differs by leukemia type, summarize the data separately for AML patients and for ALL patients, then make comparisons. For example, how does the expression of Gene A differ between individuals with AML versus ALL? Among the 5 individuals with AML, the mean expression for Gene A is 45,186; among the 5 ALL individuals, mean expression for Gene A is 36,103.

Figure 1.57 shows mean expression values for each gene among AML patients and Figure 1.58 among ALL patients.

AML	A	B	C	D	E	F	G	H	I	J
	46178	6189	12557	34485	-331	-776	-551	-48	4074	-578
	43791	33662	38380	29758	-47	124	1118	3425	7018	1133
	53420	26109	31427	23810	396	108	1040	1915	4095	-709
	41242	37590	47326	30099	15	-429	784	-532	1085	-1912
	41301	49198	66026	56249	-418	-948	-340	-905	877	745
Mean	45186	30550	39143	34880	-77	-384	410	771	3430	-264

Figure 1.57: Expression data for AML patients, where the last row contains mean expression value for each gene among the 5 AML patients. The first five rows are duplicated from the last five rows in Figure 1.56.

ALL	A	B	C	D	E	F	G	H	I	J
	39308	35232	41171	35793	-593	-1053	-513	-537	1702	1120
	32282	41432	59329	49608	-123	-511	265	-272	3567	-489
	47430	35569	56075	42858	-208	-712	32	-313	433	400
	25534	16984	28057	32694	89	-534	-24	195	3355	990
	35961	24192	27638	22241	-274	-632	-488	20	2259	348
Mean	36103	30682	42454	36639	-222	-689	-146	-181	2263	474

Figure 1.58: Expression data for ALL patients, where the last row contains mean expression value for each gene among the 5 ALL patients. The first five rows are duplicated from the first five rows in Figure 1.56.

EXAMPLE 1.30

On average, which genes are more highly expressed in AML patients? Which genes are more highly expressed in ALL patients?

For each gene, compare the mean expression value among ALL patients to the mean among AML patients. For example, the difference in mean expression levels for Gene A is

$$\overline{x}_{AML} - \overline{x}_{ALL} = 45186 - 36103 = 9083.$$

The differences in means for each gene are shown in Figure 1.59. Due to the order of subtraction used, genes with a positive difference value are more highly expressed in AML patients: A, E, F, G, H, and I. Genes B, C, D, and J are more highly expressed in ALL patients.

	A	B	C	D	E	F	G	H	I	J
AML mean	45186	30550	39143	34880	-77	-384	410	771	3430	-264
ALL mean	36103	30682	42454	36639	-222	-689	-146	-181	2263	474
Difference	9083	-132	-3310	-1758	145	304	556	952	1167	-738

Figure 1.59: The difference in mean expression levels by leukemia type for each gene in golub.small.

The most informative genes for predicting leukemia type are ones for which the difference in means seems relatively large, compared to the entire distribution of differences. Figure 1.60 visually displays the distribution of differences; the boxplot indicates that there is one large outlier and one small outlier.

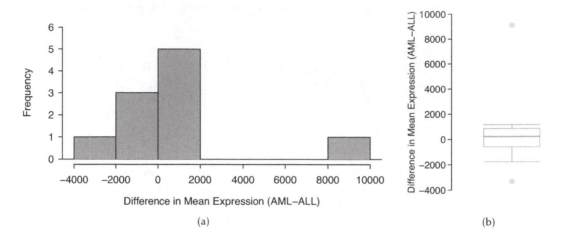

Figure 1.60: A histogram and boxplot of the differences in mean expression level between AML and ALL in the golub.small data.

It is possible to identify the outliers from simply looking at the list of differences, since the list is short: Genes A and C, with differences of 9,083 and -3,310, respectively.[58] It is important to remember that Genes A and C are only outliers out of the specific 10 genes in golub.small, where mean expression has been calculated using data from 10 patients; these genes do not necessarily show outlier levels of expression relative to the complete dataset.

With the use of computing software, the same process of calculating means, differences of means, and identifying outliers can easily be applied to the complete version of the data. Figure 1.61 shows the distribution of differences in mean expression level between AML and ALL patients for all 7,129 genes in the dataset, from 62 patients. The vast majority of genes are expressed at similar levels in AML and ALL patients; most genes have a difference in mean expression within -5,000 to 5,000. However, there are many genes that show extreme differences, as much as higher by 20,000 in AML or lower by 30,000 in ALL. These genes may be useful for differentiating between AML and ALL. The corresponding R lab illustrates the details of using R to identify these genes.[59]

Note how Figure 1.61 uses data from only 62 patients out of the 72 in the Golub dataset; this subset is called golub.train. The remaining 10 patients have been set aside as a "test" dataset (golub.test). Based on what has been learned about expression patterns from the 62 patients in golub.train, how well can the leukemia type of the 10 patients in golub.test be predicted?[60]

---

[58]For a numerical approach, calculate the outlier boundaries defined by $1.5 \times IQR$.

[59]Lab 3, Chapter 1.

[60]The original analysis used data from 38 patients to identify informative genes, then tested predictions on an independent collection of data from 34 patients.

(a)

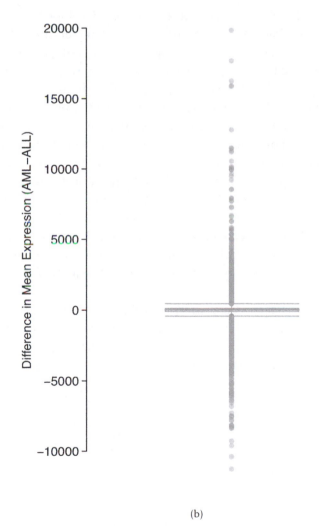

(b)

Figure 1.61: A histogram and boxplot of the differences in mean expression level between AML and ALL, using information from 7,129 genes and 62 patients in the Golub data (golub.train).

Figure 1.62: Schematic of the prediction strategy used by the Golub team, reproduced with modifications from Fig. 1B of the original paper.

Figure 1.62 illustrates the main ideas behind the strategy developed by the Golub team to predict leukemia type from expression data. The vertical orange bars represent the gene expression levels of a patient for each gene, relative to the mean expression for AML patients and ALL patients from the training dataset (vertical blue bars). A gene will "vote" for either AML or ALL, depending on whether the patient's expression level is closer to μ_{AML} or μ_{ALL}. In the example shown, three of the genes are considered to have ALL-like expression, versus the other two that are more AML-like. The votes are also weighted to account for how far an observation is from the midpoint between the two means (horizontal dotted blue line), i.e. the length of the dotted line shows the deviation from the midpoint. For example, the observed expression value for gene 2 is not as strong an indicator of ALL as the expression value for gene 1. The magnitude of the deviations (v_1, v_2, ...) are summed to obtain V_{AML} and V_{ALL}, and a higher value indicates a prediction of either AML or ALL, respectively.

The published analysis chose to use 50 informative genes; a decision about how many genes to use in a diagnostic panel typically involves considering factors such as the number of genes practical for a clinical setting. For simplicity, a smaller number of genes will be used in the analysis shown here.

Suppose that 10 genes are selected as predictors—the 5 largest outliers and 5 smallest outliers for the difference in mean expression between AML and ALL. Figure 1.63 shows expression data for these 10 genes from the 10 patients in golub.test, while Figure 1.64 contains the mean expression value for each gene among AML and ALL patients in golub.train.

	M19507_at	M27891_at	M11147_at	M96326_rna1_at	Y00787_s_at	M14483_rna1_s_at	X82240_rna1_at	X58529_at	M33680_at	U05259_rna1_at
1	4481	47532	56261	1785	-77	7824	-231	9520	7181	2757
2	11513	2839	42469	5018	20831	27407	-1116	-221	6978	-187
3	21294	6439	30239	61951	-187	19692	-540	216	1741	-84
4	-399	26023	40910	1271	26842	30092	-1247	19033	13117	-188
5	-147	29609	37606	20053	12745	26985	-1104	-273	8701	-168
6	-1229	-1206	16932	2250	360	38058	20951	12406	9927	8378
7	-238	-610	21798	-991	-348	23986	6500	20451	8500	7005
8	-1021	-792	17732	730	5102	17893	158	9287	7924	9221
9	432	-1099	9683	-576	-804	14386	7097	5556	9915	5594
10	-518	-862	26386	-2971	-1032	30100	32706	21007	23932	14841

Figure 1.63: Expression data from the 10 patients in golub.test, for the 10 genes selected as predictors. Each row represents a patient; the five right-most columns are the 5 largest outliers and the five left-most columns are the 5 smallest outliers.

Probe	AML Mean	ALL Mean	Midpoint
M19507_at	20143	322	10232
M27891_at	17395	-262	8567
M11147_at	32554	16318	24436
M96326_rna1_at	16745	830	8787
Y00787_s_at	16847	1002	8924
M14483_rna1_s_at	22268	33561	27914
X82240_rna1_at	-917	9499	4291
X58529_at	598	10227	5413
M33680_at	4151	13447	8799
U05259_rna1_at	74	8458	4266

Table 1.64: Mean expression value for each gene among AML patients and ALL patients in golub.train, and the midpoint between the means.

EXAMPLE 1.31

Consider the expression data for the patient in the first row of Figure 1.63. For each gene, identify whether the expression level is more AML-like or more ALL-like.

For the gene represented by the M19507_at probe, the patient has a recorded expression level of 4,481, which is closer to the ALL mean of 322 than the AML mean of 20,143. However, for the gene represented by the M27891_at probe, the expression level of 47,532 is closer to the AML mean of 17,395 than the ALL mean of -262.

Expression at genes represented by M19507_at, M96326_rna1_at, Y00787_s_at, and X58529_at are more ALL-like than AML-like. All other expression levels are closer to μ_{AML}.

EXAMPLE 1.32

Use the information in Figures 1.63 and 1.64 to calculate the magnitude of the deviations v_1 and v_{10} for the first patient.

For the gene represented by the M19507_at probe, the magnitude of the deviation is $v_1 = |4,481 - 10,232| = 5,751$.

For the gene represented by the U05259_rna1_at probe, the magnitude of the deviation is $v_{10} = |2,757 - 4,266| = 1,509$.

	M19507_at	M27891_at	M11147_at	M96326_rna1_at	Y00787_s_at	M14483_rna1_s_at	X82240_rna1_at	X58529_at	M33680_at	U05259_rna1_at
1	5751	38966	31825	7003	9001	20090	4522	4108	1618	1509
2	1281	5727	18032	3769	11906	507	5408	5634	1821	4453
3	11061	2128	5803	53164	9111	8222	4831	5196	7058	4350
4	10632	17457	16474	7516	17918	2178	5538	13621	4318	4454
5	10379	21042	13169	11265	3820	929	5395	5685	98	4434
6	11461	9773	7504	6537	8564	10144	16660	6994	1128	4112
7	10470	9176	2638	9778	9272	3928	2209	15038	300	2739
8	11254	9358	6704	8057	3823	10021		3875	875	4955
9	9800	9666	14754	9363	9728	13529	2806	144	1116	1328
10	10750	9428	1949	11759	9956	2186	28415	15594	15133	10575

Table 1.65: The magnitude of deviations from the midpoints. Cells for which the expression level is more ALL-like (closer to μ_{ALL} than μ_{AML}) are highlighted in blue.

EXAMPLE 1.33

Using the information in Figure 1.65, make a prediction for the leukemia status of Patient 1.

Calculate the total weighted votes for each category:

$$V_{AML} = 38,966 + 31,825 + 20,090 + 4,522 + 1,618 + 1,509 = 98,530$$

$$V_{ALL} = 5,571 + 7,003 + 9,001 + 4,108 = 25,863$$

Since $V_{AML} > V_{ALL}$, Patient 1 is predicted to have AML.

GUIDED PRACTICE 1.34

Make a prediction for the leukemia status of Patient 10.[61]

Figure 1.66 shows the comparison between actual leukemia status and predicted leukemia status based on the described prediction strategy. The prediction matches patient leukemia status for all patients.

	Actual	Prediction
1	aml	aml
2	aml	aml
3	aml	aml
4	aml	aml
5	aml	aml
6	allB	all
7	allB	all
8	allB	all
9	allB	all
10	allB	all

Figure 1.66: Actual leukemia status versus predicted leukemia status for the patients in golub.test

The analysis presented here is meant to illustrate how basic statistical concepts such as the definition of an outlier can be leveraged to address a relatively complex scientific question. There are entirely different approaches possible for analyzing these data, and many other considerations that have not been discussed. For example, this method of summing the weighted votes for each gene assumes that each gene is equally informative; the analysis in the published paper incorporates an additional weighting factor when calculating V_{AML} and V_{ALL} that accounts for how correlated each gene is with leukemia type. The published analysis also calculates prediction strength based on the values of V_{AML} and V_{AML} in order to provide a measure of how reliable each prediction is.

---

[61]Since $V_{AML} = 1,949$ and $V_{ALL} = 113,796$, Patient 10 is predicted to have ALL.

Finally, it is important to remember that the Golub analysis represented one of the earliest investigations into the use of gene expression data for diagnostic purposes. While the overall logical goals remain the same—identifying informative genes and developing a prediction strategy—the means of accomplishing them have become far more sophisticated. A modern study would have the benefit of referencing established, well-defined techniques for analyzing microarray data.

1.7.3 Case study: cold-responsive genes in the plant *Arabidopsis arenosa*

In contrast to hybridization-based approaches, RNA sequencing (RNA-Seq) allows for the entire transcriptome to be surveyed in a high-throughput, quantitative manner.[62] Microarrays require gene-specific probes, which limits microarray experiments to detecting transcripts that correspond to known gene sequences. In contrast, RNA-Seq can still be used when genome sequence information is not available, such as for non-model organisms. RNA-Seq is an especially powerful tool for researchers interested in studying small-scale genetic variation, such as single nucleotide polymorphisms, which microarrays are not capable of detecting.[63] Compared to microarrays, RNA-Seq technology offers increased sensitivity for detecting genes expressed at either low or very high levels.

This section introduces the concepts behind RNA-Seq technology and discusses a study that used RNA-Seq to explore the genetic basis of cold response in the plant *Arabidopsis arenosa*.

RNA sequencing (RNA-Seq)

The first step in an RNA-Seq experiment is to prepare cDNA sequence libraries for each RNA sample being sequenced. RNA is converted into cDNA and sheared into short fragments; sequencing adapters and barcodes are added to each fragment that initiate the sequencing reaction and identify sequences that originate from different samples. Once all the cDNA fragments are sequenced, the resulting short sequence reads must be re-constructed to produce the transcriptome. At this point, even the simplest RNA-Seq experiment has generated a relatively large amount of data; the complexity involved in processing and analyzing RNA-Seq data represents a significant challenge to widespread adoption of RNA-Seq technology. While a number of programs are available to help researchers process RNA-Seq data, improving computational methods for working with RNA-Seq data remains an active area of research.

A transcriptome can be assembled from the short sequence reads by either *de novo* assembly or genome mapping. In *de novo* assembly, sequencing data are run through computer algorithms that identify overlapping regions in the short sequence reads to gradually piece together longer stretches of continuous sequence. Alternatively, the reads can be aligned to a reference genome, a genome sequence which functions as a representative template for a given species; in cases where a species has not been sequenced, the genome of a close relative can also function as a reference genome. By mapping reads against a genome, it is possible to identify the position (and thus, the gene) from which a given RNA transcript originated. It is also possible to use a combination of these two strategies, an approach that is especially advantageous when genomes have experienced major rearrangements, such as in the case of cancer cells.[64] Once the transcripts have been assembled, information stored in sequence databases such as those hosted by the National Center for Biotechnology (NCBI) can be used to identify gene sequences (i.e., annotate the transcripts).

---

[62]Wang, et al. RNA-Seq: a revolutionary tool for transcriptomics. *Nature Genetics* 2009; **10**: 57-63.

[63]A single nucleotide polymorphism (SNP) represents variation at a single position in DNA sequence among individuals.

[64]Garber, et al. Computational methods for transcriptome annotation and quantification using RNA-seq. *Nature Methods* 2011; **8**: 469-477.

Quantifying gene expression levels from RNA-Seq data is based on counting the number of sequence reads per gene. If a particular gene is highly expressed, there will be a relatively high number of RNA transcripts originating from that gene; thus, the probability that transcripts from this gene are sequenced multiple times is also relatively high, and the gene will have a high number of sequencing reads associated with it. The number of read counts for a given gene provides a measure of gene expression level, when normalized for transcript length. If a short transcript and long transcript are present in equal amounts, the long transcript will have more sequencing reads associated with it due to the fragmentation step in library construction. Additional normalization steps are necessary when comparing data between samples to account for factors such as differences in the starting amount of RNA or the total number of sequencing reads generated (sequencing depth, in the language of genomics). A variety of strategies have been developed to carry out such normalization procedures.

Cold-responsive genes in *A. arenosa*

Arabidopsis arenosa populations exist in different habitats, and exhibit a range of differences in flowering time, cold sensitivity, and perenniality. Sensitivity to cold is an important trait for perennials, plants that live longer than one year. It is common for perennials to require a period of prolonged cold in order to flower. This mechanism, known as vernalization, allows perennials to synchronize their life cycle with the seasons such that they flower only once winter is over. Plant response to low temperatures is under genetic control, and mediated by a specific set of cold-responsive genes.

In a recent study, researchers used RNA-Seq to investigate how cold responsiveness differs in two populations of *A. arenosa*: TBG (collected from Triberg, Germany) and KA (collected from Kasparstein, Austria).[65] TBG grows in and around railway tracks, while KA is found on shaded limestone outcrops in wooded forests. As an annual, TBG has lost the vernalization response and does not required extended cold in order to flower; in the wild, TBG plants usually die before the onset of winter. In contrast, KA is a perennial plant, in which vernalization is known to greatly accelerate the onset of flowering.

Winter conditions can be simulated by incubating plants at 4 °C for several weeks; a plant that has undergone cold treatment is considered vernalized, while plants that have not been exposed to cold treatment are non-vernalized. Expression data were collected for 1,088 genes known to be cold-responsive in TBG and KA plants that were either vernalized or non-vernalized.

Figure 1.67 shows the data collected for the KA plants analyzed in the study, while Figure 1.68 shows the TBG expression data. Each row corresponds to a gene; the first column indicates gene name, while the rest correspond to expression measured in a plant sample. Three individuals of each population were exposed to cold (vernalized, denoted by V), and three were not (non-vernalized, denoted by NV). Expression was measured in gene counts (i.e. the number of RNA transcripts present in a sample); the data were then normalized between samples to allow for comparisons between gene counts. For example, a value of 288.20 for the *PUX4* gene in KA NV 1 indicates that in one of the non-vernalized KA individuals, about 288 copies of *PUX4* were detected.

A high number of transcripts indicates a high level of gene expression. As seen by comparing the expression levels across the first rows of Figures 1.67 and 1.68, the expression levels of *PUX4* are higher in vernalized plants than non-vernalized plants.

---

[65]Baduel P, et al. Habitat-Associated Life History and Stress-Tolerance Variation in *Arabidopsis arenosa*. *Plant Physiology* 2016; **171**: 437-451.

	Gene Name	KA NV 1	KA NV 2	KA NV 3	KA V 1	KA V 2	KA V 3
1	PUX4	288.20	322.55	305.35	1429.29	1408.25	1487.08
2	TZP	79.36	93.34	73.44	1203.40	1230.49	1214.03
3	GAD2	590.59	492.69	458.02	2639.42	2645.05	2705.32
4	GAUT6	86.88	99.25	57.98	586.24	590.03	579.71
5	FB	791.08	912.12	746.94	3430.03	3680.12	3467.06

Figure 1.67: Five rows and seven columns from the arenosa dataset, showing expression levels in KA plants.

	Gene Name	TBG NV 1	TBG NV 2	TBG NV 3	TBG V 1	TBG V 2	TBG V 3
1	PUX4	365.23	288.13	365.01	601.39	800.64	698.73
2	TZP	493.23	210.27	335.33	939.72	974.36	993.14
3	GAD2	1429.14	1339.50	2215.27	1630.77	1500.36	1621.28
4	GAUT6	129.63	76.40	135.02	320.57	298.91	399.27
5	FB	1472.35	1120.49	1313.14	3092.37	3230.72	3173.00

Figure 1.68: Five rows and seven columns from the arenosa dataset, showing expression levels in TBG plants.

The three measured individuals in a particular group represent biological replicates, individuals of the same type grown under identical conditions; collecting data from multiple individuals of the same group captures the inherent biological variability between organisms. Averaging expression levels across these replicates provides an estimate of the typical expression level in the larger population. Figure 1.69 shows the mean expression levels for five genes.

	Gene Name	KA NV	KA V	TBG NV	TBG V
1	PUX4	305.36	1441.54	339.46	700.25
2	TZP	82.05	1215.97	346.28	969.07
3	GAD2	513.77	2663.26	1661.30	1584.14
4	GAUT6	81.37	585.33	113.68	339.58
5	FB	816.71	3525.74	1301.99	3165.36

Figure 1.69: Mean gene expression levels of five cold-responsive genes, for non-vernalized and vernalized KA and TBG.

Figure 1.70(a) plots the mean gene expression levels of all 1,088 genes for each group. The expression levels are heavily right-skewed, with many genes present at unusually high levels relative to other genes. This is an example of a situation in which a transformation can be useful for clarifying the features of a distribution. In Figure 1.70(b), it is easier to see that expression levels of vernalized plants are shifted upward relative to non-vernalized plants. Additionally, while median expression is slightly higher in non-vernalized TBG than non-vernalized KA, median expression in vernalized KA is higher than in vernalized TBG. Vernalization appears to trigger a stronger change in expression of cold-responsive genes in KA plants than in TBG plants.

Figure 1.70 is only a starting point for exploring how expression of cold-responsive genes differs between KA and TBG plants. Consider a gene-level approach, in which the responsiveness of a gene to vernalization is quantified as the ratio of expression in a vernalized sample to expression in a non-vernalized sample.

Figure 1.70: (a) Mean gene expression levels for non-vernalized KA, vernalized KA, non-vernalized TBG, and vernalized TBG plants. (b) Log-transformed mean gene expression levels.

Figure 1.71(a) shows responsiveness for five genes, calculated separately between V and NV TBG and V and NV KA, using the means in Figure 1.69. The ratios provide a measure of how much expression differs between vernalized and non-vernalized individuals. For example, the gene *TZP* is expressed almost 15 times as much in vernalized KA than it is in non-vernalized KA. In contrast, the gene *GAD2* is expressed slightly less in vernalized TBG than in non-vernalized TBG.

As with the mean gene expression levels, it is useful to apply a log transformation (Figure 1.71(b)). On the log scale, values close to 0 are indicative of low responsiveness, while large values in either direction correspond to high responsiveness. Figure 1.72 shows the log2-transformed expression ratios as a side-by-side boxplot.[66]

	Gene Name	TBG	KA		Gene Name	TBG	KA
1	PUX4	2.06	4.72	1	PUX4	1.04	2.24
2	TZP	2.80	14.82	2	TZP	1.48	3.89
3	GAD2	0.95	5.18	3	GAD2	-0.07	2.37
4	GAUT6	2.99	7.19	4	GAUT6	1.58	2.85
5	FB	2.43	4.32	5	FB	1.28	2.11
	(a)				(b)		

Figure 1.71: (a) Ratio of mean expression in vernalized individuals to mean expression in non-vernalized individuals. (b) Log2-transformation of expression ratios in Figure 1.71(a).

---

[66]One gene is omitted because the expression ratio in KA is 0, and the logarithm of 0 is undefined.

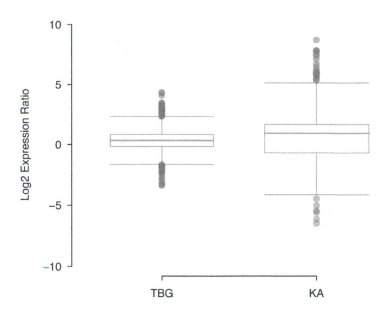

Figure 1.72: Responsiveness for 1,087 genes in arenosa, calculated as the log2 ratio of vernalized over non-vernalized expression levels.

Figure 1.72 directly illustrates how the magnitude of response to vernalization in TBG is smaller than in KA. The spread of responsiveness in KA is larger than for TBG, as indicated by the larger IQR and range of values; this indicates that more genes in KA are differentially expressed between vernalized and non-vernalized samples. Additionally, the median responsiveness in KA is higher than in TBG.

There are several outliers for both KA and TBG, with large outliers representing genes that were much more highly expressed in vernalized plants than non-vernalized plants, and vice versa for low outliers. These highly cold-responsive genes likely play a role in how plants cope with colder temperatures; they could be involved in regulating freezing tolerance, or controlling how plants detect cold temperatures. With the help of computing software, it is a simple matter to identify the outliers and address questions such as whether particular genes are highly vernalization-responsive in both KA and TBG.

Advanced data visualization

There are many ways to numerically and graphically summarize data that are not explicitly introduced in this chapter. Presentation-style graphics in published manuscripts can be especially complex, and may feature techniques specific to a certain field as well as novel approaches designed to highlight particular features of a dataset. This section discusses the figures generated by the Baduel, et al. research team to visualize the differences in vernalization response between KA and TBG *A. arenosa* plants.

Figure 1.73: Figure 4 from the original manuscript. Plot A compares mean expression levels between non-vernalized KA and TBG; Plot B compares mean expression levels between vernalized KA and TBG.

Each dot in Figure 1.73 represents a gene; each gene is plotted by its mean expression level in KA against its mean expression level in TBG. The overall trend can be summarized by a line fit to the points.[67] For the slope of the line to equal 1, each gene would have to be equally expressed in KA and TBG. In the upper plot, the slope of the line is less than 1, which indicates that for non-vernalized plants, cold-responsive genes have a higher expression in TBG than in KA. In the lower plot, the slope is greater than 1, indicating that the trend is reversed in vernalized plants: cold-responsive genes are more highly expressed in KA. This trend is also discernible from the side-by-side boxplot in Figure 1.70. Using a scatterplot, however, makes it possible to directly compare expression in KA versus TBG on a gene-by-gene basis, and also locate particular genes of interest that are known from previous research (e.g., the labeled genes in Figure 1.73.)[68] The colors in the plot signify plot density, with warmer colors representing a higher concentration of points.

---

[67] Lines of best fit are discussed in Chapter 6.

[68] Only a subset of the 1,088 genes are plotted in Figure 1.73.

Figure 1.74: Figure 3 from the original manuscript. Each gene is plotted based on the values of the log2 expression ratio in KA versus TBG.

Figure 1.74, like Figure 1.72, compares the cold-responsiveness in KA versus TBG, calculating responsiveness as the log2 ratio of vernalized over non-vernalized expression levels. As in Figure 1.73, each dot represents a single gene. The slope of the best fitting line is greater than 1, indicating that the assayed genes typically show greater responsiveness in KA than in TBG.[69]

While presentation-style graphics may use relatively sophisticated approaches to displaying data that seem far removed from the simple plots discussed in this chapter, the end goal remains the same – to effectively highlight key features of data.

---

[69]These 608 genes are a subset of the ones plotted in Figure 1.73; genes with expression ratio 0 are not included.

1.8   Notes

Introductory treatments of statistics often emphasize the value of formal methods of probability and inference, topics which are covered in the remaining chapters of this text. However, numerical and graphical summaries are essential for understanding the features of a dataset and should be applied before the process of inference begins. It is inadvisable to begin conducting tests or constructing models without a careful understanding of the strengths and weaknesses of a dataset. For example, are some measurements out of range, or the result of errors in data recording?

The tools of descriptive statistics form the basis of exploratory data analysis; having the intuition for exploring and interpreting data in the context of a research question is an essential statistical skill. With computing software, it is a relatively simple matter to produce numerical and graphical summaries, even with large datasets. The challenge lies instead in understanding how to wade through a dataset, disentangle complex relationships between variables, and piece together the underlying story.

It is important to note that the graphical methods illustrated in the text are relatively simple, static graphs that, for instance, do not show changes dynamically over time. They will be surprisingly useful in the later chapters. But there has been considerable progress in the visual display of data in the last decade, and many wonderful displays exist that show complex, time dependent data. For examples of sophisticated graphical displays, we especially recommend the bubble charts available at the Gapminder web site (`https://www.gapminder.org`) that show international trends in public health outcomes and the graphical displays of data in the Upshot section of the New York Times (`https://www.nytimes.com/section/upshot`).

There are four labs associated with Chapter 1. The first lab introduces basic commands for working with data in R, and shows how to produce the graphical and numerical summaries discussed in this chapter. The exercises in Lab 1 rely heavily on the introduction to R and *R Studio* in Lab 00 (Getting Started). The Lab Notes corresponding to Lab 1 provide a systematic introduction to R functions useful for getting started with applied data analysis.

The remaining three labs explore the data presented in the case studies in Section 1.7, outlining analyses driven by questions similar to what one might encounter in practice. Does the state of California discriminate in its distribution of funds for developmental disability support (Lab 2)? Are particular genes associated with a subtype of pediatric leukemia (Lab 3)? Is there a genetic basis to the cold weather response in the plant *Arabidopsis arenosa* (Lab 4)? Labs 3 and 4 demonstrate how computing is essential for data analysis; even though the two datasets are relatively small by modern standards, they are already too large to feasibly analyze without statistical computing software. All three labs illustrate how important questions can be examined even with relatively simple statistical concepts.

1.9 Exercises

1.9.1 Case study

1.1 Migraine and acupuncture, Part I. A migraine is a particularly painful type of headache, which patients sometimes wish to treat with acupuncture. To determine whether acupuncture relieves migraine pain, researchers conducted a randomized controlled study where 89 females diagnosed with migraine headaches were randomly assigned to one of two groups: treatment or control. 43 patients in the treatment group received acupuncture that is specifically designed to treat migraines. 46 patients in the control group received placebo acupuncture (needle insertion at non-acupoint locations). 24 hours after patients received acupuncture, they were asked if they were pain free. Results are summarized in the contingency table below.[70]

<table>
<tr><td></td><td></td><td colspan="3">Pain free</td></tr>
<tr><td></td><td></td><td>Yes</td><td>No</td><td>Total</td></tr>
<tr><td rowspan="2">Group</td><td>Treatment</td><td>10</td><td>33</td><td>43</td></tr>
<tr><td>Control</td><td>2</td><td>44</td><td>46</td></tr>
<tr><td></td><td>Total</td><td>12</td><td>77</td><td>89</td></tr>
</table>

Figure from the original paper displaying the appropriate area (M) versus the inappropriate area (S) used in the treatment of migraine attacks.

(a) What percent of patients in the treatment group were pain free 24 hours after receiving acupuncture?

(b) What percent were pain free in the control group?

(c) In which group did a higher percent of patients become pain free 24 hours after receiving acupuncture?

(d) Your findings so far might suggest that acupuncture is an effective treatment for migraines for all people who suffer from migraines. However this is not the only possible conclusion that can be drawn based on your findings so far. What is one other possible explanation for the observed difference between the percentages of patients that are pain free 24 hours after receiving acupuncture in the two groups?

1.2 Sinusitis and antibiotics, Part I. Researchers studying the effect of antibiotic treatment for acute sinusitis compared to symptomatic treatments randomly assigned 166 adults diagnosed with acute sinusitis to one of two groups: treatment or control. Study participants received either a 10-day course of amoxicillin (an antibiotic) or a placebo similar in appearance and taste. The placebo consisted of symptomatic treatments such as acetaminophen, nasal decongestants, etc. At the end of the 10-day period, patients were asked if they experienced improvement in symptoms. The distribution of responses is summarized below.[71]

<table>
<tr><td></td><td></td><td colspan="2">Self-reported improvement
in symptoms</td><td></td></tr>
<tr><td></td><td></td><td>Yes</td><td>No</td><td>Total</td></tr>
<tr><td rowspan="2">Group</td><td>Treatment</td><td>66</td><td>19</td><td>85</td></tr>
<tr><td>Control</td><td>65</td><td>16</td><td>81</td></tr>
<tr><td></td><td>Total</td><td>131</td><td>35</td><td>166</td></tr>
</table>

(a) What percent of patients in the treatment group experienced improvement in symptoms?

(b) What percent experienced improvement in symptoms in the control group?

(c) In which group did a higher percentage of patients experience improvement in symptoms?

(d) Your findings so far might suggest a real difference in effectiveness of antibiotic and placebo treatments for improving symptoms of sinusitis. However, this is not the only possible conclusion that can be drawn based on your findings so far. What is one other possible explanation for the observed difference between the percentages of patients in the antibiotic and placebo treatment groups that experience improvement in symptoms of sinusitis?

[70]G. Allais et al. "Ear acupuncture in the treatment of migraine attacks: a randomized trial on the efficacy of appropriate versus inappropriate acupoints". In: *Neurological Sci.* 32.1 (2011), pp. 173–175.

[71]J.M. Garbutt et al. "Amoxicillin for Acute Rhinosinusitis: A Randomized Controlled Trial". In: *JAMA: The Journal of the American Medical Association* 307.7 (2012), pp. 685–692.

1.9.2 Data basics

1.3 Air pollution and birth outcomes, study components. Researchers collected data to examine the relationship between air pollutants and preterm births in Southern California. During the study air pollution levels were measured by air quality monitoring stations. Specifically, levels of carbon monoxide were recorded in parts per million, nitrogen dioxide and ozone in parts per hundred million, and coarse particulate matter (PM_{10}) in $\mu g/m^3$. Length of gestation data were collected on 143,196 births between the years 1989 and 1993, and air pollution exposure during gestation was calculated for each birth. The analysis suggested that increased ambient PM_{10} and, to a lesser degree, CO concentrations may be associated with the occurrence of preterm births.[72]

(a) Identify the main research question of the study.

(b) Who are the subjects in this study, and how many are included?

(c) What are the variables in the study? Identify each variable as numerical or categorical. If numerical, state whether the variable is discrete or continuous. If categorical, state whether the variable is ordinal.

1.4 Buteyko method, study components. The Buteyko method is a shallow breathing technique developed by Konstantin Buteyko, a Russian doctor, in 1952. Anecdotal evidence suggests that the Buteyko method can reduce asthma symptoms and improve quality of life. In a scientific study to determine the effectiveness of this method, researchers recruited 600 asthma patients aged 18-69 who relied on medication for asthma treatment. These patients were randomly split into two research groups: one practiced the Buteyko method and the other did not. Patients were scored on quality of life, activity, asthma symptoms, and medication reduction on a scale from 0 to 10. On average, the participants in the Buteyko group experienced a significant reduction in asthma symptoms and an improvement in quality of life.[73]

(a) Identify the main research question of the study.

(b) Who are the subjects in this study, and how many are included?

(c) What are the variables in the study? Identify each variable as numerical or categorical. If numerical, state whether the variable is discrete or continuous. If categorical, state whether the variable is ordinal.

1.5 Cheaters, study components. Researchers studying the relationship between honesty, age and self-control conducted an experiment on 160 children between the ages of 5 and 15. Participants reported their age, sex, and whether they were an only child or not. The researchers asked each child to toss a fair coin in private and to record the outcome (white or black) on a paper sheet, and said they would only reward children who report white. The study's findings can be summarized as follows: "Half the students were explicitly told not to cheat and the others were not given any explicit instructions. In the no instruction group probability of cheating was found to be uniform across groups based on child's characteristics. In the group that was explicitly told to not cheat, girls were less likely to cheat, and while rate of cheating didn't vary by age for boys, it decreased with age for girls."[74]

(a) Identify the main research question of the study.

(b) Who are the subjects in this study, and how many are included?

(c) How many variables were recorded for each subject in the study in order to conclude these findings? State the variables and their types.

---

[72]B. Ritz et al. "Effect of air pollution on preterm birth among children born in Southern California between 1989 and 1993". In: *Epidemiology* 11.5 (2000), pp. 502–511.

[73]J. McGowan. "Health Education: Does the Buteyko Institute Method make a difference?" In: *Thorax* 58 (2003).

[74]Alessandro Bucciol and Marco Piovesan. "Luck or cheating? A field experiment on honesty with children". In: *Journal of Economic Psychology* 32.1 (2011), pp. 73–78.

1.6 Hummingbird taste behavior, study components. Researchers hypothesized that a particular taste receptor in hummingbirds, T1R1-T1R3, played a primary role in dictating taste behavior; specifically, in determining which compounds hummingbirds detect as sweet. In a series of field tests, hummingbirds were presented simultaneously with two filled containers, one containing test stimuli and a second containing sucrose. The test stimuli included aspartame, erythritol, water, and sucrose. Aspartame is an artificial sweetener that tastes sweet to humans, but is not detected by hummingbird T1R1-T1R3 , while erythritol is an artificial sweetener known to activate T1R1-T1R3.

Data were collected on how long a hummingbird drank from a particular container for a given trial, measured in seconds. For example, in one field test comparing aspartame and sucrose, a hummingbird drank from the aspartame container for 0.54 seconds and from the sucrose container for 3.21 seconds.

(a) Which tests are controls? Which tests are treatments?

(b) Identify the response variable(s) in the study. Are they numerical or categorical?

(c) Describe the main research question.

1.7 Egg coloration. The evolutionary significance of variation in egg coloration among birds is not fully understood. One hypothesis suggests that egg coloration may be an indication of female quality, with healthier females being capable of depositing blue-green pigment into eggshells instead of using it for themselves as an antioxidant. In a study conducted on 32 collared flycatchers, half of the females were given supplementary diets before and during egg laying. Eggs were measured for darkness of blue color using spectrophotometry; for example, the mean amount of blue-green chroma was 0.594 absorbance units. Egg mass was also recorded.

(a) Identify the control and treatment groups.

(b) Describe the main research question.

(c) Identify the primary response variable of interest, and whether it is numerical or categorical.

1.8 Smoking habits of UK residents. A survey was conducted to study the smoking habits of UK residents. Below is a data matrix displaying a portion of the data collected in this survey. Note that "£" stands for British Pounds Sterling, "cig" stands for cigarettes, and "N/A" refers to a missing component of the data.[75]

	sex	age	marital	grossIncome	smoke	amtWeekends	amtWeekdays
1	Female	42	Single	Under £2,600	Yes	12 cig/day	12 cig/day
2	Male	44	Single	£10,400 to £15,600	No	N/A	N/A
3	Male	53	Married	Above £36,400	Yes	6 cig/day	6 cig/day
⋮	⋮	⋮	⋮	⋮	⋮	⋮	⋮
1691	Male	40	Single	£2,600 to £5,200	Yes	8 cig/day	8 cig/day

(a) What does each row of the data matrix represent?

(b) How many participants were included in the survey?

(c) For each variable, indicate whether it is numerical or categorical. If numerical, identify the variable as continuous or discrete. If categorical, indicate if the variable is ordinal.

[75]National STEM Centre, Large Datasets from stats4schools.

1.9 The microbiome and colon cancer. A study was conducted to assess whether the abundance of particular bacterial species in the gastrointestinal system is associated with the development of colon cancer. The following data matrix shows a subset of the data collected in the study. Cancer stage is coded 1-4, with larger values indicating cancer that is more difficult to treat. The abundance levels are given for five bacterial species; abundance is calculated as the frequency of that species divided by the total number of bacteria from all species.

	age	gender	stage	bug 1	bug 2	bug 3	bug 4	bug 5
1	71	Female	2	0.03	0.09	0.52	0.00	0.00
2	53	Female	4	0.16	0.08	0.08	0.00	0.00
3	55	Female	2	0.00	0.01	0.31	0.00	0.00
4	44	Male	2	0.11	0.14	0.00	0.07	0.05
⋮	⋮	⋮	⋮	⋮	⋮	⋮	⋮	⋮
73	48	Female	3	0.21	0.05	0.00	0.00	0.04

(a) What does each row of the data matrix represent?

(b) Identify explanatory and response variables.

(c) For each variable, indicate whether it is numerical or categorical.

1.9.3   Data collection principles

1.10 Cheaters, scope of inference. Exercise 1.5 introduces a study where researchers studying the relationship between honesty, age, and self-control conducted an experiment on 160 children between the ages of 5 and 15. The researchers asked each child to toss a fair coin in private and to record the outcome (white or black) on a paper sheet, and said they would only reward children who report white. Half the students were explicitly told not to cheat and the others were not given any explicit instructions. Differences were observed in the cheating rates in the instruction and no instruction groups, as well as some differences across children's characteristics within each group.

(a) Identify the population of interest and the sample in this study.

(b) Comment on whether or not the results of the study can be generalized to the population, and if the findings of the study can be used to establish causal relationships.

1.11 Air pollution and birth outcomes, scope of inference. Exercise 1.3 introduces a study where researchers collected data to examine the relationship between air pollutants and preterm births in Southern California. During the study, air pollution levels were measured by air quality monitoring stations. Length of gestation data were collected on 143,196 births between the years 1989 and 1993, and air pollution exposure during gestation was calculated for each birth. It can be assumed that the 143,196 births are effectively the entire population of births during this time period.

(a) Identify the population of interest and the sample in this study.

(b) Comment on whether or not the results of the study can be generalized to the population, and if the findings of the study can be used to establish causal relationships.

1.12 Herbal remedies. Echinacea has been widely used as an herbal remedy for the common cold, but previous studies evaluating its efficacy as a remedy have produced conflicting results. In a new study, researchers randomly assigned 437 volunteers to receive either a placebo or echinacea treatment before being infected with rhinovirus. Healthy young adult volunteers were recruited for the study from the University of Virginia community.

(a) Identify the population of interest and the sample in this study.

(b) Comment on whether or not the results of the study can be generalized to a larger population.

(c) Can the findings of the study be used to establish causal relationships? Justify your answer.

1.13 Buteyko method, scope of inference. Exercise 1.4 introduces a study on using the Buteyko shallow breathing technique to reduce asthma symptoms and improve quality of life. As part of this study 600 asthma patients aged 18-69 who relied on medication for asthma treatment were recruited and randomly assigned to two groups: one practiced the Buteyko method and the other did not. Those in the Buteyko group experienced, on average, a significant reduction in asthma symptoms and an improvement in quality of life.

(a) Identify the population of interest and the sample in this study.

(b) Comment on whether or not the results of the study can be generalized to the population, and if the findings of the study can be used to establish causal relationships.

1.14 Vitamin supplements. In order to assess the effectiveness of taking large doses of vitamin C in reducing the duration of the common cold, researchers recruited 400 healthy volunteers from staff and students at a university. A quarter of the patients were randomly assigned a placebo, and the rest were randomly allocated between 1g Vitamin C, 3g Vitamin C, or 3g Vitamin C plus additives to be taken at onset of a cold for the following two days. All tablets had identical appearance and packaging. No significant differences were observed in any measure of cold duration or severity between the four medication groups, and the placebo group had the shortest duration of symptoms.[76]

(a) Was this an experiment or an observational study? Why?

(b) What are the explanatory and response variables in this study?

(c) Participants are ultimately able to choose whether or not to use the pills prescribed to them. We might expect that not all of them will adhere and take their pills. Does this introduce a confounding variable to the study? Explain your reasoning.

1.15 Chicks and antioxidants. Environmental factors early in life can have long-lasting effects on an organism. In one study, researchers examined whether dietary supplementation with vitamins C and E influences body mass and corticosterone level in yellow-legged gull chicks. Chicks were randomly assigned to either the nonsupplemented group or the vitamin supplement experimental group. The initial study group consisted of 108 nests, with 3 eggs per nest. Chicks were assessed at age 7 days.

(a) What type of study is this?

(b) What are the experimental and control treatments in this study?

(c) Explain why randomization is an important feature of this experiment.

1.16 Exercise and mental health. A researcher is interested in the effects of exercise on mental health and he proposes the following study: Use stratified random sampling to recruit 18-30, 31-40 and 41-55 year olds from the population. Next, randomly assign half the subjects from each age group to exercise twice a week, and instruct the rest not to exercise. Conduct a mental health exam at the beginning and at the end of the study, and compare the results.

(a) What type of study is this?

(b) What are the treatment and control groups in this study?

(c) Does this study make use of blocking? If so, what is the blocking variable?

(d) Comment on whether or not the results of the study can be used to establish a causal relationship between exercise and mental health, and indicate whether or not the conclusions can be generalized to the population at large.

(e) Suppose you are given the task of determining if this proposed study should get funding. Would you have any reservations about the study proposal?

---

[76]C. Audera et al. "Mega-dose vitamin C in treatment of the common cold: a randomised controlled trial". In: *Medical Journal of Australia* 175.7 (2001), pp. 359–362.

1.17 Internet use and life expectancy. The following scatterplot was created as part of a study evaluating the relationship between estimated life expectancy at birth (as of 2014) and percentage of internet users (as of 2009) in 208 countries for which such data were available.[77]

(a) Describe the relationship between life expectancy and percentage of internet users.

(b) What type of study is this?

(c) State a possible confounding variable that might explain this relationship and describe its potential effect.

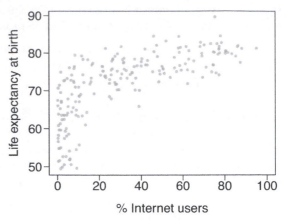

1.18 Stressed out. A study that surveyed a random sample of otherwise healthy high school students found that they are more likely to get muscle cramps when they are stressed. The study also noted that students drink more coffee and sleep less when they are stressed.

(a) What type of study is this?

(b) Can this study be used to conclude a causal relationship between increased stress and muscle cramps?

(c) State possible confounding variables that might explain the observed relationship between increased stress and muscle cramps.

1.19 Evaluate sampling methods. A university wants to assess how many hours of sleep students are getting per night. For each proposed method below, discuss whether the method is reasonable or not.

(a) Survey a simple random sample of 500 students.

(b) Stratify students by their field of study, then sample 10% of students from each stratum.

(c) Cluster students by their class year (e.g. freshmen in one cluster, sophomores in one cluster, etc.), then randomly sample three clusters and survey all students in those clusters.

1.20 City council survey. A city council has requested a household survey be conducted in a suburban area of their city. The area is broken into many distinct and unique neighborhoods, some including large homes, some with only apartments, and others a diverse mixture of housing structures. Identify the sampling methods described below, and comment on whether or not you think they would be effective in this setting.

(a) Randomly sample 50 households from the city.

(b) Divide the city into neighborhoods, and sample 20 households from each neighborhood.

(c) Divide the city into neighborhoods, randomly sample 10 neighborhoods, and sample all households from those neighborhoods.

(d) Divide the city into neighborhoods, randomly sample 10 neighborhoods, and then randomly sample 20 households from those neighborhoods.

(e) Sample the 200 households closest to the city council offices.

---

[77]CIA Factbook, Country Comparisons, 2014.

1.21 Flawed reasoning. Identify the flaw(s) in reasoning in the following scenarios. Explain what the individuals in the study should have done differently if they wanted to make such conclusions.

(a) Students at an elementary school are given a questionnaire that they are asked to return after their parents have completed it. One of the questions asked is, "Do you find that your work schedule makes it difficult for you to spend time with your kids after school?" Of the parents who replied, 85% said "no". Based on these results, the school officials conclude that a great majority of the parents have no difficulty spending time with their kids after school.

(b) A survey is conducted on a simple random sample of 1,000 women who recently gave birth, asking them about whether or not they smoked during pregnancy. A follow-up survey asking if the children have respiratory problems is conducted 3 years later, however, only 567 of these women are reached at the same address. The researcher reports that these 567 women are representative of all mothers.

(c) An orthopedist administers a questionnaire to 30 of his patients who do not have any joint problems and finds that 20 of them regularly go running. He concludes that running decreases the risk of joint problems.

1.22 Reading the paper. Below are excerpts from two articles published in the *NY Times*:

(a) An article titled *Risks: Smokers Found More Prone to Dementia* states the following:[78]

> "Researchers analyzed data from 23,123 health plan members who participated in a voluntary exam and health behavior survey from 1978 to 1985, when they were 50-60 years old. 23 years later, about 25% of the group had dementia, including 1,136 with Alzheimer's disease and 416 with vascular dementia. After adjusting for other factors, the researchers concluded that pack-a-day smokers were 37% more likely than nonsmokers to develop dementia, and the risks went up with increased smoking; 44% for one to two packs a day; and twice the risk for more than two packs."

Based on this study, can it be concluded that smoking causes dementia later in life? Explain your reasoning.

(b) Another article titled *The School Bully Is Sleepy* states the following:[79]

> "The University of Michigan study, collected survey data from parents on each child's sleep habits and asked both parents and teachers to assess behavioral concerns. About a third of the students studied were identified by parents or teachers as having problems with disruptive behavior or bullying. The researchers found that children who had behavioral issues and those who were identified as bullies were twice as likely to have shown symptoms of sleep disorders."

A friend of yours who read the article says, "The study shows that sleep disorders lead to bullying in school children." Is this statement justified? If not, how best can you describe the conclusion that can be drawn from this study?

1.23 Alcohol consumption and STIs. An observational study published last year in *The American Journal of Preventive Medicine* investigated the effects of an increased alcohol sales tax in Maryland on the rates of gonorrhea and chlamydia.[80] After a tax increase from 6% to 9% in 2011, the statewide gonorrhea rate declined by 24%, the equivalent of 1,600 cases per year. In a statement to the *New York Times*, the lead author of the paper was quoted saying, "Policy makers should consider raising liquor taxes if they're looking for ways to prevent sexually transmitted infections. In the year and a half following the alcohol tax rise in Maryland, this prevented 2,400 cases of gonorrhea and saved half a million dollars in health care costs." Explain whether the lead author's statement is accurate.

[78] R.C. Rabin. "Risks: Smokers Found More Prone to Dementia". In: *New York Times* (2010).

[79] T. Parker-Pope. "The School Bully Is Sleepy". In: *New York Times* (2011).

[80] S. Staras, et al., 2015. Maryland Alcohol Sales Tax and Sexually Transmitted Infections. The American Journal of Preventive Medicine 50: e73-e80.

1.24 Income and education in US counties. The scatterplot below shows the relationship between per capita income (in thousands of dollars) and percent of population with a bachelor's degree in 3,143 counties in the US in 2010.

(a) What are the explanatory and response variables?

(b) Describe the relationship between the two variables. Make sure to discuss unusual observations, if any.

(c) Can we conclude that having a bachelor's degree increases one's income?

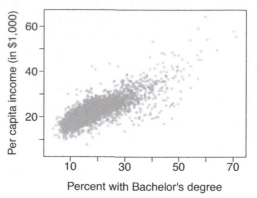

1.25 Eat better, feel better. In a public health study on the effects of consumption of fruits and vegetables on psychological well-being in young adults, participants were randomly assigned to three groups: (1) diet-as-usual, (2) an ecological momentary intervention involving text message reminders to increase their fruits and vegetable consumption plus a voucher to purchase them, or (3) a fruit and vegetable intervention in which participants were given two additional daily servings of fresh fruits and vegetables to consume on top of their normal diet. Participants were asked to take a nightly survey on their smartphones. Participants were student volunteers at the University of Otago, New Zealand. At the end of the 14-day study, only participants in the third group showed improvements to their psychological well-being across the 14-days relative to the other groups.[81]

(a) What type of study is this?

(b) Identify the explanatory and response variables.

(c) Comment on whether the results of the study can be generalized to the population.

(d) Comment on whether the results of the study can be used to establish causal relationships.

(e) A newspaper article reporting on the study states, "The results of this study provide proof that giving young adults fresh fruits and vegetables to eat can have psychological benefits, even over a brief period of time." How would you suggest revising this statement so that it can be supported by the study?

1.9.4 Numerical data

1.26 Means and SDs. For each part, compare distributions (1) and (2) based on their means and standard deviations. You do not need to calculate these statistics; simply state how the means and the standard deviations compare. Make sure to explain your reasoning. *Hint:* It may be useful to sketch dot plots of the distributions.

(a) (1) 3, 5, 5, 5, 8, 11, 11, 11, 13
    (2) 3, 5, 5, 5, 8, 11, 11, 11, 20

(b) (1) -20, 0, 0, 0, 15, 25, 30, 30
    (2) -40, 0, 0, 0, 15, 25, 30, 30

(c) (1) 0, 2, 4, 6, 8, 10
    (2) 20, 22, 24, 26, 28, 30

(d) (1) 100, 200, 300, 400, 500
    (2) 0, 50, 300, 550, 600

---

[81] Tamlin S Conner et al. "Let them eat fruit! The effect of fruit and vegetable consumption on psychological well-being in young adults: A randomized controlled trial". In: *PloS one* 12.2 (2017), e0171206.

1.27 Medians and IQRs. For each part, compare distributions (1) and (2) based on their medians and IQRs. You do not need to calculate these statistics; simply state how the medians and IQRs compare. Make sure to explain your reasoning.

(a) (1) 3, 5, 6, 7, 9
     (2) 3, 5, 6, 7, 20

(b) (1) 3, 5, 6, 7, 9
     (2) 3, 5, 8, 7, 9

(c) (1) 1, 2, 3, 4, 5
     (2) 6, 7, 8, 9, 10

(d) (1) 0, 10, 50, 60, 100
     (2) 0, 100, 500, 600, 1000

1.28 Mix-and-match. Describe the distribution in the histograms below and match them to the box plots.

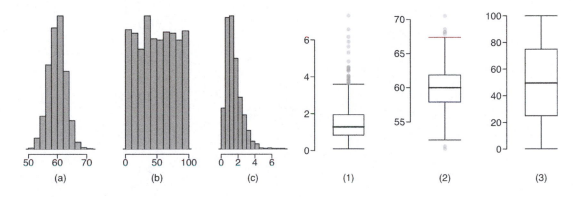

1.29 Air quality. Daily air quality is measured by the air quality index (AQI) reported by the Environmental Protection Agency. This index reports the pollution level and what associated health effects might be a concern. The index is calculated for five major air pollutants regulated by the Clean Air Act and takes values from 0 to 300, where a higher value indicates lower air quality. AQI was reported for a sample of 91 days in 2011 in Durham, NC. The relative frequency histogram below shows the distribution of the AQI values on these days.[82]

(a) Based on the histogram, describe the distribution of daily AQI.

(b) Estimate the median AQI value of this sample.

(c) Would you expect the mean AQI value of this sample to be higher or lower than the median? Explain your reasoning.

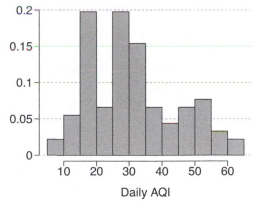

---

[82]US Environmental Protection Agency, AirData, 2011.

1.30 Nursing home residents. Since states with larger numbers of elderly residents would naturally have more nursing home residents, the number of nursing home residents in a state is often adjusted for the number of people 65 years of age or order (65+). That adjustment is usually given as the number of nursing home residents age 65+ per 1,000 members of the population age 65+. For example, a hypothetical state with 200 nursing home residents age 65+ and 50,000 people age 65+ would have the same adjusted number of residents as a state with 400 residents and a total age 65+ population of 100,000 residents: 4 residents per 1,000.

Use the two plots below to answer the following questions. Both plots show the distribution of the number of nursing home residents per 1,000 members of the population 65+ (in each state).

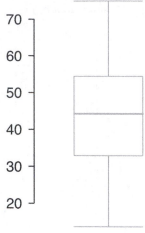

(a) Is the distribution of adjusted number of nursing home residents symmetric or skewed? Are there any states that could be considered outliers?

(b) Which plot is more informative: the histogram or the boxplot? Explain your answer.

(c) What factors might influence the substantial amount of variability among different states? This question cannot be answered from the data; speculate using what you know about the demographics of the United States.

1.31 Income at the coffee shop. The first histogram below shows the distribution of the yearly incomes of 40 patrons at a college coffee shop. Suppose two new people walk into the coffee shop: one making $225,000 and the other $250,000. The second histogram shows the new income distribution. Summary statistics are also provided.

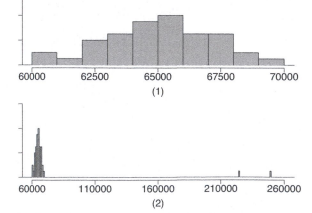

	(1)	(2)
n	40	42
Min.	60,680	60,680
1st Qu.	63,620	63,710
Median	65,240	65,350
Mean	65,090	73,300
3rd Qu.	66,160	66,540
Max.	69,890	250,000
SD	2,122	37,321

(a) Would the mean or the median best represent what we might think of as a typical income for the 42 patrons at this coffee shop? What does this say about the robustness of the two measures?

(b) Would the standard deviation or the IQR best represent the amount of variability in the incomes of the 42 patrons at this coffee shop? What does this say about the robustness of the two measures?

1.32 Midrange. The *midrange* of a distribution is defined as the average of the maximum and the minimum of that distribution. Is this statistic robust to outliers and extreme skew? Explain your reasoning.

1.9.5 Categorical data

1.33 Flossing habits. Suppose that an anonymous questionnaire is given to patients at a dentist's office once they arrive for an appointment. One of the questions asks "How often do you floss?", and four answer options are provided: a) at least twice a day, b) at least once a day, c) a few times a week, and d) a few times a month. At the end of a week, the answers are tabulated: 31 individuals chose answer a), 55 chose b), 39 chose c), and 12 chose d).

(a) Describe how these data could be numerically and graphically summarized.

(b) Assess whether the results of this survey can be generalized to provide information about flossing habits in the general population.

1.34 Views on immigration. 910 randomly sampled registered voters from Tampa, FL were asked if they thought workers who have illegally entered the US should be (i) allowed to keep their jobs and apply for US citizenship, (ii) allowed to keep their jobs as temporary guest workers but not allowed to apply for US citizenship, or (iii) lose their jobs and have to leave the country. The results of the survey by political ideology are shown below.[83]

		Political ideology			
		Conservative	Moderate	Liberal	Total
	(i) Apply for citizenship	57	120	101	278
	(ii) Guest worker	121	113	28	262
Response	(iii) Leave the country	179	126	45	350
	(iv) Not sure	15	4	1	20
	Total	372	363	175	910

(a) What percent of these Tampa, FL voters identify themselves as conservatives?

(b) What percent of these Tampa, FL voters are in favor of the citizenship option?

(c) What percent of these Tampa, FL voters identify themselves as conservatives and are in favor of the citizenship option?

(d) What percent of these Tampa, FL voters who identify themselves as conservatives are also in favor of the citizenship option? What percent of moderates share this view? What percent of liberals share this view?

1.9.6 Relationships between two variables

1.35 Mammal life spans. Data were collected on life spans (in years) and gestation lengths (in days) for 62 mammals. A scatterplot of life span versus length of gestation is shown below.[84]

(a) Does there seem to be an association between length of gestation and life span? If so, what type of association? Explain your reasoning.

(b) What type of an association would you expect to see if the axes of the plot were reversed, i.e. if we plotted length of gestation versus life span?

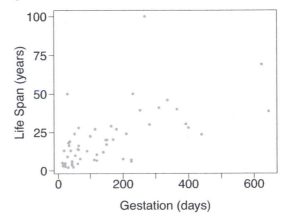

[83]SurveyUSA, News Poll #18927, data collected Jan 27-29, 2012.

[84]T. Allison and D.V. Cicchetti. "Sleep in mammals: ecological and constitutional correlates". In: *Arch. Hydrobiol* 75 (1975), p. 442.

1.36 Associations. Indicate which of the plots show a

(a) positive association

(b) negative association

(c) no association

Also determine if the positive and negative associations are linear or nonlinear. Each part may refer to more than one plot.

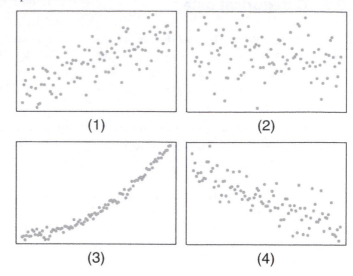

(1)   (2)

(3)   (4)

1.37 Adolescent fertility. Data are available on the number of children born to women aged 15-19 from 189 countries in the world for the years 1997, 2000, 2002, 2005, and 2006. The data are defined using a scaling similar to that used for the nursing home data in Exercise 1.30. The values for the annual adolescent fertility rates represent the number of live births among women aged 15-19 per 1,000 female members of the population of that age.

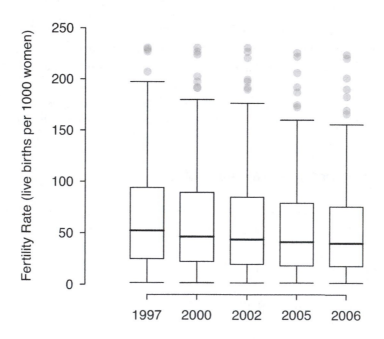

(a) In 2006, the standard deviation of the distribution of adolescent fertility is 75.73. Write a sentence explaining the 75th percentile in the context of this data.

(b) For the years 2000-2006, data are not available for Iraq. Why might those observations be missing? Would the five-number summary have been affected very much if the values had been available?

(c) From the side-by-side boxplots shown above, describe how the distribution of fertility rates changes over time. Is there a trend?

1.38 Smoking and stenosis. Researchers collected data from an observational study to investigate the association between smoking status and the presence of aortic stenosis, a narrowing of the aorta that impedes blood flow to the body.

		Smoking Status		
		Non-smoker	Smoker	Total
Disease Status	Absent	67	43	110
	Present	54	51	105
	Total	121	94	215

(a) What percentage of the 215 participants were both smokers and had aortic stenosis? This percentage is one component of the *joint distribution* of smoking and stenosis; what are the other three numbers of the joint distribution?

(b) Among the smokers, what proportion have aortic stenosis? This number is a component of the conditional distribution of stenosis for the two categories of smokers. What proportion of non-smokers have aortic stenosis?

(c) In this context, relative risk is the ratio of the proportion of smokers with stenosis to the proportion of non-smokers with stenosis. Relative risks greater than 1 indicate that smokers are at a higher risk for aortic stenosis than non-smokers; relative risks of 1.2 or higher are generally considered cause for alarm. Calculate the relative risk for the 215 participants, comparing smokers to non-smokers. Does there seem to be evidence that smoking is associated with an increased probability of stenosis?

1.39 Anger and cardiovascular health. Trait anger is defined as a relatively stable personality trait that is manifested in the frequency, intensity, and duration of feelings associated with anger. People with high trait anger have rage and fury more often, more intensely, and with long-laster episodes than people with low trait anger. It is thought that people with high trait anger might be particularly susceptible to coronary heart disease; 12,986 participants were recruited for a study examining this hypothesis. Participants were followed for five years. The following table shows data for the participants identified as having normal blood pressure (normotensives).

		Trait Anger Score			
		Low	Moderate	High	Total
CHD Event	Yes	53	110	27	190
	No	3057	4704	606	8284
	Total	3110	4731	633	8474

(a) What percentage of participants have moderate anger scores?

(b) What percentage of individuals who experienced a CHD event have moderate anger scores?

(c) What percentage of participants with high trait anger scores experienced a CHD event (i.e., heart attack)?

(d) What percentage of participants with low trait anger scores experienced a CHD event?

(e) Are individuals with high trait anger more likely to experience a CHD event than individuals with low trait anger? Calculate the relative risk of a CHD event for individuals with high trait anger compared to low trait anger.

(f) Researchers also collected data on various participant traits, such as level of blood cholesterol (measured in mg/dL). What graphical summary might be useful for examining how blood cholesterol level differs between anger groups?

1.9.7 Exploratory data analysis

Since exploratory data analysis relies heavily on the use of computation, refer to the labs for exercises related to this section, which are free and may be found at **openintro.org/book/biostat**.

Chapter 2

Probability

What are the chances that a woman with an abnormal mammogram has breast cancer? What is the probability that a woman with an abnormal mammogram has breast cancer, given that she is in her 40's? What is the likelihood that out of 100 women who undergo a mammogram and test positive for breast cancer, at least one of the women has received a false positive result?

These questions use the language of probability to express statements about outcomes that may or may not occur. More specifically, probability is used to quantify the level of uncertainty about each outcome. Like all mathematical tools, probability becomes easier to understand and work with once important concepts and terminology have been formalized.

This chapter introduces that formalization, using two types of examples. One set of examples uses settings familiar to most people – rolling dice or picking cards from a deck. The other set of examples draws from medicine, biology, and public health, reflecting the contexts and language specific to those fields. The approaches to solving these two types of problems are surprisingly similar, and in both cases, seemingly difficult problems can be solved in a series of reliable steps.

For labs, slides, and other resources, please visit
www.openintro.org/book/biostat

2.1   Defining probability

2.1.1   Some examples

The rules of probability can easily be modeled with classic scenarios, such as flipping coins or rolling dice. When a coin is flipped, there are only two possible outcomes, heads or tails. With a fair coin, each outcome is equally likely; thus, the chance of flipping heads is 1/2, and likewise for tails. The following examples deal with rolling a die or multiple dice; a die is a cube with six faces numbered 1, 2, 3, 4, 5, and 6.

EXAMPLE 2.1

What is the chance of getting 1 when rolling a die?

If the die is fair, then there must be an equal chance of rolling a 1 as any other possible number. Since there are six outcomes, the chance must be 1-in-6 or, equivalently, 1/6.

EXAMPLE 2.2

What is the chance of not rolling a 2?

Not rolling a 2 is the same as getting a 1, 3, 4, 5, or 6, which makes up five of the six equally likely outcomes and has probability 5/6.

EXAMPLE 2.3

Consider rolling two fair dice. What is the chance of getting two 1s?

If $1/6^{th}$ of the time the first die is a 1 and $1/6^{th}$ of *those* times the second die is also a 1, then the chance that both dice are 1 is (1/6)(1/6) or 1/36.

Probability can also be used to model less artificial contexts, such as to predict the inheritance of genetic disease. Cystic fibrosis (CF) is a life-threatening genetic disorder caused by mutations in the *CFTR* gene located on chromosome 7. Defective copies of *CFTR* can result in the reduced quantity and function of the CFTR protein, which leads to the buildup of thick mucus in the lungs and pancreas.[1] CF is an autosomal recessive disorder; an individual only develops CF if they have inherited two affected copies of *CFTR*. Individuals with one normal (wild-type) copy and one defective (mutated) copy are known as carriers; they do not develop CF, but may pass the disease-causing mutation onto their offspring.

---

[1]The CFTR protein is responsible for transporting sodium and chloride ions across cell membranes.

EXAMPLE 2.4

Suppose that both members of a couple are CF carriers. What is the probability that a child of this couple will be affected by CF? Assume that a parent has an equal chance of passing either gene copy (i.e., allele) to a child.

Solution 1: Enumerate all of the possible outcomes and exploit the fact that the outcomes are equally likely, as in Example 2.1. Figure 2.1 shows the four possible genotypes for a child of these parents. The paternal chromosome is in blue and the maternal chromosome in green, while chromosomes with the wild-type and mutated versions of *CFTR* are marked with + and −, respectively. The child is only affected if they have genotype (−/−), with two mutated copies of *CFTR*. Each of the four outcomes occurs with equal likelihood, so the child will be affected with probability 1-in-4, or 1/4. It is important to recognize that the child being an unaffected carrier (+/−) consists of two distinct outcomes, not one.

Solution 2: Calculate the proportion of outcomes that produce an affected child, as in Example 2.3. During reproduction, one parent will pass along an affected copy half of the time. When the child receives an affected allele from one parent, half of the those times, they will also receive an affected allele from the other parent. Thus, the proportion of times the child will have two affected copies is $(1/2) \times (1/2) = 1/4$.

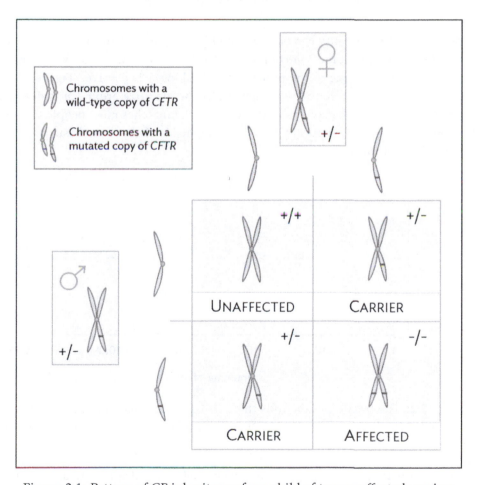

Figure 2.1: Pattern of CF inheritance for a child of two unaffected carriers

Suppose the father has CF and the mother is an unaffected carrier. What is the probability that their child will be affected by the disease?[2]

2.1.2 Probability

Probability is used to assign a level of uncertainty to the outcomes of phenomena that either happen randomly (e.g. rolling dice, inheriting of disease alleles), or appear random because of a lack of understanding about exactly how the phenomenon occurs (e.g. a woman in her 40's developing breast cancer). Modeling these complex phenomena as random can be useful, and in either case, the interpretation of probability is the same: the chance that some event will occur.

Mathematicians and philosophers have struggled for centuries to arrive at a clear statement of how probability is defined, or what it means. The most common definition is used in this text.

PROBABILITY

The **probability** of an outcome is the proportion of times the outcome would occur if the random phenomenon could be observed an infinite number of times.

This definition of probability can be illustrated by simulation. Suppose a die is rolled many times. Let $\hat{p}_n$ be the proportion of outcomes that are 1 after the first n rolls. As the number of rolls increases, $\hat{p}_n$ will converge to the probability of rolling a 1, $p = 1/6$. Figure 2.2 shows this convergence for 100,000 die rolls. The tendency of $\hat{p}_n$ to stabilize around p is described by the **Law of Large Numbers**. The behavior shown in Figure 2.2 matches most people's intuition about probability, but proving mathematically that the behavior is always true is surprisingly difficult and beyond the level of this text.

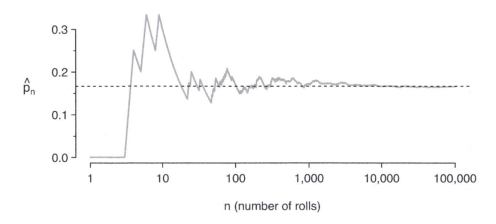

Figure 2.2: The fraction of die rolls that are 1 at each stage in a simulation. The proportion tends to get closer to the probability $1/6 \approx 0.167$ as the number of rolls increases.

Occasionally the proportion veers off from the probability and appear to defy the Law of Large Numbers, as $\hat{p}_n$ does many times in Figure 2.2. However, the likelihood of these large deviations becomes smaller as the number of rolls increases.

---

[2]Since the father has CF, he must have two affected copies; he will always pass along a defective copy of the gene. Since the mother will pass along a defective copy half of the time, the child will be affected half of the time, or with probability $(1) \times (1/2) = 1/2$.

LAW OF LARGE NUMBERS

As more observations are collected, the proportion $\hat{p}_n$ of occurrences with a particular outcome converges to the probability p of that outcome.

Probability is defined as a proportion, and it always takes values between 0 and 1 (inclusively). It may also be expressed as a percentage between 0% and 100%. The probability of rolling a 1, p, can also be written as $P(\text{rolling a 1})$.

This notation can be further abbreviated. For instance, if it is clear that the process is "rolling a die", $P(\text{rolling a 1})$ can be written as $P(1)$. There also exists a notation for an event itself; the event A of rolling a 1 can be written as $A = \{\text{rolling a 1}\}$, with associated probability $P(A)$.

$P(A)$
Probability of outcome A

2.1.3 Disjoint or mutually exclusive outcomes

Two outcomes are **disjoint** or **mutually exclusive** if they cannot both happen at the same time. When rolling a die, the outcomes 1 and 2 are disjoint since they cannot both occur. However, the outcomes 1 and "rolling an odd number" are not disjoint since both occur if the outcome of the roll is a 1.[3]

What is the probability of rolling a 1 or a 2? When rolling a die, the outcomes 1 and 2 are disjoint. The probability that one of these outcomes will occur is computed by adding their separate probabilities:

$$P(1 \text{ or } 2) = P(1) + P(2) = 1/6 + 1/6 = 1/3.$$

What about the probability of rolling a 1, 2, 3, 4, 5, or 6? Here again, all of the outcomes are disjoint, so add the individual probabilities:

$$P(1 \text{ or } 2 \text{ or } 3 \text{ or } 4 \text{ or } 5 \text{ or } 6)$$
$$= P(1) + P(2) + P(3) + P(4) + P(5) + P(6)$$
$$= 1/6 + 1/6 + 1/6 + 1/6 + 1/6 + 1/6 = 1.$$

ADDITION RULE OF DISJOINT OUTCOMES

If A_1 and A_2 represent two disjoint outcomes, then the probability that either one of them occurs is given by

$$P(A_1 \text{ or } A_2) = P(A_1) + P(A_2).$$

If there are k disjoint outcomes A_1, ..., A_k, then the probability that either one of these outcomes will occur is

$$P(A_1) + P(A_2) + \cdots + P(A_k). \tag{2.6}$$

---

[3]The terms *disjoint* and *mutually exclusive* are equivalent and interchangeable.

GUIDED PRACTICE 2.7

Ⓖ

Consider the CF example. Is the event that two carriers of CF have a child that is also a carrier represented by mutually exclusive outcomes? Calculate the probability of this event.[4]

Probability problems often deal with *sets* or *collections* of outcomes. Let A represent the event in which a die roll results in 1 or 2 and B represent the event that the die roll is a 4 or a 6. We write A as the set of outcomes {1, 2} and B = {4, 6}. These sets are commonly called **events**. Because A and B have no elements in common, they are disjoint events. A and B are represented in Figure 2.3.

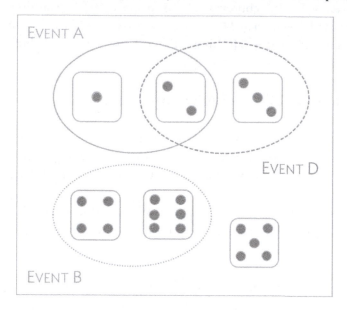

Figure 2.3: Three events, A, B, and D, consist of outcomes from rolling a die. A and B are disjoint since they do not have any outcomes in common.

The Addition Rule applies to both disjoint outcomes and disjoint events. The probability that one of the disjoint events A or B occurs is the sum of the separate probabilities:

$$P(A \text{ or } B) = P(A) + P(B) = 1/3 + 1/3 = 2/3.$$

GUIDED PRACTICE 2.8

Ⓖ

(a) Verify the probability of event A, P(A), is 1/3 using the Addition Rule. (b) Do the same for event B.[5]

GUIDED PRACTICE 2.9

Ⓖ

(a) Using Figure 2.3 as a reference, which outcomes are represented by event D? (b) Are events B and D disjoint? (c) Are events A and D disjoint?[6]

---

[4]Yes, there are two mutually exclusive outcomes for which a child of two carriers can also be a carrier - a child can either receive an affected copy of $CFTR$ from the mother and a normal copy from the father, or vice versa (since each parent can only contribute one allele). Thus, the probability that a child will be a carrier is $1/4 + 1/4 = 1/2$.

[5](a) $P(A) = P(1 \text{ or } 2) = P(1) + P(2) = \frac{1}{6} + \frac{1}{6} = \frac{2}{6} = \frac{1}{3}$. (b) Similarly, $P(B) = 1/3$.

[6](a) Outcomes 2 and 3. (b) Yes, events B and D are disjoint because they share no outcomes. (c) The events A and D share an outcome in common, 2, and so are not disjoint.

GUIDED PRACTICE 2.10

In Guided Practice 2.9, you confirmed B and D from Figure 2.3 are disjoint. Compute the probability that event B or event D occurs.[7]

2.1.4 Probabilities when events are not disjoint

Venn diagrams are useful when outcomes can be categorized as "in" or "out" for two or three variables, attributes, or random processes. The Venn diagram in Figure 2.5 uses one oval to represent diamonds and another to represent face cards (the cards labeled jacks, queens, and kings); if a card is both a diamond and a face card, it falls into the intersection of the ovals.

2♣	3♣	4♣	5♣	6♣	7♣	8♣	9♣	10♣	J♣	Q♣	K♣	A♣
2◇	3◇	4◇	5◇	6◇	7◇	8◇	9◇	10◇	J◇	Q◇	K◇	A◇
2♡	3♡	4♡	5♡	6♡	7♡	8♡	9♡	10♡	J♡	Q♡	K♡	A♡
2♠	3♠	4♠	5♠	6♠	7♠	8♠	9♠	10♠	J♠	Q♠	K♠	A♠

Figure 2.4: A regular deck of 52 cards is split into four suits: ♣ (club), ◇ (diamond), ♡ (heart), ♠ (spade). Each suit has 13 labeled cards: 2, 3, ..., 10, J (jack), Q (queen), K (king), and A (ace). Thus, each card is a unique combination of a suit and a label, e.g. 4♡ and J♣.

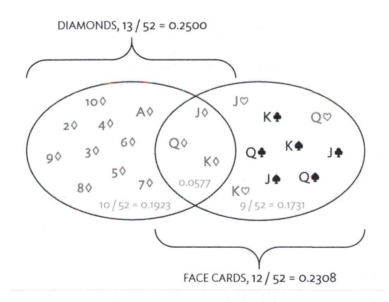

Figure 2.5: A Venn diagram for diamonds and face cards.

GUIDED PRACTICE 2.11

(a) What is the probability that a randomly selected card is a diamond? (b) What is the probability that a randomly selected card is a face card?[8]

---

[7]Since B and D are disjoint events, use the Addition Rule: $P(B \text{ or } D) = P(B) + P(D) = \frac{1}{3} + \frac{1}{3} = \frac{2}{3}$.

[8](a) There are 52 cards and 13 diamonds. If the cards are thoroughly shuffled, each card has an equal chance of being drawn, so the probability that a randomly selected card is a diamond is $P(◇) = \frac{13}{52} = 0.250$. (b) Likewise, there are 12 face cards, so $P(\text{face card}) = \frac{12}{52} = \frac{3}{13} = 0.231$.

Let A represent the event that a randomly selected card is a diamond and B represent the event that it is a face card. Events A and B are not disjoint – the cards J$\diamond$, Q$\diamond$, and K$\diamond$ fall into both categories.

As a result, adding the probabilities of the two events together is not sufficient to calculate $P(A \text{ or } B)$:

$$P(A) + P(B) = P(\diamond) + P(\text{face card}) = 12/52 + 13/52.$$

Instead, a small modification is necessary. The three cards that are in both events were counted twice. To correct the double counting, subtract the probability that both events occur:

$$
\begin{aligned}
P(A \text{ or } B) &= P(\text{face card or } \diamond) \\
&= P(\text{face card}) + P(\diamond) - P(\text{face card and } \diamond) \quad\quad (2.12) \\
&= 13/52 + 12/52 - 3/52 \\
&= 22/52 = 11/26.
\end{aligned}
$$

Equation (2.12) is an example of the **General Addition Rule**.

GENERAL ADDITION RULE

If A and B are any two events, disjoint or not, then the probability that at least one of them will occur is

$$P(A \text{ or } B) = P(A) + P(B) - P(A \text{ and } B), \quad\quad (2.13)$$

where $P(A \text{ and } B)$ is the probability that both events occur.

Note that in the language of statistics, "or" is inclusive such that A or B occurs means A, B, or both A and B occur.

GUIDED PRACTICE 2.14

(a) If A and B are disjoint, describe why this implies $P(A \text{ and } B) = 0$. (b) Using part (a), verify that the General Addition Rule simplifies to the Addition Rule for disjoint events if A and B are disjoint.[9]

GUIDED PRACTICE 2.15

Human immunodeficiency virus (HIV) and tuberculosis (TB) affect substantial proportions of the population in certain areas of the developing world. Individuals sometimes are co-infected (i.e., have both diseases). Children of HIV-infected mothers may have HIV and TB can spread from one family member to another. In a mother-child pair, let A = {the mother has HIV}, B = {the mother has TB}, C = {the child has HIV}, D = {the child has TB}. Write out the definitions of the events A or B, A and B, A and C, A or D.[10]

---

[9](a) If A and B are disjoint, A and B can never occur simultaneously. (b) If A and B are disjoint, then the last term of Equation (2.13) is 0 (see part (a)) and we are left with the Addition Rule for disjoint events.

[10]Events A or B: the mother has HIV, the mother has TB, or the mother has both HIV and TB. Events A and B: the mother has both HIV and TB. Events A and C: The mother has HIV and the child has HIV. A or D: The mother has HIV, the child has TB, or the mother has HIV and the child has TB.

2.1.5 Probability distributions

A **probability distribution** consists of all disjoint outcomes and their associated probabilities. Figure 2.6 shows the probability distribution for the sum of two dice.

Dice sum	2	3	4	5	6	7	8	9	10	11	12
Probability	$\frac{1}{36}$	$\frac{2}{36}$	$\frac{3}{36}$	$\frac{4}{36}$	$\frac{5}{36}$	$\frac{6}{36}$	$\frac{5}{36}$	$\frac{4}{36}$	$\frac{3}{36}$	$\frac{2}{36}$	$\frac{1}{36}$

Figure 2.6: Probability distribution for the sum of two dice.

RULES FOR A PROBABILITY DISTRIBUTION

A probability distribution is a list of all possible outcomes and their associated probabilities that satisfies three rules:

1. The outcomes listed must be disjoint.

2. Each probability must be between 0 and 1.

3. The probabilities must total to 1.

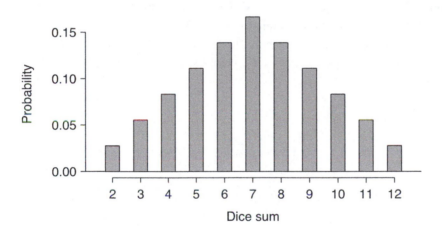

Figure 2.7: The probability distribution of the sum of two dice.

Probability distributions can be summarized in a bar plot. The probability distribution for the sum of two dice is shown in Figure 2.7, with the bar heights representing the probabilities of outcomes.

Figure 2.8 shows a bar plot of the birth weight data for 3,999,386 live births in the United States in 2010, for which total counts have been converted to proportions. Since birth weight trends do not change much between years, it is valid to consider the plot as a representation of the probability distribution of birth weights for upcoming years, such as 2017. The data are available as part of the US CDC National Vital Statistics System.[11]

The graph shows that while most babies born weighed between 2000 and 5000 grams (2 to 5 kg), there were both small (less than 1000 grams) and large (greater than 5000 grams) babies. Pediatricians consider birth weights between 2.5 and 5 kg as normal.[12] A probability distribution gives a sense of which outcomes can be considered unusual (i.e., outcomes with low probability).

---

[11]http://205.207.175.93/vitalstats/ReportFolders/reportFolders.aspx

[12]https://www.nlm.nih.gov/medlineplus/birthweight.html

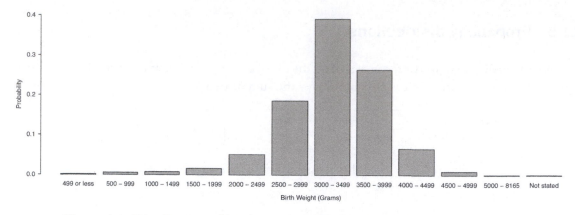

Figure 2.8: Distribution of birth weights (in grams) of babies born in the US in 2010.

Continuous probability distributions

Probability distributions for events that take on a finite number of possible outcomes, such as the sum of two dice rolls, are referred to as **discrete probability distributions**.

Consider how the probability distribution for adult heights in the US might best be represented. Unlike the sum of two dice rolls, height can occupy any value over a continuous range. Thus, height has a **continuous probability distribution**, which is specified by a **probability density function** rather than a table; Figure 2.9 shows a histogram of the height for 3 million US adults from the mid-1990's, with an overlaid density curve.[13]

Just as in the discrete case, the probabilities of all possible outcomes must still sum to 1; the total area under a probability density function equals 1.

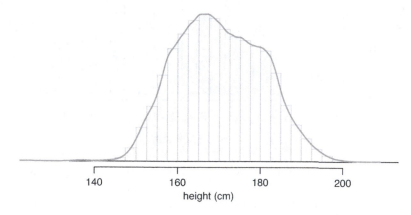

Figure 2.9: The continuous probability distribution of heights for US adults.

---

[13]This sample can be considered a simple random sample from the US population. It relies on the USDA Food Commodity Intake Database.

EXAMPLE 2.16

Estimate the probability that a randomly selected adult from the US population has height between 180 and 185 centimeters. In Figure 2.10(a), the two bins between 180 and 185 centimeters have counts of 195,307 and 156,239 people.

Find the proportion of the histogram's area that falls in the range 180 cm and 185: add the heights of the bins in the range and divide by the sample size:

$$\frac{195,307 + 156,239}{3,000,000} = 0.1172.$$

The probability can be calculated precisely with the use of computing software, by finding the area of the shaded region under the curve between 180 and 185:

$$P(\text{height between 180 and 185}) = \text{area between 180 and 185} = 0.1157.$$

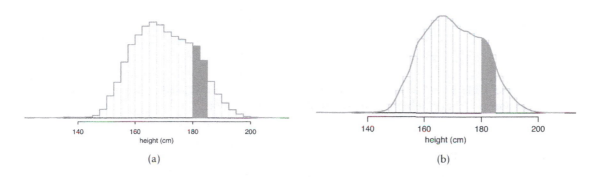

(a)   (b)

Figure 2.10: (a) A histogram with bin sizes of 2.5 cm, with bars between 180 and 185 cm shaded. (b) Density for heights in the US adult population with the area between 180 and 185 cm shaded.

EXAMPLE 2.17

What is the probability that a randomly selected person is **exactly** 180 cm? Assume that height can be measured perfectly.

This probability is zero. A person might be close to 180 cm, but not exactly 180 cm tall. This also coheres with the definition of probability as an area under the density curve; there is no area captured between 180 cm and 180 cm.

GUIDED PRACTICE 2.18

Suppose a person's height is rounded to the nearest centimeter. Is there a chance that a random person's **measured** height will be 180 cm?[14]

---

[14]This has positive probability. Anyone between 179.5 cm and 180.5 cm will have a *measured* height of 180 cm. This a more realistic scenario to encounter in practice versus Example 2.17.

2.1.6 Complement of an event

S

Sample space

A^c

Complement
of outcome A

Rolling a die produces a value in the set {1, 2, 3, 4, 5, 6}. This set of all possible outcomes is called the **sample space** (S) for rolling a die.

Let $D = \{2, 3\}$ represent the event that the outcome of a die roll is 2 or 3. The **complement** of D represents all outcomes in the sample space that are not in D, which is denoted by $D^c = \{1, 4, 5, 6\}$. That is, D^c is the set of all possible outcomes not already included in D. Figure 2.11 shows the relationship between D, D^c, and the sample space S.

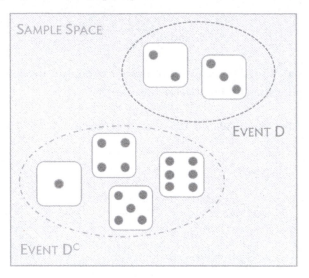

Figure 2.11: Event $D = \{2, 3\}$ and its complement, $D^c = \{1, 4, 5, 6\}$. S represents the sample space, which is the set of all possible events.

GUIDED PRACTICE 2.19

(a) Compute $P(D^c) = P(\text{rolling a 1, 4, 5, or 6})$. (b) What is $P(D) + P(D^c)$?[15]

GUIDED PRACTICE 2.20

Events $A = \{1, 2\}$ and $B = \{4, 6\}$ are shown in Figure 2.3 on page 94. (a) Write out what A^c and B^c represent. (b) Compute $P(A^c)$ and $P(B^c)$. (c) Compute $P(A) + P(A^c)$ and $P(B) + P(B^c)$.[16]

A complement of an event A is constructed to have two very important properties: every possible outcome not in A is in A^c, and A and A^c are disjoint. If every possible outcome not in A is in A^c, this implies that

$$P(A \text{ or } A^c) = 1. \tag{2.21}$$

Then, by Addition Rule for disjoint events,

$$P(A \text{ or } A^c) = P(A) + P(A^c). \tag{2.22}$$

Combining Equations (2.21) and (2.22) yields a useful relationship between the probability of an event and its complement.

---

[15](a) The outcomes are disjoint and each has probability 1/6, so the total probability is 4/6 = 2/3. (b) We can also see that $P(D) = \frac{1}{6} + \frac{1}{6} = 1/3$. Since D and D^c are disjoint, $P(D) + P(D^c) = 1$.

[16]Brief solutions: (a) $A^c = \{3, 4, 5, 6\}$ and $B^c = \{1, 2, 3, 5\}$. (b) Noting that each outcome is disjoint, add the individual outcome probabilities to get $P(A^c) = 2/3$ and $P(B^c) = 2/3$. (c) A and A^c are disjoint, and the same is true of B and B^c. Therefore, $P(A) + P(A^c) = 1$ and $P(B) + P(B^c) = 1$.

COMPLEMENT

The complement of event A is denoted A^c, and A^c represents all outcomes not in A. A and A^c are mathematically related:

$$P(A) + P(A^c) = 1, \quad \text{i.e.} \quad P(A) = 1 - P(A^c). \tag{2.23}$$

In simple examples, computing either A or A^c is feasible in a few steps. However, as problems grow in complexity, using the relationship between an event and its complement can be a useful strategy.

GUIDED PRACTICE 2.24

Let A represent the event of selecting an adult from the US population with height between 180 and 185 cm, as calculated in Example 2.16. What is $P(A^c)$?[17]

GUIDED PRACTICE 2.25

Let A represent the event in which two dice are rolled and their total is less than 12. (a) What does the event A^c represent? (b) Determine $P(A^c)$ from Figure 2.6 on page 97. (c) Determine $P(A)$.[18]

GUIDED PRACTICE 2.26

Consider again the probabilities from Figure 2.6 and rolling two dice. Find the following probabilities: (a) The sum of the dice is *not* 6. (b) The sum is at least 4. That is, determine the probability of the event $B = \{4, 5, ..., 12\}$. (c) The sum is no more than 10. That is, determine the probability of the event $D = \{2, 3, ..., 10\}$.[19]

2.1.7 Independence

Just as variables and observations can be independent, random phenomena can also be independent. Two processes are **independent** if knowing the outcome of one provides no information about the outcome of the other. For instance, flipping a coin and rolling a die are two independent processes – knowing that the coin lands heads up does not help determine the outcome of the die roll. On the other hand, stock prices usually move up or down together, so they are not independent.

---

[17] $P(A^c) = 1 - P(A) = 1 - 0.1157 = 0.8843$.

[18] (a) The complement of A: when the total is equal to 12. (b) $P(A^c) = 1/36$. (c) Use the probability of the complement from part (b), $P(A^c) = 1/36$, and Equation (2.23): $P(\text{less than } 12) = 1 - P(12) = 1 - 1/36 = 35/36$.

[19] (a) First find $P(6) = 5/36$, then use the complement: $P(\text{not } 6) = 1 - P(6) = 31/36$.
 (b) First find the complement, which requires much less effort: $P(2 \text{ or } 3) = 1/36 + 2/36 = 1/12$. Then calculate $P(B) = 1 - P(B^c) = 1 - 1/12 = 11/12$.
 (c) As before, finding the complement is the more direct way to determine $P(D)$. First find $P(D^c) = P(11 \text{ or } 12) = 2/36 + 1/36 = 1/12$. Then calculate $P(D) = 1 - P(D^c) = 11/12$.

Example 2.3 provides a basic example of two independent processes: rolling two dice. What is the probability that both will be 1? Suppose one of the dice is blue and the other green. If the outcome of the blue die is a 1, it provides no information about the outcome of the green die. This question was first encountered in Example 2.3: $1/6^{th}$ of the time the blue die is a 1, and $1/6^{th}$ of *those* times the green die will also be 1. This is illustrated in Figure 2.12. Because the rolls are independent, the probabilities of the corresponding outcomes can be multiplied to obtain the final answer: $(1/6)(1/6) = 1/36$. This can be generalized to many independent processes.

Figure 2.12: $1/6^{th}$ of the time, the first roll is a 1. Then $1/6^{th}$ of *those* times, the second roll will also be a 1.

Complicated probability problems, such as those that arise in biology or medicine, are often solved with the simple ideas used in the dice example. For instance, independence was used implicitly in the second solution to Example 2.4, when calculating the probability that two carriers will have an affected child with cystic fibrosis. Genes are typically passed along from the mother and father independently. This allows for the assumption that, on average, half of the offspring who receive a mutated gene copy from the mother will also receive a mutated copy from the father.

GUIDED PRACTICE 2.27

What if there were also a red die independent of the other two? What is the probability of rolling the three dice and getting all 1s?[20]

GUIDED PRACTICE 2.28

Three US adults are randomly selected. The probability the height of a single adult is between 180 and 185 cm is 0.1157.[21]

(a) What is the probability that all three are between 180 and 185 cm tall?

(b) What is the probability that none are between 180 and 185 cm tall?

---

[20]The same logic applies from Example 2.3. If $1/36^{th}$ of the time the blue and green dice are both 1, then $1/6^{th}$ of *those* times the red die will also be 1, so multiply:

$$P(blue = 1 \text{ and } green = 1 \text{ and } red = 1) = P(blue = 1)P(green = 1)P(red = 1)$$
$$= (1/6)(1/6)(1/6) = 1/216.$$

[21]Brief answers: (a) $0.1157 \times 0.1157 \times 0.1157 = 0.0015$. (b) $(1 - 0.1157)^3 = 0.692$.

MULTIPLICATION RULE FOR INDEPENDENT PROCESSES

If A and B represent events from two different and independent processes, then the probability that both A and B occur is given by:

$$P(A \text{ and } B) = P(A)P(B). \tag{2.29}$$

Similarly, if there are k events $A_1, ..., A_k$ from k independent processes, then the probability they all occur is

$$P(A_1)P(A_2)\cdots P(A_k).$$

EXAMPLE 2.30

Mandatory drug testing. Mandatory drug testing in the workplace is common practice for certain professions, such as air traffic controllers and transportation workers. A false positive in a drug screening test occurs when the test incorrectly indicates that a screened person is an illegal drug user. Suppose a mandatory drug test has a false positive rate of 1.2% (i.e., has probability 0.012 of indicating that an employee is using illegal drugs when that is not the case). Given 150 employees who are in reality drug free, what is the probability that at least one will (falsely) test positive? Assume that the outcome of one drug test has no effect on the others.

First, note that the complement of at least 1 person testing positive is that no one tests positive (i.e., all employees test negative). The multiplication rule can then be used to calculate the probability of 150 negative tests.

$$\begin{aligned}
P(\text{At least 1 "+"}) &= P(1 \text{ or } 2 \text{ or } 3 \dots \text{ or } 150 \text{ are "+"}) \\
&= 1 - P(\text{None are "+"}) \\
&= 1 - P(150 \text{ are "-"}) \\
&= 1 - P(\text{"-"})^{150} \\
&= 1 - (0.988)^{150} = 1 - 0.16 = 0.84.
\end{aligned}$$

Even when using a test with a small probability of a false positive, the company is more than 80% likely to incorrectly claim at least one employee is an illegal drug user!

GUIDED PRACTICE 2.31

Because of the high likelihood of at least one false positive in company wide drug screening programs, an individual with a positive test is almost always re-tested with a different screening test: one that is more expensive than the first, but has a lower false positive probability. Suppose the second test has a false positive rate of 0.8%. What is the probability that an employee who is not using illegal drugs will test positive on both tests?[22]

---

[22]The outcomes of the two tests are independent of one another; $P(A \text{ and } B) = P(A) \times P(B)$, where events A and B are the results of the two tests. The probability of a false positive with the first test is 0.012 and 0.008 with the second. Thus, the probability of an employee who is not using illegal drugs testing positive on both tests is $0.012 \times 0.008 = 9.6 \times 10^{-5}$

Figure 2.13: Inheritance of ABO blood groups.

EXAMPLE 2.32

ABO blood groups. There are four different common blood types (A, B, AB, and O), which are determined by the presence of certain antigens located on cell surfaces. Antigens are substances used by the immune system to recognize self versus non-self; if the immune system encounters antigens not normally found on the body's own cells, it will attack the foreign cells. When patients receive blood transfusions, it is critical that the antigens of transfused cells match those of the patient's, or else an immune system response will be triggered.

The ABO blood group system consists of four different blood groups, which describe whether an individual's red blood cells carry the A antigen, B antigen, both, or neither. The ABO gene has three alleles: I^A, I^B, and i. The i allele is recessive to both I^A and I^B, and does not produce antigens; thus, an individual with genotype $I^A i$ is blood group A and an individual with genotype $I^B i$ is blood group B. The I^A and I^B alleles are codominant, such that individuals of $I^A I^B$ genotype are AB. Individuals homozygous for the i allele are known as blood group O, with neither A nor B antigens.

Suppose that both members of a couple have Group AB blood.

a) What is the probability that a child of this couple will have Group A blood?

b) What is the probability that they have two children with Group A blood?

---

a) An individual with Group AB blood is genotype $I^A I^B$. Two $I^A I^B$ parents can produce children with genotypes $I^A I^B$, $I^A I^A$, or $I^B I^B$. Of these possibilities, only children with genotype $I^A I^A$ have Group A blood. Each parent has 0.5 probability of passing down their I^A allele. Thus, the probability that a child of this couple will have Group A blood is P(parent 1 passes down I^A allele) × P(parent 2 passes down I^A allele) = $0.5 \times 0.5 = 0.25$.

b) Inheritance of alleles is independent between children. Thus, the probability of two children having Group A blood equals P(child 1 has Group A blood) × P(child 2 has group A blood). The probability of a child of this couple having Group A blood was previously calculated as 0.25. The answer is given by $0.25 \times 0.25 = 0.0625$.

The previous examples in this section have used independence to solve probability problems. The definition of independence can also be used to check whether two events are independent – two events A and B are independent if they satisfy Equation (2.29).

EXAMPLE 2.33

Is the event of drawing a heart from a deck of cards independent of drawing an ace?

_____

The probability the card is a heart is 1/4 (13/52 = 1/4) and the probability that it is an ace is 1/13 (4/52 = 1/13). The probability that the card is the ace of hearts (A♡) is 1/52. Check whether Equation 2.29 is satisfied:

$$P(\heartsuit)P(\text{A}) = \left(\frac{1}{4}\right)\left(\frac{1}{13}\right) = \frac{1}{52} = P(\heartsuit \text{ and A}).$$

Since the equation holds, the event that the card is a heart and the event that the card is an ace are independent events.

EXAMPLE 2.34

In the general population, about 15% of adults between 25 and 40 years of age are hypertensive. Suppose that among males of this age, hypertension occurs about 18% of the time. Is hypertension independent of sex?

_____

Assume that the population is 50% male, 50% female; it is given in the problem that hypertension occurs about 15% of the time in adults between ages 25 and 40.

$$P(\text{hypertension}) \times P(\text{male}) = (0.15)(0.50) = 0.075 \neq 0.18.$$

Equation 2.29 is not satisfied, therefore hypertension is not independent of sex. In other words, knowing whether an individual is male or female is informative as to whether they are hypertensive. If hypertension and sex were independent, then we would expect hypertension to occur at an equal rate in males as in females.

2.2   Conditional probability

While it is difficult to obtain precise estimates, the US CDC estimated that in 2012, approximately 29.1 million Americans had type 2 diabetes – about 9.3% of the population.[23] A health care practitioner seeing a new patient would expect a 9.3% chance that the patient might have diabetes.

However, this is only the case if nothing is known about the patient. The prevalence of type 2 diabetes varies with age. Between the ages of 20 and 44, only about 4% of the population have diabetes, but almost 27% of people age 65 and older have the disease. Knowing the age of a patient provides information about the chance of diabetes; age and diabetes status are not independent. While the probability of diabetes in a randomly chosen member of the population is 0.093, the *conditional* probability of diabetes in a person known to be 65 or older is 0.27.

Conditional probability is used to characterize how the probability of an outcome varies with the knowledge of another factor or condition, and is closely related to the concepts of marginal and joint probabilities.

2.2.1   Marginal and joint probabilities

Figures 2.14 and 2.15 provide additional information about the relationship between diabetes prevalence and age.[24] Figure 2.14 is a contingency table for the entire US population in 2012; the values in the table are in thousands (to make the table more readable).

	Diabetes	No Diabetes	Sum
Less than 20 years	200	86,664	86,864
20 to 44 years	4,300	98,724	103,024
45 to 64 years	13,400	68,526	81,926
Greater than 64 years	11,200	30,306	41,506
Sum	29,100	284,220	313,320

Figure 2.14: Contingency table showing type 2 diabetes status and age group, in thousands.

In the first row, for instance, Figure 2.14 shows that in the entire population of approximately 313,320,000 people, approximately 200,000 individuals were in the less than 20 years age group and diagnosed with diabetes – about 0.1%. The table also indicates that among the approximately 86,864,000 individuals less than 20 years of age, only 200,000 suffered from type 2 diabetes, approximately 0.2%. The distinction between these two statements is small but important. The first provides information about the size of the group with type 2 diabetes population that is less than 20 years of age, relative to the entire population. In contrast, the second statement is about the size of the diabetes population within the less than 20 years of age group, relative to the size of that age group.

---

[23]21 million of these cases are diagnosed, while the CDC predicts that 8.1 million cases are undiagnosed; that is, approximately 8.1 million people are living with diabetes, but they (and their physicians) are unaware that they have the condition.

[24]Because the CDC provides only approximate numbers for diabetes prevalence, the numbers in the table are approximations of actual population counts.

GUIDED PRACTICE 2.35

What fraction of the US population are 45 to 64 years of age and have diabetes? What fraction of the population age 45 to 64 have diabetes?[25]

The entries in Figure 2.15 show the proportions of the population in each of the eight categories defined by diabetes status and age, obtained by dividing each value in the cells of Figure 2.14 by the total population size.

	Diabetes	No Diabetes	Sum
Less than 20 years	0.001	0.277	0.277
20 to 44 years	0.014	0.315	0.329
45 to 64 years	0.043	0.219	0.261
Greater than 64 years	0.036	0.097	0.132
Sum	0.093	0.907	1.000

Figure 2.15: Probability table summarizing diabetes status and age group.

If these proportions are interpreted as probabilities for randomly chosen individuals from the population, the value 0.014 in the first column of the second row implies that the probability of selecting someone at random who has diabetes and whose age is between 20 and 44 is 0.014, or 1.4%. The entries in the eight main table cells (i.e., excluding the values in the margins) are **joint probabilities**, which specify the probability of two events happening at the same time – in this case, diabetes and a particular age group. In probability notation, this joint probability can be expressed as 0.014 = P(diabetes and age 20 to 44).[26]

The values in the last row and column of the table are the sums of the corresponding rows or columns. The sum of the of the probabilities of the disjoint events (diabetes, age 20 to 44) and (no diabetes, age 20 to 44), 0.329, is the probability of being in the age group 20 to 44. The row and column sums are **marginal probabilities**; they are probabilities about only one type of event, such as age. For example, the sum of the first column (0.093) is the marginal probability of a member of the population having diabetes.

MARGINAL AND JOINT PROBABILITIES

A *marginal probability* is a probability only related to a single event or process, such as P(A).
A *joint probability* is the probability that two or more events or processes occur jointly, such as P(A and B).

GUIDED PRACTICE 2.36

What is the interpretation of the value 0.907 in the last row of the table? And of the value 0.097 directly above it?[27]

---

[25]The first value is given by the intersection of "45 - 64 years of age" and "diabetes", divided by the total population number: 13,400,000/313,320,000 = 0.043. The second value is given by dividing 13,400,000 by 81,926,000, the number of individuals in that age group: 13,400,000/81,926,000 = 0.164.

[26]Alternatively, this is commonly written as as P(diabetes, age 20 to 44), with a comma replacing "and".

[27]The value 0.907 in the last row indicates the total proportion of individuals in the population who do not have diabetes. The value 0.097 indicates the joint probability of not having diabetes and being in the greater than 64 years age group.

2.2.2 Defining conditional probability

The probability that a randomly selected individual from the US has diabetes is 0.093, the sum of the first column in Figure 2.15. How does that probability change if it is known that the individual's age is 64 or greater?

The conditional probability can be calculated from Figure 2.14, which shows that 11,200,000 of the 41,506,000 people in that age group have diabetes, so the likelihood that someone from that age group has diabetes is:

$$\frac{11,200,000}{41,506,000} = 0.27,$$

or 27%. The additional information about a patient's age allows for a more accurate estimate of the probability of diabetes.

Similarly, the conditional probability can be calculated from the joint and marginal proportions in Figure 2.15. Consider the main difference between the conditional probability versus the joint and marginal probabilities. Both the joint probability and marginal probabilities are probabilities relative to the entire population. However, the conditional probability is the probability of having diabetes, *relative only to* the segment of the population greater than the age of 64.

Intuitively, the denominator in the calculation of a conditional probability must account for the fact that only a segment of the population is being considered, rather than the entire population. The conditional probability of diabetes given age 64 or older is simply the joint probability of having diabetes and being greater than 64 years of age divided by the marginal probability of being in that age group:

$$\frac{\text{prop. of population with diabetes, age 64 or greater}}{\text{prop. of population greater than age 64}} = \frac{11,200,000/313,320,000}{41,506,000/313,320,000}$$
$$= \frac{0.036}{0.132}$$
$$= 0.270.$$

This leads to the mathematical definition of conditional probability.

CONDITIONAL PROBABILITY

The conditional probability of an event A given an event or condition B is:

$$P(A|B) = \frac{P(A \text{ and } B)}{P(B)}. \tag{2.37}$$

GUIDED PRACTICE 2.38

Calculate the probability that a randomly selected person has diabetes, given that their age is between 45 and 64.[28]

---

[28]Let A be the event a person has diabetes, and B the event that their age is between 45 and 64. Use the information in Figure 2.15 to calculate $P(A|B)$. $P(A|B) = \frac{P(A \text{ and } B)}{P(B)} = \frac{0.043}{0.261} = 0.165.$

> **GUIDED PRACTICE 2.39**
>
> (G) Calculate the probability that a randomly selected person is between 45 and 64 years old, given that the person has diabetes.[29]

Conditional probabilities have similar properties to regular (unconditional) probabilities.

SUM OF CONDITIONAL PROBABILITIES

Let $A_1, ..., A_k$ represent all the disjoint outcomes for a variable or process. Then if B is an event, possibly for another variable or process, we have:

$$P(A_1|B) + \cdots + P(A_k|B) = 1.$$

The rule for complements also holds when an event and its complement are conditioned on the same information:

$$P(A|B) = 1 - P(A^c|B).$$

> **GUIDED PRACTICE 2.40**
>
> (G) Calculate the probability a randomly selected person is older than 20 years of age, given that the person has diabetes.[30]

---

[29] Again, let A be the event a person has diabetes, and B the event that their age is between 45 and 64. Find $P(B|A)$. $P(B|A) = \frac{P(A \text{ and } B)}{P(A)} = \frac{0.043}{0.093} = 0.462$.

[30] Let A be the event that a person has diabetes, and B be the event that their age is less than 20 years. The desired probability is $P(B^c|A) = 1 - P(B|A) = 1 - \frac{0.001}{0.093} = 0.989$.

2.2.3   General multiplication rule

Section 2.1.7 introduced the Multiplication Rule for independent processes. Here, the **General Multiplication Rule** is introduced for events that might not be independent.

GENERAL MULTIPLICATION RULE

If A and B represent two outcomes or events, then

$$P(A \text{ and } B) = P(A|B)P(B).$$

It is useful to think of A as the outcome of interest and B as the condition.

This General Multiplication Rule is simply a rearrangement of the definition for conditional probability in Equation (2.37) on page 108.

EXAMPLE 2.41

Suppose that among male adults between 25 and 40 years of age, hypertension occurs about 18% of the time. Assume that the population is 50% male, 50% female. What is the probability of randomly selecting a male with hypertension from the population of individuals 25-40 years of age?

Let A be the event that a person has hypertension, and B the event that they are a male adult between 25 and 40 years of age. $P(A|B)$, the probability of hypertension given male sex, is 0.18. Thus, $P(A \text{ and } B) = (0.18)(0.50) = 0.09$.

2.2.4   Independence and conditional probability

If two events are independent, knowing the outcome of one should provide no information about the other.

EXAMPLE 2.42

Let X and Y represent the outcomes of rolling two dice. Use the formula for conditional probability to compute $P(Y = 1 \mid X = 1)$. What is $P(Y = 1)$? Is this different from $P(Y = 1 \mid X = 1)$?

$$\frac{P(Y = 1 \text{ and } X = 1)}{P(X = 1)} = \frac{1/36}{1/6} = 1/6.$$

The probability $P(Y = 1) = 1/6$ is the same as the conditional probability. The probability that $Y = 1$ was unchanged by knowledge about X, since the events X and Y are independent.

Using the Multiplication Rule for independent events allows for a mathematical illustration of why the condition information has no influence in Example 2.42:

$$
\begin{aligned}
P(Y = 1 \mid X = 1) &= \frac{P(Y = 1 \text{ and } X = 1)}{P(X = 1)} \\
&= \frac{P(Y = 1)P(X = 1)}{P(X = 1)} \\
&= P(Y = 1).
\end{aligned}
$$

This is a specific instance of the more general result that if two events A and B are independent, $P(A|B) = P(A)$ as long as $P(B) > 0$:

$$
\begin{aligned}
P(A|B) &= \frac{P(A \text{ and } B)}{P(B)} \\
&= \frac{P(A)P(B)}{P(B)} \\
&= P(A).
\end{aligned}
$$

GUIDED PRACTICE 2.43

In the US population, about 45% of people are blood group O. Suppose that 40% of Asian people living in the US are blood group O, and that the Asian population in the United States is approximately 4%. Do these data suggest that blood group is independent of ethnicity?[31]

2.2.5 Bayes' Theorem

This chapter began with a straightforward question – what are the chances that a woman with an abnormal (i.e., positive) mammogram has breast cancer? For a clinician, this question can be rephrased as the conditional probability that a woman has breast cancer, given that her mammogram is abnormal. This conditional probability is called the **positive predictive value** (PPV) of a mammogram. More concisely, if $A = \{$a woman has breast cancer$\}$, and $B = \{$a mammogram is positive$\}$, the PPV of a mammogram is $P(A|B)$.

The characteristics of a mammogram (and other diagnostic tests) are given with the reverse conditional probabilities—the probability that the mammogram correctly returns a positive result if a woman has breast cancer, as well as the probability that the mammogram correctly returns a negative result if a woman does not have breast cancer. These are the probabilities $P(B|A)$ and $P(B^c|A^c)$, respectively.

Given the probabilities $P(B|A)$ and $P(B^c|A^c)$, as well as the marginal probability of disease $P(A)$, how can the positive predictive value $P(A|B)$ be calculated?

There are several possible strategies for approaching this type of problem—1) constructing tree diagrams, 2) using a purely algebraic approach using Bayes' Theorem, and 3) creating contingency tables based on calculating conditional probabilities from a large, hypothetical population.

---

[31] Let A represent blood group O, and B represent Asian ethnicity. Since $P(A|B) = 0.40$ does not equal $P(A) = 0.45$, the two events are not independent. Blood group does not seem to be independent of ethnicity.

EXAMPLE 2.44

In Canada, about 0.35% of women over 40 will develop breast cancer in any given year. A common screening test for cancer is the mammogram, but it is not perfect. In about 11% of patients with breast cancer, the test gives a **false negative**: it indicates a woman does not have breast cancer when she does have breast cancer. Similarly, the test gives a **false positive** in 7% of patients who do not have breast cancer: it indicates these patients have breast cancer when they actually do not.[32] If a randomly selected woman over 40 is tested for breast cancer using a mammogram and the test is positive – that is, the test suggests the woman has cancer – what is the probability she has breast cancer?

Read on in the text for three solutions to this example.

Example 2.44 Solution 1. Tree Diagram.

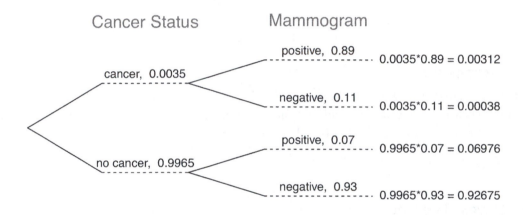

Figure 2.16: A tree diagram for breast cancer screening.

A **tree diagram** is a tool to organize outcomes and probabilities around the structure of data, and is especially useful when two or more processes occur in a sequence, with each process conditioned on its predecessors.

In Figure 2.16, the primary branches split the population by cancer status, and show the marginal probabilities 0.0035 and 0.9965 of having cancer or not, respectively. The secondary branches are conditioned on the primary branch and show conditional probabilities; for example, the top branch is the probability that a mammogram is positive given that an individual has cancer. The problem provides enough information to compute the probability of testing positive if breast cancer is present, since this probability is the complement of the probability of a false negative: $1 - 0.11 = 0.89$.

Joint probabilities can be constructed at the end of each branch by multiplying the numbers from right to left, such as the probability that a woman tests positive given that she has breast cancer (abbreviated as BC), times the probability she has breast cancer:

$$P(\text{BC and mammogram}^+) = P(\text{mammogram}^+ \mid \text{BC}) \times P(\text{BC})$$
$$= (0.89)(0.0035) = 0.00312.$$

---

[32] The probabilities reported here were obtained using studies reported at www.breastcancer.org and www.ncbi.nlm.nih.gov/pmc/articles/PMC1173421.

Using the tree diagram allows for the information in the problem to be mapped out in a way that makes it easier to calculate the desired conditional probability. In this case, the diagram makes it clear that there are two scenarios in which someone can test positive: either testing positive when having breast cancer or by testing positive in the absence of breast cancer. To find the probability that a woman has breast cancer given that she tests positive, apply the conditional probability formula: divide the probability of testing positive when having breast cancer by the probability of testing positive.

The probability of a positive test result is the sum of the two corresponding scenarios:

$$P(\text{mammogram}^+) = P(\text{mammogram}^+ \text{ and has BC}) + P(\text{mammogram}^+ \text{ and no BC})$$
$$= [P(\text{mammogram}^+ \mid \text{has BC}) \times P(\text{has BC})] + [P(\text{mammogram}^+ \mid \text{no BC}) \times P(\text{no BC})]$$
$$= (0.0035)(0.89) + (0.9965)(0.07) = 0.07288.$$

Thus, if the mammogram screening is positive for a patient, the probability that the patient has breast cancer is given by:

$$P(\text{has BC} \mid \text{mammogram}^+) = \frac{P(\text{has BC and mammogram}^+)}{P(\text{mammogram}^+)}$$
$$= \frac{0.00312}{0.07288} \approx 0.0428.$$

Even with a positive mammogram, there is still only a 4% chance of breast cancer! It may seem surprising that even when the false negative and false positive probabilities of the test are small (0.11 and 0.07, respectively), the conditional probability of disease given a positive test could also be so small. In this population, the probability that a woman does not have breast cancer is high (1 - 0.0035 = 0.9965), which results in a relatively high number of false positives in comparison to true positives.

Calculating probabilities for diagnostic tests is done so often in medicine that the topic has some specialized terminology. The **sensitivity** of a test is the probability of a positive test result when disease is present, such as a positive mammogram when a patient has breast cancer. The **specificity** of a test is the probability of a negative test result when disease is absent.[33] The probability of disease in a population is referred to as the **prevalence**. With specificity and sensitivity information for a particular test, along with disease prevalence, the **positive predictive value** (PPV) can be calculated: the probability that disease is present when a test result is positive. Similarly, the **negative predictive value** is the probability that disease is absent when test results are negative. These terms are used for nearly all diagnostic tests used to screen for diseases.

GUIDED PRACTICE 2.45

Identify the prevalence, sensitivity, specificity, and PPV from the scenario in Example 2.44.[34]

---

[33]The sensitivity and specificity are, respectively, the probability of a true positive test result and the probability of a true negative test result.

[34]The prevalence of breast cancer is 0.0035. The sensitivity is the probability of a positive test result when disease is present, which is the complement of a false negative: 1 − 0.11 = 0.89. The specificity is the probability of a negative test result when disease is absent, which is the complement of a false positive: 1 − 0.07 = 0.93. The PPV is 0.04, the probability of breast cancer given a positive mammogram.

Example 2.44 Solution 2. Bayes' Rule.

The process used to solve the problem via the tree diagram can be condensed into a single algebraic expression by substituting the original probability expressions into the numerator and denominator:

$$P(\text{has BC} \mid \text{mammogram}^+) = \frac{P(\text{has BC and mammogram}^+)}{P(\text{mammogram}^+)}$$

$$= \frac{P(\text{mammogram}^+ \mid \text{has BC}) \times P(\text{has BC})}{[P(\text{mammogram}^+ \mid \text{has BC}) \times P(\text{has BC})] + [P(\text{mammogram}^+ \mid \text{no BC}) \times P(\text{no BC})]}.$$

The expression can also be written in terms of diagnostic testing language, where $D = \{\text{has disease}\}$, $D^c = \{\text{does not have disease}\}$, $T^+ = \{\text{positive test result}\}$, and $T^- = \{\text{negative test result}\}$.

$$P(D|T^+) = \frac{P(D \text{ and } T^+)}{P(T^+)}$$

$$= \frac{P(T^+|D) \times P(D)}{[P(T^+|D) \times P(D)] + [P(T^+|D^c) \times P(D^c)]}$$

$$\text{PPV} = \frac{\text{sensitivity} \times \text{prevalence}}{[\text{sensitivity} \times \text{prevalence}] + [(1 - \text{specificity}) \times (1 - \text{prevalence})]}.$$

The generalization of this formula is known as Bayes' Theorem or Bayes' Rule.

BAYES' THEOREM

Consider the following conditional probability for variable 1 and variable 2:

$P(\text{outcome } A_1 \text{ of variable 1} \mid \text{outcome } B \text{ of variable 2}).$

Bayes' Theorem states that this conditional probability can be identified as the following fraction:

$$\frac{P(B|A_1)P(A_1)}{P(B|A_1)P(A_1) + P(B|A_2)P(A_2) + \cdots + P(B|A_k)P(A_k)}, \qquad (2.46)$$

where $A_2, A_3, ...,$ and A_k represent all other possible outcomes of the first variable.

The numerator identifies the probability of getting both A_1 and B. The denominator is the marginal probability of getting B. This bottom component of the fraction describes the adding of probabilities from the different ways to get B.

To apply Bayes' Theorem correctly, there are two preparatory steps:

(1) First identify the marginal probabilities of each possible outcome of the first variable: $P(A_1)$, $P(A_2), ..., P(A_k)$.

(2) Then identify the probability of the outcome B, conditioned on each possible scenario for the first variable: $P(B|A_1), P(B|A_2), ..., P(B|A_k)$.

Once these probabilities are identified, they can be applied directly within the formula.

Example 2.44 Solution 3. Contingency Table.

The positive predictive value (PPV) of a diagnostic test can be calculated by constructing a two-way contingency table for a large, hypothetical population and calculating conditional probabilities by conditioning on rows or columns. Using a large enough hypothetical population results in an empirical estimate of PPV that is very close to the exact value obtained via using the previously discussed approaches.

Begin by constructing an empty 2×2 table, with the possible outcomes of the diagnostic test as the rows, and the possible disease statuses as the columns (Figure 2.17). Include cells for the row and column sums.

Choose a large number N, for the hypothetical population size. Typically, N of 100,000 is sufficient for an accurate estimate.

	Breast Cancer Present	Breast Cancer Absent	Sum
Mammogram Positive	–	–	–
Mammogram Negative	–	–	–
Sum	–	–	100,000

Figure 2.17: A 2×2 table for the mammogram example, with hypothetical population size N of 100,000.

Continue populating the table, using the provided information about the prevalence of breast cancer in this population (0.35%), the chance of a false negative mammogram (11%), and the chance of a false positive (7%):

1. Calculate the two column totals (the number of women with and without breast cancer) from $P(\text{BC})$, the disease prevalence:

$$N \times P(\text{BC}) = 100,000 \times .0035 = 350 \text{ women with BC}$$

$$N \times [1 - P(\text{BC})] = 100,000 \times [1 - .0035] = 99,650 \text{ women without BC}$$

Alternatively, the number of women without breast cancer can be calculated by subtracting the number of women with breast cancer from N.

2. Calculate the two numbers in the first column: the number of women who have breast cancer and tested either negative (false negative) or positive (true positive).

$$\text{women with BC} \times P(\text{false "-"}) = 350 \times .11 = 38.5 \text{ false "-" results}$$

$$\text{women with BC} \times [1 - P(\text{false "-"})] = 350 \times [1 - .11] = 311.5 \text{ true "+" results}$$

3. Calculate the two numbers in the second column: the number of women who do not have breast cancer and tested either positive (false positive) or negative (true negative).

$$\text{women without BC} \times P(\text{false "+"}) = 99,650 \times .07 = 6,975.5 \text{ false "+" results}$$

$$\text{women without BC} \times [1 - P(\text{false "+"})] = 99,650 \times [1 - .07] = 92,674.5 \text{ true "-" results}$$

4. Complete the table by calculating the two row totals: the number of positive and negative mammograms out of 100,000.

$$(\text{true "+" results}) + (\text{false "+" results}) = 311.5 + 6,975.5 = 7,287 \text{ "+" mammograms}$$

$$(\text{true "-" results}) + (\text{false "-" results}) = 38.5 + 92,674.5 = 92,713 \text{ "-" mammograms}$$

5. Finally, calculate the PPV of the mammogram by using the ratio of the number of true positives to the total number of positive mammograms. This estimate is more than accurate enough, with the calculated value differing only in the third decimal place from the exact calculation,

$$\frac{\text{true "+" results}}{\text{"+" mammograms}} = \frac{311.5}{7,287} = 0.0427.$$

	Breast Cancer Present	Breast Cancer Absent	Sum
Mammogram Positive	311.5	6,975.5	7,287
Mammogram Negative	38.5	92,674.5	92,713
Sum	350	99,650	100,000

Figure 2.18: Completed table for the mammogram example. The table shows again why the PPV of the mammogram is low: almost 7,300 women will have a positive mammogram result in this hypothetical population, but only ~312 of those women actually have breast cancer.

GUIDED PRACTICE 2.47

Some congenital disorders are caused by errors that occur during cell division, resulting in the presence of additional chromosome copies. Trisomy 21 occurs in approximately 1 out of 800 births. Cell-free fetal DNA (cfDNA) testing is one commonly used way to screen fetuses for trisomy 21. The test sensitivity is 0.98 and the specificity is 0.995. Calculate the PPV and NPV of the test.[35]

---

[35]$PPV = \dfrac{P(T^+|D) \times P(D)}{[P(T^+|D) \times P(D)] + [P(T^+|D^c) \times P(D^c)]} = \dfrac{(0.98)(1/800)}{(0.98)(1/800) + (1-0.995)(799/800)} = 0.197.$

$NPV = \dfrac{P(T^-|D^c) \times P(D^c)}{[P(T^-|D) \times P(D)] + [P(T^-|D^c) \times P(D^c)]} = \dfrac{(0.995)(799/800)}{(1-0.98)(1/800) + (0.995)(799/800)} = 0.999975.$

2.3 Extended example: cat genetics

So far, the principles of probability have only been illustrated with short examples. In a more complex setting, it can be surprisingly difficult to accurately translate a problem scenario into the language of probability. This section demonstrates how the rules of probability can be applied to work through a relatively sophisticated conditioning problem.

Problem statement

The gene that controls white coat color in cats, KIT, is known to be responsible for multiple phenotypes such as deafness and blue eye color. A dominant allele W at one location in the gene has complete penetrance for white coat color; all cats with the W allele have white coats. There is incomplete penetrance for blue eyes and deafness; not all white cats will have blue eyes and not all white cats will be deaf. However, deafness and blue eye color are strongly linked, such that white cats with blue eyes are much more likely to be deaf. The variation in penetrance for eye color and deafness may be due to other genes as well as environmental factors.

Suppose that 30% of white cats have one blue eye, while 10% of white cats have two blue eyes. About 73% of white cats with two blue eyes are deaf and 40% of white cats with one blue eye are deaf. Only 19% of white cats with other eye colors are deaf.

a) Calculate the prevalence of deafness among white cats.

b) Given that a white cat is deaf, what is the probability that it has two blue eyes?

c) Suppose that deaf, white cats have an increased chance of being blind, but that the prevalence of blindness differs according to eye color. While deaf, white cats with two blue eyes or two non-blue eyes have probability 0.20 of developing blindness, deaf and white cats with one blue eye have probability 0.40 of developing blindness. White cats that are not deaf have probability 0.10 of developing blindness, regardless of their eye color.

   i. What is the prevalence of blindness among deaf, white cats?

   ii. What is the prevalence of blindness among white cats?

   iii. Given that a cat is white and blind, what is the probability that it has two blue eyes?

Defining notation

Before beginning any calculations, it is essential to clearly define any notation that will be used. For this problem, there are several events of interest: deafness, number of blue eyes (either 0, 1, or 2), and blindness.

   – Let D represent the event that a white cat is deaf.

   – Let $B_0 = \{$zero blue eyes$\}$, $B_1 = \{$one blue eye$\}$, and $B_2 = \{$two blue eyes$\}$.

   – Let L represent the event that a white cat is blind.

Note that since all cats mentioned in the problem are white, it is not necessary to define whiteness as an event; white cats represent the sample space.

Part a) Deafness

The prevalence of deafness among white cats is the proportion of white cats that are deaf; i.e., the probability of deafness among white cats. In the notation of probability, this question asks for the value of $P(D)$.

EXAMPLE 2.48

The following information has been given in the problem. Re-write the information using the notation defined earlier.

> Suppose that 30% of white cats have one blue eye, while 10% of white cats have two blue eyes. About 73% of white cats with two blue eyes are deaf and 40% of white cats with one blue eye are deaf. Only 19% of white cats with other eye colors are deaf.

The first sentence provides information about the prevalence of white cats with one blue eye and white cats with two blue eyes: $P(B_1) = 0.30$ and $P(B_2) = 0.10$. The only other possible eye color combination is zero blue eyes (i.e., two non-blue eyes); i.e., since $P(B_0) + P(B_1) + P(B_2) = 1$, $P(B_0) = 1 - P(B_1) - P(B_2) = 0.60$. 60% of white cats have two non-blue eyes.

While it is not difficult to recognize that the second and third sentences provide information about deafness in relation to eye color, it can be easy to miss that these probabilities are conditional probabilities. A close reading should focus on the language—"About 73% *of* white cats with two blue eyes are deaf...": i.e., out of the white cats that have two blue eyes, 73% are deaf. Thus, these are probabilities of deafness conditioned on eye color. From these sentences, $P(D|B_2) = 0.73$, $P(D|B_1) = 0.40$, and $P(D|B_0) = 0.19$.

Consider that there are three possible ways to partition the event D, that a white cat is deaf: a cat could be deaf and have two blue eyes, be deaf and have one blue eye (and one non-blue eye), or be deaf and have two non-blue eyes. Thus, by the addition rule of disjoint outcomes:

$$P(D) = P(D \text{ and } B_2) + P(D \text{ and } B_1) + P(D \text{ and } B_0).$$

Although the joint probabilities of being deaf and having particular eye colors are not given in the problem, these can be solved for based on the given information. The definition of conditional probability $P(A|B)$ relates the joint probability $P(A \text{ and } B)$ with the marginal probability $P(B)$.[36]

$$P(A|B) = \frac{P(A \text{ and } B)}{P(B)} \qquad P(A \text{ and } B) = P(A|B)P(B).$$

Thus, the probability $P(D)$ is given by:

$$
\begin{aligned}
P(D) &= P(D \text{ and } B_2) + P(D \text{ and } B_1) + P(D \text{ and } B_0) \\
&= P(D|B_2)P(B_2) + P(D|B_1)P(B_1) + P(D|B_0)P(B_0) \\
&= (0.73)(0.10) + (0.40)(0.30) + (0.19)(0.60) \\
&= 0.307.
\end{aligned}
$$

The prevalence of deafness among white cats is 0.307.

---

[36]This rearrangement of the definition of conditional probability, $P(A \text{ and } B) = P(A|B)P(B)$, is also known as the general multiplication rule.

Part b) Deafness and eye color

The probability that a white cat has two blue eyes, given that it is deaf, can be expressed as $P(B_2|D)$.

EXAMPLE 2.49

Using the definition of conditional probability, solve for $P(B_2|D)$.

$$P(B_2|D) = \frac{P(D \text{ and } B_2)}{P(D)} = \frac{P(D|B_2)P(B_2)}{P(D)} = \frac{(0.73)(0.10)}{0.307} = 0.238.$$

The probability that a white cat has two blue eyes, given that it is deaf, is 0.238.

It is also possible to think of this as a Bayes' Rule problem, where there are three possible partitions of the event of deafness, D. In this problem, it is possible to directly solve from the definition of conditional probability since $P(D)$ was solved for in part a); note that the expanded denominator below matches the earlier work to calculate $P(D)$.

$$P(B_2|D) = \frac{P(D \text{ and } B_2)}{P(D)} = \frac{P(D|B_2)P(B_2)}{P(D|B_2)P(B_2) + P(D|B_1)P(B_1) + P(D|B_0)P(B_0)}.$$

Part c) Blindness, deafness, and eye color

EXAMPLE 2.50

The following information has been given in the problem. Re-write the information using the notation defined earlier.

> Suppose that deaf, white cats have an increased chance of being blind, but that the prevalence of blindness differs according to eye color. While deaf, white cats with two blue eyes or two non-blue eyes have probability 0.20 of developing blindness, deaf and white cats with one blue eye have probability 0.40 of developing blindness. White cats that are not deaf have probability 0.10 of developing blindness, regardless of their eye color.

The second sentence gives probabilities of blindness, conditional on eye color and being deaf: $P(L|B_2, D) = P(L|B_0, D) = 0.20$, and $P(L|B_1, D) = 0.40$. The third sentence gives the probability that a white cat is blind, given that it is not deaf: $P(L|D^C) = 0.10$.

Part i. asks for the prevalence of blindness among deaf, white cats: $P(L|D)$. As in part a), the event of blindness given deafness can be partitioned by eye color:

$$P(L|D) = P(L \text{ and } B_0|D) + P(L \text{ and } B_1|D) + P(L \text{ and } |D).$$

EXAMPLE 2.51

Expand the previous expression using the general multiplication rule, $P(A \text{ and } B) = P(A|B)P(B)$.

The general multiplication rule may seem difficult to apply when conditioning is present, but the principle remains the same. Think of the conditioning as a way to restrict the sample space; in this context, conditioning on deafness implies that for this part of the problem, all the cats being considered are deaf (and white).

For instance, consider the first term, $P(L \text{ and } B_0|D)$, the probability of being blind and having two non-blue eyes, given deafness. How could this be rewritten if the probability were simply $P(L \text{ and } B_0)$?

$$P(L \text{ and } B_0) = P(L|B_0)P(B_0)$$

Now, recall that for this part of the problem, the sample space is restricted to deaf (and white) cats. Thus, all of the terms in the expansion should include conditioning on deafness:

$$P(L \text{ and } B_0|D) = P(L|D, B_0)P(B_0|D).$$

Thus,

$$P(L|D) = P(L|D, B_0)P(B_0|D) + P(L|D, B_1)P(B_1|D) + P(L|D, B_2)P(B_2|D).$$

Although $P(L|D, B_0)$, $P(L|D, B_1)$, and $P(L|D, B_2)$ are given from the problem statement, $P(B_0|D$, $P(B_1|D)$, and $P(B_2|D)$ are not. However, note that the probability that a white cat has two blue eyes given that it is deaf, $P(B_2|D)$, was calculated in part b).

GUIDED PRACTICE 2.52

Calculate $P(B_0|D)$ and $P(B_1|D)$.[37]

There is now sufficient information to calculate $P(L|D)$:

$$\begin{aligned}
P(L|D) &= P(L \text{ and } B_0|D) + P(L \text{ and } B_1|D) + P(L \text{ and } B_2|D) \\
&= P(L|D, B_0)P(B_0|D) + P(L|D, B_1)P(B_1|D) + P(L|D, B_2)P(B_2|D) \\
&= (0.20)(0.371) + (0.40)(0.391) + (0.20)(0.238) \\
&= 0.278.
\end{aligned}$$

The prevalence of blindness among deaf, white cats is 0.278.

Part ii. asks for the prevalence of blindness among white cats, $P(L)$. Again, partitioning is an effective strategy. Instead of partitioning by eye color, however, partition by deafness.

---

[37]
$$P(B_0|D) = \frac{P(D \text{ and } B_0)}{P(D)} = \frac{P(D|B_0)P(B_0)}{P(D)} = \frac{(0.19)(0.60)}{0.307} = 0.371.$$
$$P(B_1|D) = \frac{P(D \text{ and } B_1)}{P(D)} = \frac{P(D|B_1)P(B_1)}{P(D)} = \frac{(0.40)(0.30)}{0.307} = 0.391.$$

EXAMPLE 2.53

Calculate the prevalence of blindness among white cats, $P(L)$.

⎯⎯⎯⎯⎯⎯

$$
\begin{aligned}
P(L) &= P(L \text{ and } D) + P(L \text{ and } D^C) \\
&= P(L|D)P(D) + P(L|D^C)P(D^C) \\
&= (0.278)(0.307) + (0.10)(1 - 0.307) \\
&= 0.155.
\end{aligned}
$$

$P(D)$ was calculated in part a), while $P(L|D)$ was calculated in part c, i. The conditioning probability of blindness given a white cat is not deaf is 0.10, as given in the question statement. By the definition of the complement, $P(D^C) = 1 - P(D)$.

The prevalence of blindness among white cats is 0.155.

Part iii. asks for the probability that a cat has two blue eyes, given that it is white and blind. This probability can be expressed as $P(B_2|L)$. Recall that since all cats being discussed in the problem are white, it is not necessary to condition on coat color.

Start out with the definition of conditional probability:

$$
P(B_2|L) = \frac{P(B_2 \text{ and } L)}{P(L)}.
$$

The key to calculating $P(B_2|L)$ relies on recognizing that the event a cat is blind and has two blue eyes can be partitioned by whether or not the cat is also deaf:

$$
P(B_2|L) = \frac{P(B_2 \text{ and } L \text{ and } D) + P(B_2 \text{ and } L \text{ and } D^C)}{P(L)}. \tag{2.54}
$$

EXAMPLE 2.55

Draw a tree diagram to organize the events involved in this problem. Identify the branches that represent the possible paths for a white cat to both have two blue eyes and be blind.

⎯⎯⎯⎯⎯⎯

When drawing a tree diagram, remember that each branch is conditioned on the previous branches. While there are various possible trees, the goal is to construct a tree for which as many of the branches as possible have known probabilities.

The tree for this problem will have three branch points, corresponding to either deafness, blindness, or eye color. The first set of branches contain unconditional probabilities, the second set contains conditional probabilities given one event, and the third set contains conditional probabilities given two events.

Recall that the probabilities $P(L|D, B_0)$, $P(L|D, B_1)$, and $P(L|D, B_2)$ were provided in the problem statement. These are the only probabilities conditioned on two events that have previously appeared in the problem, so blindness is the most convenient choice of third branch point.

It is not immediately obvious whether it will be more efficient to start with deafness or eye color, since unconditional and conditional probabilities related to both have appeared in the problem. Figure 2.19 shows two trees, one starting with deafness and the other starting with eye color. The two possible paths for a white cat to both have two blue eyes and be blind are shown in green.

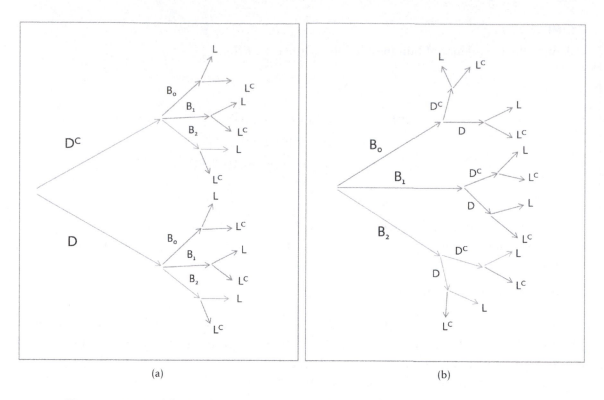

(a)   (b)

Figure 2.19: In (a), the first branch is based on deafness, while in (b), the first branch is based on eye color.

EXAMPLE 2.56

Expand Equation 2.54 according to the tree shown in Figure 2.19(a), and solve for $P(B_2|L)$.

$$P(B_2|L) = \frac{P(B_2 \text{ and } L \text{ and } D) + P(B_2 \text{ and } L \text{ and } D^C)}{P(L)}$$

$$= \frac{P(L|B_2, D)P(B_2|D)P(D) + P(L|B_2, D^C)P(B_2|D^C)P(D^C)}{P(L)}$$

$$= \frac{(0.20)(0.238)(0.307) + (0.10)P(B_2|D^C)P(D^C)}{0.155}.$$

Two of the probabilities have not been calculated previously: $P(B_2|D^C)$ and $P(D^C)$. From the definition of the complement, $P(D^C) = 1 - P(D) = 0.693$; $P(D)$ was calculated in part a). To calculate $P(B_2|D^C)$, apply the definition of conditional probability as in part b), where $P(B_2|D)$ was calculated:

$$P(B_2|D^C) = \frac{P(D^C \text{ and } B_2)}{P(D^C)} = \frac{P(D^C|B_2)(P(B_2))}{P(D^C)} = \frac{(1 - 0.73)(0.10)}{0.693} = 0.0390.$$

$$P(B_2|L) = \frac{(0.20)(0.238)(0.307) + (0.10)(0.0390)(0.693)}{0.155}$$

$$= 0.112$$

The probability that a white cat has two blue eyes, given that it is blind, is 0.112.

GUIDED PRACTICE 2.57

Expand Equation 2.54 according to the tree shown in Figure 2.19(b), and solve for $P(B_2|L)$.[38]

A tree diagram is useful for visualizing the different possible ways that a certain set of outcomes can occur. Although conditional probabilities can certainly be calculated without the help of tree diagrams, it is often easy to make errors with a strictly algebraic approach. Once a tree is constructed, it can be used to solve for several probabilities of interest. The following example shows how one of the previous trees can be applied to answer a different question than the one posed in part c), iii.

---

[38]

$$P(B_2|L) = \frac{P(L|B_2,D)P(D|B_2)P(B_2) + P(L|B_2,D^C)P(D^C|B_2)P(B_2)}{P(L)} = \frac{(0.20)(0.73)(0.10) + (0.10)(1-0.73)(0.10)}{0.155} = 0.112.$$

EXAMPLE 2.58

What is the probability that a white cat has one blue eye and one non-blue eye, given that it is not blind?

———————

Calculate $P(B_1|L^C)$. Start with the definition of conditional probability, then expand.

$$P(B_1|L^C) = \frac{P(B_1 \text{ and } L^C)}{P(L^C)} = \frac{P(B_1 \text{ and } L^C \text{ and } D) + P(B_1 \text{ and } L^C \text{ and } D^C)}{P(L^C)}.$$

Figure 2.20 is a reproduction of the earlier tree diagram (Figure 2.19(b)), with yellow arrows showing the two paths of interest.

As before, expand the numerator and fill in the known values.

$$P(B_1|L^C) = \frac{P(B_1 \text{ and } L^C \text{ and } D) + P(B_1 \text{ and } L^C \text{ and } D^C)}{P(L^C)}$$

$$= \frac{P(L^C|D, B_1)P(D|B_1)P(B_1) + P(L^C|D^C, B_1)P(D^C|B_1)P(B_1)}{P(L^C)}$$

$$= \frac{P(L^C|D, B_1)(0.40)(0.30) + P(L^C|D^C, B_1)P(D^C|B_1)(0.30)}{P(L^C)}.$$

The probabilities in bold are not known. Apply the definition of the complement; recall that the rule for complements holds when an event and its complement are conditioned on the same information: $P(A|B) = 1 - P(A^C|B)$.

– $P(L^C) = 1 - P(L) = 1 - 0.155 = 0.845$

– $P(D^C|B_1) = 1 - P(D|B_1) = 1 - 0.40 = 0.60$

– $P(L^C|D, B_1) = 1 - P(L|D, B_1) = 1 - 0.40 = 0.60$

The definition of the complement can also be applied to calculate $P(L^C|D^C, B_1)$. The problem statement originally specified that white cats that are not deaf have probability 0.10 of developing blindness regardless of eye color: $P(L|D^C) = 0.10$. Thus, $P(L^C|D^C, B_1) = P(L^C|D^C)$. By the definition of the complement, $P(L^C|D^C) = 1 - P(L|D^C) = 1 - 0.10 = 0.90$.

$$P(B_1|L^C) = \frac{P(B_1 \text{ and } L^C \text{ and } D) + P(B_1 \text{ and } L^C \text{ and } D^C)}{P(L^C)}$$

$$= \frac{P(L^C|D, B_1)P(D|B_1)P(B_1) + P(L^C|D^C, B_1)P(D^C|B_1)P(B_1)}{P(L^C)}$$

$$= \frac{(0.60)(0.40)(0.30) + (0.90)(0.60)(0.30)}{0.845}$$

$$= 0.277.$$

The probability that a white cat has one blue eye and one non-blue eye, given that it is not blind, is 0.277.

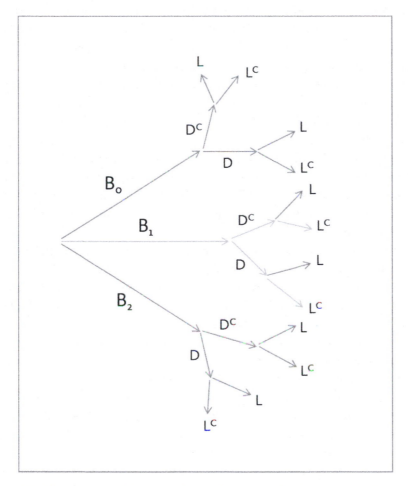

Figure 2.20: The two possible paths for a white cat to both have one blue eye (and one non-blue eye) and to not be blind are shown in yellow.

2.4 Notes

Probability is a powerful framework for quantifying uncertainty and randomness. In particular, conditional probability represents a way to update the uncertainty associated with an event given that specific information has been observed. For example, the probability that a person has a particular disease can be adjusted based on observed information, such as age, sex, or the results of a diagnostic test.

As discussed in the text, there are several possible approaches to solving conditional probability problems, including the use of tree diagrams or contingency tables. It can also be intuitive to use a simulation approach in computing software; refer to the labs for details about this method. Regardless of the specific approach that will be used for calculation, it is always advisable to start any problem by understanding the problem context (i.e., the sample space, given information, probabilities of interest) and reading the problem carefully, in order to avoid mistakes when translating between words and probability notation. A common mistake is to confuse joint and conditional probabilities.

Probability distributions were briefly introduced in Section 2.1.5. This topic will be discussed in greater detail in the next chapter.

Probability forms the foundation for data analysis and statistical inference, since nearly every conclusion to a study should be accompanied by a measure of uncertainty. For example, the publication reporting the results of the LEAP study discussed in Chapter 1 included the probability that the observed results could have been due to chance variation. This aspect of probability will be discussed in later chapters.

The four labs for Chapter 2 cover basic principles of probability, conditional probability, positive predictive value of a diagnostic test (via Bayes' Theorem), and the calculation of probabilities conditional on several events in the context of genetic inheritance. Probabilities can be calculated algebraically, using formulas given in this and other texts, but can also be calculated with simple simulations, since a probability represents a proportion of times an event happens when an experiment is repeated many times. Computers are particularly good at keeping track of events during many replications of an experiment. The labs for this chapter use both algebraic and simulation methods, and are particularly useful for building programming skills with the R language.

In medicine, the positive predictive value of a diagnostic test may be one of the most important applications of probability theory. It is certainly the most common. The positive predictive value of a test is the conditional probability of the presence of a disease or condition, given a positive test for the condition, and is often used when counseling patients about their risk for being diagnosed with a disease in the future. The lab on positive predictive value examines the conditional probability of a trisomy 21 genetic mutation (Down syndrome) given that a test based on cell-free DNA suggests its presence.

2.5 Exercises

2.5.1 Defining probability

2.1 True or false. Determine if the statements below are true or false, and explain your reasoning.

(a) Assume that a couple has an equal chance of having a boy or a girl. If a couple's previous three children have all been boys, then the chance that their next child is a boy is somewhat less than 50%.

(b) Drawing a face card (jack, queen, or king) and drawing a red card from a full deck of playing cards are mutually exclusive events.

(c) Drawing a face card and drawing an ace from a full deck of playing cards are mutually exclusive events.

2.2 Dice rolls. If you roll a pair of fair dice, what is the probability of

(a) getting a sum of 1?

(b) getting a sum of 5?

(c) getting a sum of 12?

2.3 Colorblindness. Red-green colorblindness is a commonly inherited form of colorblindness; the gene involved is transmitted on the X chromosome in a recessive manner. If a male inherits an affected X chromosome, he is necessarily colorblind (genotype X^-Y). However, a female can only be colorblind if she inherits two defective copies (genotype X^-X^-); heterozygous females are not colorblind. Suppose that a couple consists of a genotype X^+Y male and a genotype X^+X^- female.

(a) What is the probability of the couple producing a colorblind male?

(b) True or false: Among the couple's offspring, colorblindness and female sex are mutually exclusive events.

2.4 Diabetes and hypertension. Diabetes and hypertension are two of the most common diseases in Western, industrialized nations. In the United States, approximately 9% of the population have diabetes, while about 30% of adults have high blood pressure. The two diseases frequently occur together: an estimated 6% of the population have both diabetes and hypertension.

(a) Are having diabetes and having hypertension disjoint?

(b) Draw a Venn diagram summarizing the variables and their associated probabilities.

(c) Let A represent the event of having diabetes, and B the event of having hypertension. Calculate $P(A \text{ or } B)$.

(d) What percent of Americans have neither hypertension nor diabetes?

(e) Is the event of someone being hypertensive independent of the event that someone has diabetes?

2.5 Educational attainment by gender. The table below shows the distribution of education level attained by US residents by gender based on data collected during the 2010 American Community Survey.[39]

		Gender	
		Male	Female
	Less than 9th grade	0.07	0.13
	9th to 12th grade, no diploma	0.10	0.09
Highest	HS graduate (or equivalent)	0.30	0.20
education	Some college, no degree	0.22	0.24
attained	Associate's degree	0.06	0.08
	Bachelor's degree	0.16	0.17
	Graduate or professional degree	0.09	0.09
	Total	1.00	1.00

(a) What is the probability that a randomly chosen individual is a high school graduate? Assume that there is an equal proportion of males and females in the population.

(b) Define Event A as having a graduate or professional degree. Calculate the probability of the complement, A^c.

(c) What is the probability that a randomly chosen man has at least a Bachelor's degree?

(d) What is the probability that a randomly chosen woman has at least a Bachelor's degree?

(e) What is the probability that a man and a woman getting married both have at least a Bachelor's degree? Note any assumptions made – are they reasonable?

2.6 Poverty and language. The American Community Survey is an ongoing survey that provides data every year to give communities the current information they need to plan investments and services. The 2010 American Community Survey estimates that 14.6% of Americans live below the poverty line, 20.7% speak a language other than English (foreign language) at home, and 4.2% fall into both categories.[40]

(a) Are living below the poverty line and speaking a foreign language at home disjoint?

(b) Draw a Venn diagram summarizing the variables and their associated probabilities.

(c) What percent of Americans live below the poverty line and only speak English at home?

(d) What percent of Americans live below the poverty line or speak a foreign language at home?

(e) What percent of Americans live above the poverty line and only speak English at home?

(f) Is the event that someone lives below the poverty line independent of the event that the person speaks a foreign language at home?

2.7 Urgent care visits. Urgent care centers are open beyond typical office hours and provide a broader range of services than that of many primary care offices. A study conducted to collect information about urgent care centers in the United States reported that in one week, 15.8% of centers saw 0-149 patients, 33.7% saw 150-299 patients, 28.8% saw 300-449 patients, and 21.7% saw 450 or more patients. Assume that the data can be treated as a probability distribution of patient visits for any given week.

(a) What is the probability that three random urgent care centers in a county all see between 300-449 patients in a week? Note any assumptions made. Are the assumptions reasonable?

(b) What is the probability that ten random urgent care centers throughout a state all see 450 or more patients in a week? Note any assumptions made. Are the assumptions reasonable?

(c) With the information provided, is it possible to compute the probability that one urgent care center sees between 150-299 patients in one week and 300-449 patients in the next week? Explain why or why not.

---

[39]U.S. Census Bureau, 2010 American Community Survey 1-Year Estimates, Educational Attainment.

[40]U.S. Census Bureau, 2010 American Community Survey 1-Year Estimates, Characteristics of People by Language Spoken at Home.

2.8 School absences. Data collected at elementary schools in DeKalb County, GA suggest that each year roughly 25% of students miss exactly one day of school, 15% miss 2 days, and 28% miss 3 or more days due to sickness.[41]

(a) What is the probability that a student chosen at random doesn't miss any days of school due to sickness this year?

(b) What is the probability that a student chosen at random misses no more than one day?

(c) What is the probability that a student chosen at random misses at least one day?

(d) If a parent has two kids at a DeKalb County elementary school, what is the probability that neither kid will miss any school? Note any assumptions made and evaluate how reasonable they are.

(e) If a parent has two kids at a DeKalb County elementary school, what is the probability that both kids will miss some school, i.e. at least one day? Note any assumptions made and evaluate how reasonable they are.

2.9 Disjoint vs. independent. In parts (a) and (b), identify whether the events are disjoint, independent, or neither (events cannot be both disjoint and independent).

(a) You and a randomly selected student from your class both earn A's in this course.

(b) You and your class study partner both earn A's in this course.

(c) If two events can occur at the same time, must they be dependent?

2.10 Health coverage, frequencies. The Behavioral Risk Factor Surveillance System (BRFSS) is an annual telephone survey designed to identify risk factors in the adult population and report emerging health trends. The following table summarizes two variables for the respondents: health status and health coverage, which describes whether each respondent had health insurance.[42]

| | | \multicolumn{6}{c}{Health Status} |
		Excellent	Very good	Good	Fair	Poor	Total
Health	No	459	727	854	385	99	2,524
Coverage	Yes	4,198	6,245	4,821	1,634	578	17,476
	Total	4,657	6,972	5,675	2,019	677	20,000

(a) If one individual is drawn at random, what is the probability that the respondent has excellent health and doesn't have health coverage?

(b) If one individual is drawn at random, what is the probability that the respondent has excellent health or doesn't have health coverage?

---

[41]S.S. Mizan et al. "Absence, Extended Absence, and Repeat Tardiness Related to Asthma Status among Elementary School Children". In: *Journal of Asthma* 48.3 (2011), pp. 228–234.

[42]Office of Surveillance, Epidemiology, and Laboratory Services Behavioral Risk Factor Surveillance System, BRFSS 2010 Survey Data.

2.5.2 Conditional probability

2.11 Global warming. A Pew Research poll asked 1,306 Americans "From what you've read and heard, is there solid evidence that the average temperature on earth has been getting warmer over the past few decades, or not?". The table below shows the distribution of responses by party and ideology, where the counts have been replaced with relative frequencies.[43]

		Response			
		Earth is warming	Not warming	Don't Know Refuse	Total
	Conservative Republican	0.11	0.20	0.02	0.33
Party and	Mod/Lib Republican	0.06	0.06	0.01	0.13
Ideology	Mod/Cons Democrat	0.25	0.07	0.02	0.34
	Liberal Democrat	0.18	0.01	0.01	0.20
	Total	0.60	0.34	0.06	1.00

(a) Are believing that the earth is warming and being a liberal Democrat mutually exclusive?

(b) What is the probability that a randomly chosen respondent believes the earth is warming or is a liberal Democrat?

(c) What is the probability that a randomly chosen respondent believes the earth is warming given that he is a liberal Democrat?

(d) What is the probability that a randomly chosen respondent believes the earth is warming given that he is a conservative Republican?

(e) Does it appear that whether or not a respondent believes the earth is warming is independent of their party and ideology? Explain your reasoning.

(f) What is the probability that a randomly chosen respondent is a moderate/liberal Republican given that he does not believe that the earth is warming?

2.12 ABO blood groups. The ABO blood group system consists of four different blood groups, which describe whether an individual's red blood cells carry the A antigen, B antigen, both, or neither. The ABO gene has three alleles: I^A, I^B, and i. The i allele is recessive to both I^A and I^B, while the I^A and I^B allels are codominant. Individuals homozygous for the i allele are known as blood group O, with neither A nor B antigens.

Alleles inherited	Blood type
I^A and I^A	A
I^A and I^B	AB
I^A and i	A
I^B and I^B	B
I^B and i	B
i and i	O

Blood group follows the rules of Mendelian single-gene inheritance – alleles are inherited independently from either parent, with probability 0.5.

(a) Suppose that both members of a couple have Group AB blood. What is the probability that a child of this couple will have Group A blood?

(b) Suppose that one member of a couple is genotype $I^B i$ and the other is $I^A i$. What is the probability that their first child has Type O blood and the next two do not?

(c) Suppose that one member of a couple is genotype $I^B i$ and the other is $I^A i$. Given that one child has Type O blood and two do not, what is the probability of the first child having Type O blood?

---

[43]Pew Research Center, Majority of Republicans No Longer See Evidence of Global Warming, data collected on October 27, 2010.

2.13 Seat belts. Seat belt use is the most effective way to save lives and reduce injuries in motor vehicle crashes. In a 2014 survey, respondents were asked, "How often do you use seat belts when you drive or ride in a car?". The following table shows the distribution of seat belt usage by sex.

		Always	Nearly always	Sometimes	Seldom	Never	Total
Sex	Male	146,018	19,492	7,614	3,145	4,719	180,988
	Female	229,246	16,695	5,549	1,815	2,675	255,980
	Total	375,264	36,187	13,163	4,960	7,394	436,968

Seat Belt Usage

(a) Calculate the marginal probability that a randomly chosen individual always wears seatbelts.

(b) What is the probability that a randomly chosen female always wears seatbelts?

(c) What is the conditional probability of a randomly chosen individual always wearing seatbelts, given that they are female?

(d) What is the conditional probability of a randomly chosen individual always wearing seatbelts, given that they are male?

(e) Calculate the probability that an individual who never wears seatbelts is male.

(f) Does gender seem independent of seat belt usage?

2.14 Health coverage, relative frequencies. The Behavioral Risk Factor Surveillance System (BRFSS) is an annual telephone survey designed to identify risk factors in the adult population and report emerging health trends. The following table displays the distribution of health status of respondents to this survey (excellent, very good, good, fair, poor) conditional on whether or not they have health insurance.

		Excellent	Very good	Good	Fair	Poor	Total
Health	No	0.0230	0.0364	0.0427	0.0192	0.0050	0.1262
Coverage	Yes	0.2099	0.3123	0.2410	0.0817	0.0289	0.8738
	Total	0.2329	0.3486	0.2838	0.1009	0.0338	1.0000

Health Status

(a) Are being in excellent health and having health coverage mutually exclusive?

(b) What is the probability that a randomly chosen individual has excellent health?

(c) What is the probability that a randomly chosen individual has excellent health given that he has health coverage?

(d) What is the probability that a randomly chosen individual has excellent health given that he doesn't have health coverage?

(e) Do having excellent health and having health coverage appear to be independent?

2.15 HIV in Swaziland. Swaziland has the highest HIV prevalence in the world: 25.9% of this country's population is infected with HIV.[44] The ELISA test is one of the first and most accurate tests for HIV. For those who carry HIV, the ELISA test is 99.7% accurate. For those who do not carry HIV, the test is 92.6% accurate. Calculate the PPV and NPV of the test.

---

[44]Source: CIA Factbook, Country Comparison: HIV/AIDS - Adult Prevalence Rate.

2.16 Assortative mating. Assortative mating is a nonrandom mating pattern where individuals with similar genotypes and/or phenotypes mate with one another more frequently than what would be expected under a random mating pattern. Researchers studying this topic collected data on eye colors of 204 Scandinavian men and their female partners. The table below summarizes the results. For simplicity, we only include heterosexual relationships in this exercise.[45]

		Partner (female)			
		Blue	Brown	Green	Total
	Blue	78	23	13	114
Self (male)	Brown	19	23	12	54
	Green	11	9	16	36
	Total	108	55	41	204

(a) What is the probability that a randomly chosen male respondent or his partner has blue eyes?

(b) What is the probability that a randomly chosen male respondent with blue eyes has a partner with blue eyes?

(c) What is the probability that a randomly chosen male respondent with brown eyes has a partner with blue eyes? What about the probability of a randomly chosen male respondent with green eyes having a partner with blue eyes?

(d) Does it appear that the eye colors of male respondents and their partners are independent? Explain your reasoning.

2.17 It's never lupus. Lupus is a medical phenomenon where antibodies that are supposed to attack foreign cells to prevent infections instead see plasma proteins as foreign bodies, leading to a high risk of blood clotting. It is believed that 2% of the population suffer from this disease. The test is 98% accurate if a person actually has the disease. The test is 74% accurate if a person does not have the disease. There is a line from the Fox television show *House* that is often used after a patient tests positive for lupus: "It's never lupus." Do you think there is truth to this statement? Use appropriate probabilities to support your answer.

2.18 Predisposition for thrombosis. A genetic test is used to determine if people have a predisposition for *thrombosis*, which is the formation of a blood clot inside a blood vessel that obstructs the flow of blood through the circulatory system. It is believed that 3% of people actually have this predisposition. The genetic test is 99% accurate if a person actually has the predisposition, meaning that the probability of a positive test result when a person actually has the predisposition is 0.99. The test is 98% accurate if a person does not have the predisposition.

(a) What is the probability that a randomly selected person who tests positive for the predisposition by the test actually has the predisposition?

(b) What is the probability that a randomly selected person who tests negative for the predisposition by the test actually does not have the predisposition?

---

[45]B. Laeng et al. "Why do blue-eyed men prefer women with the same eye color?" In: *Behavioral Ecology and Sociobiology* 61.3 (2007), pp. 371–384.

2.19 Views on evolution. A 2013 analysis conducted by the Pew Research Center found that 60% of survey respondents agree with the statement "humans and other living things have evolved over time" while 33% say that "humans and other living things have existed in their present form since the beginning of time" (7% responded "don't know"). They also found that there are differences among partisan groups in beliefs about evolution. While roughly two-thirds of Democrats (67%) and independents (65%) say that humans and other living things have evolved over time, 48% of Republicans reject the idea of evolution. Suppose that 45% of respondents identified as Democrats, 40% identified as Republicans, and 15% identified as political independents. The survey was conducted among a national sample of 1,983 adults.

(a) Suppose that a person is randomly selected from the population and found to identify as a Democrat. What is the probability that this person does not agree with the idea of evolution?

(b) Suppose that a political independent is randomly selected from the population. What is the probability that this person does not agree with the idea of evolution?

(c) Suppose that a person is randomly selected from the population and found to identify as a Republican. What is the probability that this person agrees with the idea of evolution?

(d) Suppose that a person is randomly selected from the population and found to support the idea of evolution. What is the probability that this person identifies as a Republican?

2.20 Cystic fibrosis testing. The prevalence of cystic fibrosis in the United States is approximately 1 in 3,500 births. Various screening strategies for CF exist. One strategy uses dried blood samples to check the levels of immunoreactive trypsogen (IRT); IRT levels are commonly elevated in newborns with CF. The sensitivity of the IRT screen is 87% and the specificity is 99%.

(a) In a hypothetical population of 100,000, how many individuals would be expected to test positive? Of those who test positive, how many would be true positives? Calculate the PPV of IRT.

(b) In order to account for lab error or physiological fluctuations in IRT levels, infants who tested positive on the initial IRT screen are asked to return for another IRT screen at a later time, usually two weeks after the first test. This is referred to as an IRT/IRT screening strategy. Calculate the PPV of IRT/IRT.

2.21 Mumps. Mumps is a highly contagious viral infection that most often occurs in children, but can affect adults, particularly if they are living in shared living spaces such as college dormitories. It is most recognizable by the swelling of salivary glands at the side of the face under the ears, but earlier symptoms include headaches, fever, and joint pain. Suppose a college student at a university presents to a physician with symptoms of headaches, fever, and joint pain. Let $A = \{$headaches, fever, and joint pain$\}$, and suppose that the possible disease state of the patient can be partitioned into: B_1 = normal, B_2 = common cold, B_3 = mumps. From clinical experience, the physician estimates $P(A|B_i)$: $P(A|B_1) = 0.001$, $P(A|B_2) = 0.70$, $P(A|B_3) = 0.95$. The physician, aware that some students have contracted the mumps, then estimates that for students at this university, $P(B_1) = 0.95$, $P(B_2) = 0.025$, and $P(B_3) = 0.025$. Given the previous symptoms, which of the disease states is most likely?

2.22 Twins. About 30% of human twins are identical, and the rest are fraternal. Identical twins are necessarily the same sex – half are males and the other half are females. One-quarter of fraternal twins are both male, one-quarter both female, and one-half are mixes: one male, one female. You have just become a parent of twins and are told they are both girls. Given this information, what is the probability that they are identical?

2.23 IQ testing. A psychologist conducts a study on intelligence in which participants are asked to take an IQ test consisting of n questions, each with m choices.

(a) One thing the psychologist must be careful about when analyzing the results is accounting for lucky guesses. Suppose that for a given question a particular participant either knows the answer or guesses. The participant knows the correct answer with probability p, and does not know the answer (and therefore will have to guess) with probability $1 - p$. The participant guesses completely randomly. What is the conditional probability that the participant knew the answer to a question, given that they answered it correctly?

(b) About 1 in 1,100 people have IQs over 150. If a subject receives a score of greater than some specified amount, they are considered by the psychologist to have an IQ over 150. But the psychologist's test is not perfect. Although all individuals with IQ over 150 will definitely receive such a score, individuals with IQs less than 150 can also receive such scores about 0.1% of the time due to lucky guessing. Given that a subject in the study is labeled as having an IQ over 150, what is the probability that they actually have an IQ below 150?

2.24 Breast cancer and age. The strongest risk factor for breast cancer is age; as a woman gets older, her risk of developing breast cancer increases. The following table shows the average percentage of American women in each age group who develop breast cancer, according to statistics from the National Cancer Institute. For example, approximately 3.56% of women in their 60's get breast cancer.

Age Group	Prevalence
30 - 40	0.0044
40 - 50	0.0147
50 - 60	0.0238
60 - 70	0.0356
70 - 80	0.0382

A mammogram typically identifies a breast cancer about 85% of the time, and is correct 95% of the time when a woman does not have breast cancer.

(a) Calculate the PPV for each age group. Describe any trend(s) you see in the PPV values as prevalence changes. Explain the reason for the trend(s) in language that someone who has not taken a statistics course would understand.

(b) Suppose that two new mammogram imaging technologies have been developed which can improve the PPV associated with mammograms; one improves sensitivity to 99% (but specificity remains at 95%), while the other improves specificity to 99% (while sensitivity remains at 85%). Which technology offers a higher increase in PPV? Explain why.

2.25 Prostate-specific antigen. Prostate-specific antigen (PSA) is a protein produced by the cells of the prostate gland. Blood PSA level is often elevated in men with prostate cancer, but a number of benign (not cancerous) conditions can also cause a man's PSA level to rise. The PSA test for prostate cancer is a laboratory test that measures PSA levels from a blood sample. The test measures the amount of PSA in ng/ml (nanograms per milliliter of blood).

The sensitivity and specificity of the PSA test depend on the cutoff value used to label a PSA level as abnormally high. In the last decade, 4.0 ng/ml has been considered the upper limit of normal, and values 4.1 and higher were used to classify a PSA test as positive. Using this value, the sensitivity of the PSA test is 20% and the specificity is 94%.

The likelihood that a man has undetected prostate cancer depends on his age. This likelihood is also called the prevalence of undetected cancer in the male population. The following table shows the prevalence of undetected prostate cancer by age group.

Age Group	Prevalence	PPV	NPV
< 50 years	0.001		
50 - 60 years	0.020		
61 - 70 years	0.060		
71 - 80 years	0.100		

(a) Calculate the missing PPV and NPV values.

(b) Describe any trends you see in the PPV and NPV values.

(c) Explain the reason for the trends in part b), in language that someone who has not taken a statistics course would understand.

(d) The cutoff for a positive test is somewhat controversial. Explain, in your own words, how lowering the cutoff for a positive test from 4.1 ng/ml to 2.5 ng/ml would affect sensitivity and specificity.

2.5.3   Extended example

2.26   Eye color. One of the earliest models for the genetics of eye color was developed in 1907, and proposed a single-gene inheritance model, for which brown eye color is always dominant over blue eye color. Suppose that in the population, 25% of individuals are homozygous dominant (*BB*), 50% are heterozygous (*Bb*), and 25% are homozygous recessive (*bb*).

(a) Suppose that two parents have brown eyes. What is the probability that their first child has blue eyes?

(b) Does the probability change if it is now known that the paternal grandfather had blue eyes? Justify your answer.

(c) Given that their first child has brown eyes, what is the probability that their second child has blue eyes? Ignore the condition given in part (b).

2.27   Colorblindness. The most common form of colorblindness is a recessive, sex-linked hereditary condition caused by a defect on the X chromosome. Females are XX, while males are XY. Individuals inherit one chromosome from each parent, with equal probability; for example, an individual has a 50% chance of inheriting their father's X chromosome, and a 50% chance of inheriting their father's Y chromosome. If a male has an X chromosome with the defect, he is colorblind. However, a female with only one defective X chromosome will not be colorblind. Thus, colorblindness is more common in males than females; 7% of males are colorblind but only 0.5% of females are colorblind.

(a) Assume that the X chromosome with the wild-type allele is X^+ and the one with the disease allele is X^-. What is the expected frequency of each possible female genotype: X^+X^+, X^+X^-, and X^-X^-? What is the expected frequency of each possible male genotype: X^+Y and X^-Y?

(b) Suppose that two parents are not colorblind. What is the probability that they have a colorblind child?

2.28   Rapid feathering. Sex linkage refers to the inheritance pattern that results from a mutation occurring on a gene located on a sex chromosome. A classic example of a sex-linked trait in humans is red-green color blindness; females can only be red-green colorblind if they have two copies of the mutation (one on each X chromosome), while a single copy of the mutation is sufficient to confer colorblindness in males (since males only have one X chromosome).

In birds, females are the heterogametic sex (with sex chromosomes ZW) and males are the homogametic sex (with sex chromosomes ZZ). A commonly known sex-linked trait in domestic chickens is the rapid feathering trait, which is carried on the Z chromosome. Chickens with the rapid feathering trait grow feathers at a faster rate; this difference is especially pronounced within the first few days from hatching. The wild-type allele K^- is dominant to the mutant alelle K^+; presence of the K^- allele produces slow feathering. Females can be either genotype $Z^{K^+}W$ or $Z^{K^-}W$. Males can be either heterozygous ($Z^{K^+}Z^{K^-}$), homozygous for slow feathering ($Z^{K^-}Z^{K^-}$), or homozygous for rapid feathering ($Z^{K^+}Z^{K^+}$).

In a population of chickens, 9% of males are rapid feathering and 16% of females are rapid feathering. Suppose that slow feathering chickens are mated. What is the probability that out of their 12 offspring, at least two are rapid feathering?

2.29 Genetics of Australian cattle dogs. Australian cattle dogs are known to have a high prevalence of congenital deafness. Deafness in both ears is referred to as bilateral deafness, while deafness in one ear is referred to as unilateral deafness.

Deafness in dogs is associated with the white spotting gene S that controls the expression of coat and eye pigmentation. The dominant allele S produces solid color, while the three recessive alleles contribute to increasing amounts of white in coat pigmentation: Irish spotting (s^i), piebald (s^p), and extreme white piebald (s^w). The s^p and s^w alleles are responsible for the distinctive Australian cattle dog coat pattern of white hair evenly speckled throughout either a predominantly red or black coat. The dogs are born with white coats, and the speckled pattern develops as they age.

While all Australian cattle dogs have some combination of the s^p and s^w alleles, the gene displays incomplete penetrance such that individuals show some variation in phenotype despite having the same genotype. Individuals with low penetrance of the alleles tend to have additional patterns on their coat, such as a dark "mask" around one or both eyes (in other words, a unilateral mask or a bilateral mask). High penetrance of the piebald alleles is associated with deafness.

Suppose that 40% of Australian cattle dogs have black coats; these individuals are commonly referred to as "Blue Heelers" as opposed to "Red Heelers". Among Blue Heelers, 35% of individuals have bilateral masks and 25% have unilateral masks. About 50% of Red Heelers exhibit no eye masking and 10% have bilateral masks.

Let M represent the event that an Australian cattle dog has a facial mask, where M_2 represents a bilateral mask, M_1 represents a unilateral mask, and M_0 indicates lack of a mask.

(a) Calculate the probability an Australian cattle dog has a facial mask and a black coat.

(b) Calculate the prevalence of bilateral masks in Australian cattle dogs.

(c) Among Australian cattle dogs with bilateral facial masks, what is the probability of being a Red Heeler?

(d) Unilateral deafness occurs in Red Heelers with probability 0.15, in both dogs that either lack facial masking or exhibit a unilateral mask; for both unmasked and unilaterally masked Red Heelers, 60% of dogs are not deaf. The overall prevalence of bilaterally masked Australian cattle dogs with bilateral deafness and red coats is 1.2% and the overall prevalence of bilaterally masked Australian cattle dogs with unilateral deafness and red coats is 4.5%; these prevalences are the same for Australian cattle dogs with black coats. Among Blue Heelers with either no facial masking or a unilateral mask, the probability of unilateral deafness is 0.05 and the probability of bilateral deafness is 0.01.

Let D represent the event that an Australian cattle dog is deaf (i.e., deaf in at least one ear), where D_2 represents bilateral deafness and D_1 represents unilateral deafness.

   i. What is the probability that an Australian cattle dog has a bilateral mask, no hearing deficits, and a red coat?

   ii. Calculate the proportion of bilaterally masked Blue Heelers without hearing deficits.

   iii. Compare the prevalence of deafness between Red Heelers and Blue Heelers.

   iv. If a dog is known to have no hearing deficits, what is the probability it is a Blue Heeler?

Chapter 3

Distributions of random variables

When planning clinical research studies, investigators try to anticipate the results they might see under certain hypotheses. The treatments for some forms of cancer, such as advanced lung cancer, are only effective in a small percentage of patients: typically 20% or less. Suppose that a study testing a new treatment will be conducted on 20 participants, where the working assumption is that 20% of the patients will respond to the treatment. How might the possible outcomes of the study be represented, along with their probabilities? It is possible to express various outcomes using the probability notation in the previous chapter, e.g. if A were the event that one patient responds to treatment, but this would quickly become unwieldy.

Instead, the anticipated outcome in the study can be represented as a **random variable**, which numerically summarizes the possible outcomes of a random experiment. For example, let X represent the number of patients who respond to treatment; a numerical value x can be assigned to each possible outcome, and the probabilities of $1, 2, \ldots, x$ patients having a good response can be expressed as $P(X = 1), P(X = 2), \ldots, P(X = x)$. The distribution of a random variable specifies the probability of each possible outcome associated with the random variable.

This chapter will begin by outlining general properties of random variables and their distributions. The rest of the chapter discusses specific named distributions that are commonly used throughout probability and statistics.

For labs, slides, and other resources, please visit
www.openintro.org/book/biostat

3.1 Random variables

3.1.1 Distributions of random variables

Formally, a random variable assigns numerical values to the outcome of a random phenomenon, and is usually written with a capital letter such as X, Y, or Z.

If a coin is tossed three times, the outcome is the sequence of observed heads and tails. One such outcome might be TTH: tails on the first two tosses, heads on the third. If the random variable X is the number of heads for the three tosses, $X = 1$; if Y is the number of tails, then $Y = 2$. For the sequence THT, only the order has changed, but the values of X and Y remain the same. For the sequence HHH, however, $X = 3$ and $Y = 0$. Even in this simple setting, is possible to define other random variables; for example, if Z is the toss when the first H occurs, then $Z = 3$ for the first set of tosses (TTH) and 1 for the third set (HHH).

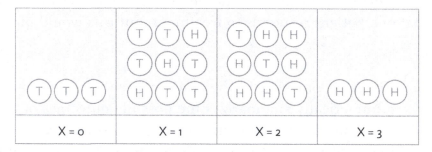

Figure 3.1: Possible outcomes for number of heads in three tosses of a coin.

If probabilities can be assigned to the outcomes in a random phenomenon or study, then those can be used to assign probabilities to values of a random variable. Using independence, $P(\text{HHH}) = (1/2)^3 = 1/8$. Since X in the above example can only be three if the three tosses are all heads, $P(X = 3) = 1/8$. The distribution of a random variable is the collection of probabilities for all of the variable's unique values. Figure 3.1 shows the eight possible outcomes when a coin is cossed three times: TTT, HTT, THT, TTH, HHT, HTH, THH, HHH. For the first set of tosses, $X = 0$; for the next three, $X = 1$, then $X = 2$ for the following three tosses and $X = 3$ for the last set (HHH).

Using independence again, each of the 8 outcomes have probability $1/8$, so $P(X = 0) = P(X = 3) = 1/8$ and $P(X = 1) = P(X = 2) = 3/8$. Figure 3.2 shows the probability distribution for X. Probability distributions for random variables follow the rules for probability; for instance, the sum of the probabilities must be 1.00. The possible outcomes of X are labeled with a corresponding lower case letter x and subscripts. The values of X are $x_1 = 0$, $x_2 = 1$, $x_3 = 2$, and $x_4 = 3$; these occur with probabilities 1/8, 3/8, 3/8 and 1/8.

i	1	2	3	4	Total
x_i	0	1	2	3	–
$P(X = x_i)$	1/8	3/8	3/8	1/8	$8/8 = 1.00$

Figure 3.2: Tabular form for the distribution of the number of heads in three coin tosses.

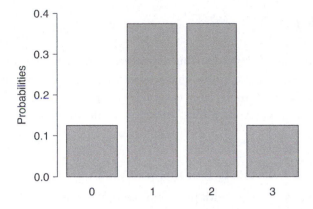

Figure 3.3: Bar plot of the distribution of the number of heads in three coin tosses.

Bar graphs can be used to show the distribution of a random variable. Figure 3.3 is a bar graph of the distribution of X in the coin tossing example. When bar graphs are used to show the distribution of a dataset, the heights of the bars show the frequency of observations; in contrast, bar heights for a probability distribution show the probabilities of possible values of a random variable.

X is an example of a **discrete random variable** since it takes on a finite number of values.[1] A **continuous random variable** can take on any real value in an interval.

In the hypothetical clinical study described at the beginning of this section, how unlikely would it be for 12 or more patients to respond to the treatment, given that only 20% of patients are expected to respond? Suppose X is a random variable that will denote the possible number of responding patients, out of a total of 20. X will have the same probability distribution as the number of heads in a 20 tosses of a weighted coin, where the probability of landing heads is 0.20. The graph of the probability distribution for X in Figure 3.4 can be used to approximate this probability. The event of 12 or more consists of nine values (12, 13, ..., 20); the graph shows that the probabilities for each value is extremely small, so the chance of 12 or more responses must be less than 0.01.[2]

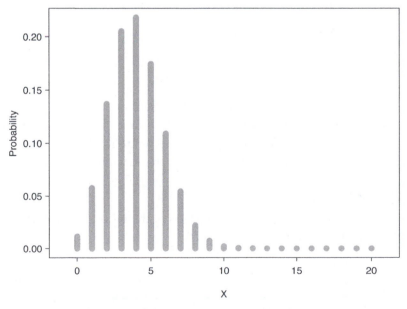

Figure 3.4: Bar plot of the distribution of the number of responses in a study with 20 participants and response probability 0.20

---

[1]Some discrete random variables have an infinite number of possible values, such as all the non-negative integers.
[2]Formulas in Section 3.2 can be used to show that the exact probability is slightly larger than 0.0001.

3.1.2 Expectation

Just like distributions of data, distributions of random variables also have means, variances, standard deviations, medians, etc.; these characteristics are computed a bit differently for random variables. The mean of a random variable is called its **expected value** and written $E(X)$. To calculate the mean of a random variable, multiply each possible value by its corresponding probability and add these products.

EXPECTED VALUE OF A DISCRETE RANDOM VARIABLE

If X takes on outcomes $x_1, ..., x_k$ with probabilities $P(X = x_1), ..., P(X = x_k)$, the expected value of X is the sum of each outcome multiplied by its corresponding probability:

$$E(X) = x_1 P(X = x_1) + \cdots + x_k P(X = x_k)$$

$$= \sum_{i=1}^{k} x_i P(X = x_i). \tag{3.1}$$

The Greek letter μ may be used in place of the notation $E(X)$.

EXAMPLE 3.2

Calculate the expected value of X, where X represents the number of heads in three tosses of a fair coin.

X can take on values 0, 1, 2, and 3. The probability of each x_k is given in Figure 3.2.

$E(X)$

Expected Value
of X

$$E(X) = x_1 P(X = x_1) + \cdots + x_k P(X = x_k)$$
$$= (0)(P(X = 0)) + (1)(P(X = 1)) + (2)(P(X = 2)) + (3)(P(X = 3))$$
$$= (0)(1/8) + (1)(3/8) + (2)(3/8) + (3)(1/8) = 12/8$$
$$= 1.5.$$

The expected value of X is 1.5.

The expected value for a random variable represents the average outcome. For example, $E(X) = 1.5$ represents the average number of heads in three tosses of a coin, if the three tosses were repeated many times.[3] It often happens with discrete random variables that the expected value is not precisely one of the possible outcomes of the variable.

GUIDED PRACTICE 3.3

Calculate the expected value of Y, where Y represents the number of heads in three tosses of an unfair coin, where the probability of heads is 0.70.[4]

---

[3]The expected value $E(X)$ can also be expressed as μ, e.g. $\mu = 1.5$

[4]First, calculate the probability distribution. $P(Y = 0) = (1 - 0.70)^3 = 0.027$ and $P(Y = 3) = (0.70)^3 = 0.343$. Note that there are three ways to obtain 1 head (HTT, THT, TTH), thus, $P(Y = 1) = (3)(0.70)(1 - 0.70)^2 = 0.189$. By the same logic, $P(Y = 2) = (3)(0.70)^2(1 - 0.70) = 0.441$. Thus, $E(Y) = (0)(0.027) + (1)(0.189) + (2)(0.441) + (3)(0.343) = 2.1$. The expected value of Y is 2.1.

3.1.3 Variability of random variables

The variability of a random variable can be described with variance and standard deviation. For data, the variance is computed by squaring deviations from the mean $(x_i - \mu)$ and then averaging over the number of values in the dataset (Section 1.4.2).

In the case of a random variable, the squared deviations from the mean of the random variable are used instead, and their sum is weighted by the corresponding probabilities. This weighted sum of squared deviations equals the variance; the standard deviation is the square root of the variance.

VARIANCE OF A DISCRETE RANDOM VARIABLE

If X takes on outcomes $x_1, ..., x_k$ with probabilities $P(X = x_1), \ldots, P(X = x_k)$ and expected value $\mu = E(X)$, then the variance of X, denoted by $\text{Var}(X)$ or σ^2, is

$$Var(X) = (x_1 - \mu)^2 P(X = x_1) + \cdots + (x_k - \mu)^2 P(X = x_k)$$

$$= \sum_{i=1}^{k} (x_i - \mu)^2 P(X = x_i). \tag{3.4}$$

The standard deviation of X, labeled $SD(X)$ or σ, is the square root of the variance.

$\text{Var}(X)$
Variance of X

The variance of a random variable can be interpreted as the expectation of the terms $(x_i - \mu)^2$; i.e., $\sigma^2 = E(X - \mu)^2$. While this compact form is not useful for direct computation, it can be helpful for understanding the concept of variability in the context of a random variable; variance is simply the average of the deviations from the mean.

EXAMPLE 3.5

Compute the variance and standard deviation of X, the number of heads in three tosses of a fair coin.

In the formula for the variance, $k = 4$ and $\mu_X = E(X) = 1.5$.

$$\sigma_X^2 = (x_1 - \mu_X)^2 P(X = x_1) + \cdots + (x_4 - \mu)^2 P(X = x_4)$$
$$= (0 - 1.5)^2 (1/8) + (1 - 1.5)^2 (3/8) + (2 - 1.5)^2 (3/8) + (3 - 1.5)^2 (1/8)$$
$$= 3/4.$$

The variance is $3/4 = 0.75$ and the standard deviation is $\sqrt{3/4} = 0.866$.

The coin tossing scenario provides a simple illustration of the mean and variance of a random variable. For the rest of this section, a more realistic example will be discussed—calculating expected health care costs.

In most typical health insurance plans in the United States, members of the plan pay annually in three categories: a monthly premium, a deductible amount that members pay each year before the insurance covers service, and "out-of-pocket" costs which include co-payments for each physician visit or prescription.[5] Picking a new health plan involves estimating costs for the next year based on a person's best guess at the type and number of services that will be needed.

---

[5]The deductible also includes care and supplies that are not covered by insurance.

In 2015, Harvard University offered several alternative plans to its employees. In the Health Maintenance Organization (HMO) plan for employees earning less than $70,000 per year, the monthly premium was $79, and the co-payment for each office visit or physical therapy session was $20. After a new employee examined her health records for the last 10 years, she noticed that in three of the 10 years, she visited the office of her primary care physician only once, for one annual physical. In four of the 10 years, she visited her physician three times: once for a physical, and twice for cases of the flu. In two of the years, she had four visits. In one of the 10 years, she experienced a knee injury that required 3 office visits and 5 physical therapy sessions.

EXAMPLE 3.6

Ignoring the cost of prescription drugs, over-the-counter medications, and the annual deductible amount, calculate the expectation and the standard deviation of the expected annual health care cost for this employee.

Let the random variable X denote annual health care costs, where x_i represents the costs in a year for i number of visits. If the last ten years are an accurate picture of annual costs for this employee, X will have four possible values.

The total cost of the monthly premiums in a single year is $12 \times \$79 = \$948$. The cost of each visit is $20, so the total visit cost for a year is $20 times the number of visits.

For example, the first column in the table contains information about the years in which the employee had one office visit. Adding the $948 for the annual premium and $20 for one visit results in $x_1 = \$968$; $P(X = x_i) = 3/10 = 0.30$.

i	1	2	3	4	Sum
Number of visits	1	3	4	8	
x_i	968	1008	1028	1108	
$P(X = x_i)$	0.30	0.40	0.20	0.10	1.00
$x_i P(X = x_i)$	290.40	403.20	205.60	110.80	1010.00

The expected cost of health care for a year, $\sum_i x_i P(X = x_i)$, is $\mu = \$1010.00$.

i	1	2	3	4	Sum
Number of visits	1	3	4	8	
x_i	968	1008	1028	1108	
$P(X = x_i)$	0.30	0.40	0.20	0.10	1.00
$(x_i)P(X = x_i)$	290.40	403.20	205.60	110.80	1010.00
$x_i - \mu$	-42.00	-2.00	18.00	98.00	
$(x_i - \mu)^2$	1764.00	4.00	324.00	9604	
$(x_i - \mu)^2 P(X = x_i)$	529.20	1.60	64.80	960.40	1556.00

The variance of X, $\sum_i (x_i - \mu)^2 P(X = x_i)$, is $\sigma^2 = 1556.00$, and the standard deviation is $\sigma = \$39.45$.[6]

---

[6]Note that the standard deviation always has the same units as the original measurements.

3.1.4   Linear combinations of random variables

Sums of random variables arise naturally in many problems. In the health insurance example, the amount spent by the employee during her next five years of employment can be represented as $X_1 + X_2 + X_3 + X_4 + X_5$, where X_1 is the cost of the first year, X_2 the second year, etc. If the employee's domestic partner has health insurance with another employer, the total annual cost to the couple would be the sum of the costs for the employee (X) and for her partner (Y), or $X + Y$. In each of these examples, it is intuitively clear that the average cost would be the sum of the average of each term.

Sums of random variables represent a special case of linear combinations of variables.

LINEAR COMBINATIONS OF RANDOM VARIABLES AND THEIR EXPECTED VALUES

If X and Y are random variables, then a linear combination of the random variables is given by

$$aX + bY,$$

where a and b are constants. The mean of a linear combination of random variables is

$$E(aX + bY) = aE(X) + bE(Y) = a\mu_X + b\mu_Y.$$

The formula easily generalizes to a sum of any number of random variables. For example, the average health care cost for 5 years, given that the cost for services remains the same, is

$$E(X_1 + X_2 + X_3 + X_4 + X_5) = E(5X_1) = 5E(X_1) = (5)(1010) = \$5,050.$$

The formula implies that for a random variable Z, $E(a + Z) = a + E(Z)$. This could have been used when calculating the average health costs for the employee by defining a as the fixed cost of the premium $(a = \$948)$ and Z as the cost of the physician visits. Thus, the total annual cost for a year could be calculated as: $E(a + Z) = a + E(Z) = \$948 + E(Z) = \$948 + .30(1 \times \$20) + .40(3 \times \$20) + .20(4 \times \$20) + 0.10(8 \times \$20) = \$1,010.00$.

GUIDED PRACTICE 3.7

Suppose the employee will begin a domestic partnership in the next year. Although she and her companion will begin living together and sharing expenses, they will each keep their existing health insurance plans; both, in fact, have the same plan from the same employer. In the last five years, her partner visited a physician only once in four of the ten years, and twice in the other six years. Calculate the expected total cost of health insurance to the couple in the next year.[7]

Calculating the variance and standard deviation of a linear combination of random variables requires more care. The formula given here requires that the random variables in the linear combination be independent, such that an observation on one of the variables provides no information about the value of the other variable.

---

[7] Let X represent the costs for the employee and Y represent the costs for her partner. $E(X) = \$1,010.00$, as previously calculated. $E(Y) = 948 + 0.4(1 \times \$20) + 0.6(2 \times \$20) = \$980.00$. Thus, $E(X + Y) = E(X) + E(Y) = \$1,010.00 + \$980.00 = \$1,990.00$.

VARIABILITY OF LINEAR COMBINATIONS OF RANDOM VARIABLES

$$\text{Var}(aX + bY) = a^2\text{Var}(X) + b^2\text{Var}(Y).$$

This equation is valid only if the random variables are independent of each other.

For the transformation $a + bZ$, the variance is $b^2\text{Var}(Z)$, since a constant a has variance 0. When $b = 1$, variance of $a + Z$ is $\text{Var}(Z)$—adding a constant to a random variable has no effect on the variability of the random variable.

EXAMPLE 3.8

Calculate the variance and standard deviation for the combined cost of next year's health care for the two partners, assuming that the costs for each person are independent.

Let X represent the sum of costs for the employee and Y the sum of costs for her partner.

First, calculate the variance of health care costs for the partner. The partner's costs are the sum of the annual fixed cost and the variable annual costs, so the variance will simply be the variance of the variable costs. If Z represents the component of the variable costs, $E(Z) = 0.4(1\times\$20)+0.6(2\times\$20) = \$8 + \$24 = \$32$. Thus, the variance of Z equals

$$\text{Var}(Z) = 0.4(20 - 32)^2 + 0.6(40 - 32)^2 = 96.$$

Under the assumption of independence, $\text{Var}(X + Y) = \text{Var}(X) + \text{Var}(Y) = 1556 + 96 = 1652$, and the standard deviation is $\sqrt{1652} = \$40.64$.

The example of health insurance costs has been simplified to make the calculations clearer. It ignores the fact that many plans have a deductible amount, and that plan members pay for services at different rates before and after the deductible has been reached. Often, insured individuals no longer need to pay for services at all once a maximum amount has been reached in a year. The example also assumes that the proportions of number of physician visits per year, estimated from the last 10 years, can be treated as probabilities measured without error. Had a different timespan been chosen, the proportions might well have been different.

It also relies on the assumption that health care costs for the two partners are independent. Two individuals living together may pass on infectious diseases like the flu, or may participate together in activities that lead to similar injuries, such as skiing or long distance running. Section 3.6 shows how to adjust a variance calculation when independence is unrealistic.

3.2 Binomial distribution

The hypothetical clinical study and coin tossing example discussed earlier in this chapter are both examples of experiments that can be modeled with a binomial distribution. The binomial distribution is a more general case of another named distribution, the Bernoulli distribution.

3.2.1 Bernoulli distribution

Psychologist Stanley Milgram began a series of experiments in 1963 to study the effect of authority on obedience. In a typical experiment, a participant would be ordered by an authority figure to give a series of increasingly severe shocks to a stranger. Milgram found that only about 35% of people would resist the authority and stop giving shocks before the maximum voltage was reached. Over the years, additional research suggested this number is approximately consistent across communities and time.[8]

Each person in Milgram's experiment can be thought of as a **trial**. Suppose that a trial is labeled a **success** if the person refuses to administer the worst shock. If the person does administer the worst shock, the trial is a **failure**. The **probability of a success** can be written as $p = 0.35$. The probability of a failure is sometimes denoted with $q = 1 - p$.

When an individual trial only has two possible outcomes, it is called a **Bernoulli random variable**. It is arbitrary as to which outcome is labeled success.

Bernoulli random variables are often denoted as 1 for a success and 0 for a failure. Suppose that ten trials are observed, of which 6 are successes and 4 are failures:

$$0\ 1\ 1\ 1\ 1\ 0\ 1\ 1\ 0\ 0.$$

The **sample proportion**, $\hat{p}$, is the sample mean of these observations:

$$\hat{p} = \frac{\text{\# of successes}}{\text{\# of trials}} = \frac{0+1+1+1+1+0+1+1+0+0}{10} = 0.6.$$

Since 0 and 1 are numerical outcomes, the mean and standard deviation of a Bernoulli random variable can be defined. If p is the true probability of a success, then the mean of a Bernoulli random variable X is given by

$$\mu = E[X] = P(X = 0) \times 0 + P(X = 1) \times 1$$
$$= (1-p) \times 0 + p \times 1 = 0 + p = p.$$

Similarly, the variance of X can be computed:

$$\sigma^2 = P(X = 0)(0 - p)^2 + P(X = 1)(1 - p)^2$$
$$= (1-p)p^2 + p(1-p)^2 = p(1-p).$$

The standard deviation is $\sigma = \sqrt{p(1-p)}$.

---

[8]Find further information on Milgram's experiment at www.cnr.berkeley.edu/ucce50/ag-labor/7article/article35.htm.

BERNOULLI RANDOM VARIABLE

If X is a random variable that takes value 1 with probability of success p and 0 with probability $1 - p$, then X is a Bernoulli random variable with mean p and standard deviation $\sqrt{p(1-p)}$.

Suppose X represents the outcome of a single toss of a fair coin, where heads is labeled success. X is a Bernoulli random variable with probability of success $p = 0.50$; this can be expressed as $X \sim \text{Bern}(p)$, or specifically, $X \sim \text{Bern}(0.50)$. It is essential to specify the probability of success when characterizing a Bernoulli random variable. For example, although the outcome of a single toss of an unfair coin can also be represented by a Bernoulli, it will have a different probability distribution since p does not equal 0.50 for an unfair coin.

Bern(p)

Bernoulli dist. with p prob. of success

The success probability p is the **parameter** of the distribution, and identifies a specific Bernoulli distribution out of the entire family of Bernoulli distributions where p can be any value between 0 and 1 (inclusive).

EXAMPLE 3.9

Suppose that four individuals are randomly selected to participate in Milgram's experiment. What is the chance that there will be exactly one successful trial, assuming independence between trials? Suppose that the probability of success remains 0.35.

Consider a scenario in which there is one success (i.e., one person refuses to give the strongest shock). Label the individuals as A, B, C, and D:

$$P(A = \text{refuse}, B = \text{shock}, C = \text{shock}, D = \text{shock})$$
$$= P(A = \text{refuse})\, P(B = \text{shock})\, P(C = \text{shock})\, P(D = \text{shock})$$
$$= (0.35)(0.65)(0.65)(0.65) = (0.35)^1 (0.65)^3 = 0.096.$$

However, there are three other possible scenarios: either B, C, or D could have been the one to refuse. In each of these cases, the probability is also $(0.35)^1(0.65)^3$. These four scenarios exhaust all the possible ways that exactly one of these four people could refuse to administer the most severe shock, so the total probability of one success is $(4)(0.35)^1(0.65)^3 = 0.38$.

3.2.2 The binomial distribution

The Bernoulli distribution is unrealistic in all but the simplest of settings. However, it is a useful building block for other distributions. The **binomial distribution** describes the probability of having exactly k successes in n independent Bernoulli trials with probability of a success p. In Example 3.9, the goal was to calculate the probability of 1 success out of 4 trials, with probability of success 0.35 ($n = 4$, $k = 1$, $p = 0.35$).

Like the Bernoulli distribution, the binomial is a discrete distribution, and can take on only a finite number of values. A binomial variable has values $0, 1, 2, \ldots, n$.

A general formula for the binomial distribution can be developed from re-examining Example 3.9. There were four individuals who could have been the one to refuse, and each of these four scenarios had the same probability. Thus, the final probability can be written as:

$$[\text{\# of scenarios}] \times P(\text{single scenario.}) \tag{3.10}$$

The first component of this equation is the number of ways to arrange the $k = 1$ successes among the $n = 4$ trials. The second component is the probability of any of the four (equally probable) scenarios.

Consider $P(\text{single scenario})$ under the general case of k successes and $n - k$ failures in the n trials. In any such scenario, the Multiplication Rule for independent events can be applied:

$$p^k(1-p)^{n-k}.$$

Secondly, there is a general formula for the number of ways to choose k successes in n trials, i.e. arrange k successes and $n - k$ failures:

$$\binom{n}{k} = \frac{n!}{k!(n-k)!}.$$

The quantity $\binom{n}{k}$ is read **n choose k**.[9] The exclamation point notation (e.g. $k!$) denotes a **factorial** expression.[10]

Using the formula, the number of ways to choose $k = 1$ successes in $n = 4$ trials can be computed as:

$$\binom{4}{1} = \frac{4!}{1!(4-1)!} = \frac{4!}{1!3!} = \frac{4 \times 3 \times 2 \times 1}{(1)(3 \times 2 \times 1)} = 4.$$

Substituting n choose k for the number of scenarios and $p^k(1-p)^{n-k}$ for the single scenario probability in Equation (3.10) yields the general binomial formula.

---

[9]Other notation for n choose k includes $_nC_k$, C_n^k, and $C(n,k)$.

[10]$0! = 1$, $1! = 1$, $2! = 2 \times 1 = 2$, $\ldots$, $n! = n \times (n-1) \times \ldots 2 \times 1$.

$\text{Bin}(n, p)$

Binomial dist.
with n trials
& p prob. of
success

BINOMIAL DISTRIBUTION

Suppose the probability of a single trial being a success is p. The probability of observing exactly k successes in n independent trials is given by

$$P(X = k) = \binom{n}{k} p^k (1 - p)^{n-k} = \frac{n!}{k!(n-k)!} p^k (1 - p)^{n-k}. \tag{3.11}$$

Additionally, the mean, variance, and standard deviation of the number of observed successes are, respectively

$$\mu = np \qquad\qquad \sigma^2 = np(1 - p) \qquad\qquad \sigma = \sqrt{np(1 - p)}. \tag{3.12}$$

A binomial random variable X can be expressed as $X \sim \text{Bin}(n, p)$.

IS IT BINOMIAL? FOUR CONDITIONS TO CHECK.

(1) The trials are independent.
(2) The number of trials, n, is fixed.
(3) Each trial outcome can be classified as a *success* or *failure*.
(4) The probability of a success, p, is the same for each trial.

EXAMPLE 3.13

What is the probability that 3 of 8 randomly selected participants will refuse to administer the worst shock?

First, check the conditions for applying the binomial model. The number of trials is fixed ($n = 8$) and each trial outcome can be classified as either success or failure. The sample is random, so the trials are independent, and the probability of success is the same for each trial.

For the outcome of interest, $k = 3$ successes occur in $n = 8$ trials, and the probability of a success is $p = 0.35$. Thus, the probability that 3 of 8 will refuse is given by

$$
\begin{aligned}
P(X = 3) = \binom{8}{3}(0.35)^3(1 - 0.35)^{8-3} &= \frac{8!}{3!(8-3)!}(0.35)^3(1 - 0.35)^{8-3} \\
&= (56)(0.35)^3(0.65)^5 \\
&= 0.28.
\end{aligned}
$$

EXAMPLE 3.14

What is the probability that at most 3 of 8 randomly selected participants will refuse to administer the worst shock?

The event of at most 3 out of 8 successes can be thought of as the combined probability of 0, 1, 2, and 3 successes. Thus, the probability that at most 3 of 8 will refuse is given by:

$$P(X \leq 3) = P(X = 0) + P(X = 1) + P(X = 2) + P(X = 3)$$

$$= \binom{8}{0}(0.35)^0(1-0.35)^{8-0} + \binom{8}{1}(0.35)^1(1-0.35)^{8-1}$$

$$+ \binom{8}{2}(0.35)^2(1-0.35)^{8-2} + \binom{8}{3}(0.35)^3(1-0.35)^{8-3}$$

$$= (1)(0.35)^0(1-0.35)^8 + (8)(0.35)^1(1-0.35)^7$$

$$+ (28)(0.35)^2(1-0.35)^6 + (56)(0.35)^3(1-0.35)^5$$

$$= 0.706.$$

EXAMPLE 3.15

If 40 individuals were randomly selected to participate in the experiment, how many individuals would be expected to refuse to administer the worst shock? What is the standard deviation of the number of people expected to refuse?

Both quantities can directly be computed from the formulas in Equation (3.12). The expected value (mean) is given by: $\mu = np = 40 \times 0.35 = 14$. The standard deviation is: $\sigma = \sqrt{np(1-p)} = \sqrt{40 \times 0.35 \times 0.65} = 3.02$.

GUIDED PRACTICE 3.16

The probability that a smoker will develop a severe lung condition in their lifetime is about 0.30. Suppose that 5 smokers are randomly selected from the population. What is the probability that (a) one will develop a severe lung condition? (b) that no more than one will develop a severe lung condition? (c) that at least one will develop a severe lung condition?[11]

---

[11]Let $p = 0.30$; $X \sim \text{Bin}(5, 0.30)$. (a) $P(X = 1) = \binom{5}{1}(0.30)^1(1-0.30)^{5-1} = 0.36$ (b) $P(X \leq 1) = P(X = 0) + P(X = 1) = \binom{5}{0}(0.30)^0(1-0.30)^{5-0} + 0.36 = 0.53$ (c) $P(X \geq 1) = 1 - P(X = 0) = 1 - 0.36 = 0.83$

3.3   Normal distribution

Among the many distributions seen in practice, one is by far the most common: the **normal distribution**, which has the shape of a symmetric, unimodal bell curve. Many variables are nearly normal, which makes the normal distribution useful for a variety of problems. For example, characteristics such as human height closely follow the normal distribution.

3.3.1   Normal distribution model

The normal distribution model always describes a symmetric, unimodal, bell-shaped curve. However, the curves can differ in center and spread; the model can be adjusted using mean and standard deviation. Changing the mean shifts the bell curve to the left or the right, while changing the standard deviation stretches or constricts the curve. Figure 3.5 shows the normal distribution with mean 0 and standard deviation 1 in the left panel and the normal distribution with mean 19 and standard deviation 4 in the right panel. Figure 3.6 shows these distributions on the same axis.

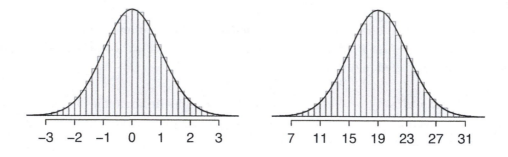

Figure 3.5: Both curves represent the normal distribution; however, they differ in their center and spread. The normal distribution with mean 0 and standard deviation 1 is called the **standard normal distribution**.

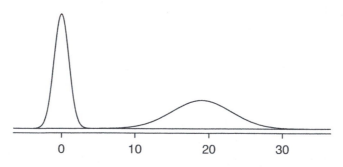

Figure 3.6: The normal models shown in Figure 3.5 but plotted together and on the same scale.

$N(\mu, \sigma)$

Normal dist.
with mean μ
& st. dev. σ

For any given normal distribution with mean μ and standard deviation σ, the distribution can be written as $N(\mu, \sigma)$; μ and σ are the parameters of the normal distribution. For example, $N(0, 1)$ refers to the standard normal distribution, as shown in Figure 3.5.

Unlike the Bernoulli and binomial distributions, the normal distribution is a continuous distribution.

3.3.2 Standardizing with Z-scores

The **Z-score** of an observation quantifies how far the observation is from the mean, in units of standard deviation(s). If x is an observation from a distribution $N(\mu, \sigma)$, the Z-score is mathematically defined as:

$$Z = \frac{x - \mu}{\sigma}.$$

Z

Z-score, the standardized observation

An observation equal to the mean has a Z-score of 0. Observations above the mean have positive Z-scores, while observations below the mean have negative Z-scores. For example, if an observation is one standard deviation above the mean, it has a Z-score of 1; if it is 1.5 standard deviations below the mean, its Z-score is -1.5.

Z-scores can be used to identify which observations are more extreme than others, and are especially useful when comparing observations from different normal distributions. One observation x_1 is said to be more unusual than another observation x_2 if the absolute value of its Z-score is larger than the absolute value of the other observation's Z-score: $|Z_1| > |Z_2|$. In other words, the further an observation is from the mean in either direction, the more extreme it is.

EXAMPLE 3.17

The SAT and the ACT are two standardized tests commonly used for college admissions in the United States. The distribution of test scores are both nearly normal. For the SAT, $N(1500, 300)$; for the ACT, $N(21, 5)$. While some colleges request that students submit scores from both tests, others allow students the choice of either the ACT or the SAT. Suppose that one student scores an 1800 on the SAT (Student A) and another scores a 24 on the ACT (Student B). A college admissions officer would like to compare the scores of the two students to determine which student performed better.

Calculate a Z-score for each student; i.e., convert x to Z.

Using $\mu_{SAT} = 1500$, $\sigma_{SAT} = 300$, and $x_A = 1800$, find Student A's Z-score:

$$Z_A = \frac{x_A - \mu_{SAT}}{\sigma_{SAT}} = \frac{1800 - 1500}{300} = 1.$$

For Student B:

$$Z_B = \frac{x_B - \mu_{ACT}}{\sigma_{ACT}} = \frac{24 - 21}{5} = 0.6.$$

Student A's score is 1 standard deviation above average on the SAT, while Student B's score is 0.6 standard deviations above the mean on the ACT. As illustrated in Figure 3.7, Student A's score is more extreme, indicating that Student A has scored higher with respect to other scores than Student B.

THE Z-SCORE

The Z-score of an observation quantifies how far the observation is from the mean, in units of standard deviation(s). The Z-score for an observation x that follows a distribution with mean μ and standard deviation σ can be calculated using

$$Z = \frac{x - \mu}{\sigma}.$$

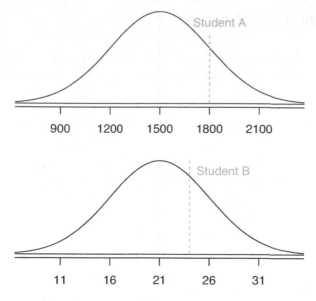

Figure 3.7: Scores of Students A and B plotted on the distributions of SAT and ACT scores.

EXAMPLE 3.18

How high would a student need to score on the ACT to have a score equivalent to Student A's score of 1800 on the SAT?

As shown in Example 3.7, a score of 1800 on the SAT is 1 standard deviation above the mean. ACT scores are normally distributed with mean 21 and standard deviation 5. To convert a value from the standard normal curve (Z) to one on a normal distribution $N(\mu, \sigma)$:

$$x = \mu + Z\sigma.$$

Thus, a student would need a score of $21 + 1(5) = 26$ on the ACT to have a score equivalent to 1800 on the SAT.

GUIDED PRACTICE 3.19

Systolic blood pressure (SBP) for adults in the United States aged 18-39 follow an approximate normal distribution, $N(115, 17.5)$. As age increases, systolic blood pressure also tends to increase. Mean systolic blood pressure for adults 60 years of age and older is 136 mm Hg, with standard deviation 40 mm Hg. Systolic blood pressure of 140 mm Hg or higher is indicative of hypertension (high blood pressure). (a) How many standard deviations away from the mean is a 30-year-old with systolic blood pressure of 125 mm Hg? (b) Compare how unusual a systolic blood pressure of 140 mm Hg is for a 65-year-old, versus a 30-year-old.[12]

---

[12](a) Calculate the Z-score: $\frac{\bar{x}-\mu}{\sigma} = \frac{125-115}{17.5} = 0.571$. A 30-year-old with systolic blood pressure of 125 mm Hg is about 0.6 standard deviations above the mean. (b) For $x_1 = 140$ mm Hg: $Z_1 = \frac{x_1-\mu}{\sigma} = \frac{140-115}{17.5} = 1.43$. For $x_2 = 140$ mm Hg: $Z_2 = \frac{x_2-\mu}{\sigma} = \frac{140-136}{40} = 0.1$. While an SBP of 140 mm Hg is almost 1.5 standard deviations above the mean for a 30-year-old, it is only 0.1 standard deviations above the mean for a 65-year-old.

3.3.3 The empirical rule

The empirical rule (also known as the 68-95-99.7 rule) states that for a normal distribution, almost all observations will fall within three standard deviations of the mean. Specifically, 68% of observations are within one standard deviation of the mean, 95% are within two SD's, and 99.7% are within three SD's.

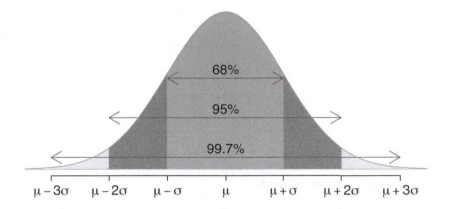

Figure 3.8: Probabilities for falling within 1, 2, and 3 standard deviations of the mean in a normal distribution.

While it is possible for a normal random variable to take on values 4, 5, or even more standard deviations from the mean, these occurrences are extremely rare if the data are nearly normal. For example, the probability of being further than 4 standard deviations from the mean is about 1-in-30,000.

3.3.4 Calculating normal probabilities

The normal distribution is a continuous probability distribution. Recall from Section 2.1.5 that the total area under the density curve is always equal to 1, and the probability that a variable has a value within a specified interval is the area under the curve over that interval. By using either statistical software or normal probability tables, the normal model can be used to identify a probability or percentile based on the corresponding Z-score (and vice versa).

Figure 3.9: The area to the left of Z represents the percentile of the observation.

Z	Second decimal place of Z									
	0.00	0.01	0.02	0.03	0.04	0.05	0.06	0.07	0.08	0.09
0.0	0.5000	0.5040	0.5080	0.5120	0.5160	0.5199	0.5239	0.5279	0.5319	0.5359
0.1	0.5398	0.5438	0.5478	0.5517	0.5557	0.5596	0.5636	0.5675	0.5714	0.5753
0.2	0.5793	0.5832	0.5871	0.5910	0.5948	0.5987	0.6026	0.6064	0.6103	0.6141
0.3	0.6179	0.6217	0.6255	0.6293	0.6331	0.6368	0.6406	0.6443	0.6480	0.6517
0.4	0.6554	0.6591	0.6628	0.6664	0.6700	0.6736	0.6772	0.6808	0.6844	0.6879
0.5	0.6915	0.6950	0.6985	0.7019	0.7054	0.7088	0.7123	0.7157	0.7190	0.7224
0.6	0.7257	0.7291	0.7324	0.7357	0.7389	0.7422	0.7454	0.7486	0.7517	0.7549
0.7	0.7580	0.7611	0.7642	0.7673	0.7704	0.7734	0.7764	0.7794	0.7823	0.7852
0.8	0.7881	0.7910	0.7939	0.7967	**0.7995**	0.8023	0.8051	0.8078	0.8106	0.8133
0.9	0.8159	0.8186	0.8212	0.8238	0.8264	0.8289	0.8315	0.8340	0.8365	0.8389
1.0	0.8413	0.8438	0.8461	0.8485	0.8508	0.8531	0.8554	0.8577	0.8599	0.8621
1.1	0.8643	0.8665	0.8686	0.8708	0.8729	0.8749	0.8770	0.8790	0.8810	0.8830
⋮	⋮	⋮	⋮	⋮	⋮	⋮	⋮	⋮	⋮	⋮	⋮

Figure 3.10: A section of the normal probability table. The percentile for a normal random variable with $Z = 0.43$ has been *highlighted*, and the percentile closest to 0.8000 has also been **highlighted**.

A **normal probability table** is given in Appendix B.1 on page 463 and abbreviated in Figure 3.10. This table can be used to identify the **percentile** corresponding to any particular Z-score; for instance, the percentile of $Z = 0.43$ is shown in row 0.4 and column 0.03 in Figure 3.10: 0.6664, or the 66.64^{th} percentile. First, find the proper row in the normal probability table up through the first decimal, and then determine the column representing the second decimal value. The intersection of this row and column is the percentile of the observation. This value also represents the probability that the standard normal variable Z takes on a value of 0.43 or less; i.e. $P(Z \leq 0.43) = 0.6664$.

The table can also be used to find the Z-score associated with a percentile. For example, to identify Z for the 80th percentile, look for the value closest to 0.8000 in the middle portion of the table: 0.7995. The Z-score for the 80th percentile is given by combining the row and column Z values: 0.84.

EXAMPLE 3.20

Student A from Example 3.17 earned a score of 1800 on the SAT, which corresponds to $Z = 1$. What percentile is this score associated with?

In this context, the **percentile** is the percentage of people who earned a lower SAT score than Student A. From the normal table, Z of 1.00 is 0.8413. Thus, the student is in the 84th percentile of test takers. This area is shaded in Figure 3.11.

GUIDED PRACTICE 3.21

Determine the proportion of SAT test takers who scored better than Student A on the SAT.[13]

---

[13]If 84% had lower scores than Student A, the number of people who had better scores must be 16%.

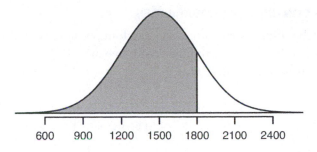

Figure 3.11: The normal model for SAT scores, with shaded area representing scores below 1800.

3.3.5 Normal probability examples

There are two main types of problems that involve the normal distribution: calculating probabilities from a given value (whether X or Z), or identifying the observation that corresponds to a particular probability.

EXAMPLE 3.22

Cumulative SAT scores are well-approximated by a normal model, $N(1500, 300)$. What is the probability that a randomly selected test taker scores at least 1630 on the SAT?

For any normal probability problem, it can be helpful to start out by drawing the normal curve and shading the area of interest.

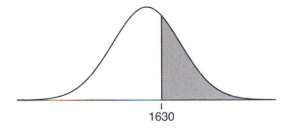

To find the shaded area under the curve, convert 1630 to a Z-score:

$$Z = \frac{x - \mu}{\sigma} = \frac{1630 - 1500}{300} = \frac{130}{300} = 0.43.$$

Look up the percentile of $Z = 0.43$ in the normal probability table shown in Figure 3.10 or in Appendix B.1 on page 463: 0.6664. However, note that the percentile describes those who had a Z-score *lower* than 0.43, or in other words, the area below 0.43. To find the area *above* $Z = 0.43$, subtract the area of the lower tail from the total area under the curve, 1:

$$1.0000 \quad - \quad 0.6664 \quad = \quad 0.3336$$

The probability that a student scores at least 1630 on the SAT is 0.3336.

DISCRETE VERSUS CONTINUOUS PROBABILITIES

Recall that the probability of a continuous random variable equaling some exact value is always 0. As a result, for a continuous random variable X, $P(X \leq x) = P(X < x)$ and $P(X \geq x) = P(X > x)$. It is valid to state that $P(X \geq x) = 1 - P(X \leq x) = 1 - P(X < x)$.

This is *not* the case for discrete random variables. For example, for a discrete random variable Y, $P(Y \geq 2) = 1 - P(Y < 2) = 1 - P(Y \leq 1)$. It would be incorrect to claim that $P(Y \geq 2) = 1 - P(Y \leq 2)$.

GUIDED PRACTICE 3.23

What is the probability of a student scoring at most 1630 on the SAT?[14]

GUIDED PRACTICE 3.24

Systolic blood pressure for adults 60 years of age and older in the United States is approximately normally distributed: $N(136, 40)$. What is the probability of an adult in this age group having systolic blood pressure of 140 mm Hg or greater?[15]

---

[14]This probability was calculated as part of Example 3.22: 0.6664. A picture for this exercise is represented by the shaded area below "0.6664" in Example 3.22.

[15]The Z-score for this observation was calculated in Exercise 3.19 as 0.1. From the table, the $P(Z \geq 0.1) = 1 - 0.54 = 0.46$.

EXAMPLE 3.25

The height of adult males in the United States between the ages of 20 and 62 is nearly normal, with mean 70 inches and standard deviation 3.3 inches.[16] What is the probability that a random adult male is between 5'9" and 6'2"?

These heights correspond to 69 inches and 74 inches. First, draw the figure. The area of interest is an interval, rather than a tail area.

To find the middle area, find the area to the left of 74; from that area, subtract the area to the left of 69.

First, convert to Z-scores:

$$Z_{74} = \frac{x-\mu}{\sigma} = \frac{74-70}{3.3} = 1.21, \qquad Z_{62} = \frac{x-\mu}{\sigma} = \frac{69-70}{3.3} = -0.30.$$

From the normal probability table, the areas are respectively, 0.8868 and 0.3821. The middle area is $0.8868 - 0.3821 = 0.5048$. The probability of being between heights 5'9" and 6'2" is 0.5048.

0.8868 $-$ 0.3821 $=$ 0.5048

GUIDED PRACTICE 3.26

What percentage of adults in the United States ages 60 and older have blood pressure between 145 and 130 mm Hg?[17]

---

[16]As based on a sample of 100 men, from the USDA Food Commodity Intake Database.

[17]First calculate Z-scores, then find the percent below 145 mm Hg and below 130 mm Hg: $Z_{145} = 0.23 \rightarrow 0.5890$, $Z_{130} = -0.15 \rightarrow 0.4404$ (area above). Final answer: $0.5890 - 0.4404 = 0.1486$.

EXAMPLE 3.27

How tall is a man with height in the 40th percentile?

First, draw a picture. The lower tail probability is 0.40, so the shaded area must start before the mean.

Determine the Z-score associated with the 40th percentile. Because the percentile is below 50%, Z will be negative. Look for the probability inside the negative part of table that is closest to 0.40: 0.40 falls in row −0.2 and between columns 0.05 and 0.06. Since it falls closer to 0.05, choose $Z = -0.25$.

Convert the Z-score to X, where $X \sim N(70, 3.3)$.

$$X = \mu + \sigma Z = 70 + (-0.25)(3.3) = 69.18.$$

A man with height in the 40th percentile is 69.18 inches tall, or about 5′ 9″.

GUIDED PRACTICE 3.28

(a) What is the 95th percentile for SAT scores? (b) What is the 97.5^{th} percentile of the male heights?[18]

---

[18](a) Look for 0.95 in the probability portion (middle part) of the normal probability table: row 1.6 and (about) column 0.05, i.e. $Z_{95} = 1.65$. Knowing $Z_{95} = 1.65$, $\mu = 1500$, and $\sigma = 300$, convert Z to x: $1500 + (1.65)(300) = 1995$. (b) Similarly, find $Z_{97.5} = 1.96$, and convert to x: $x_{97.5} = 76.5$ inches.

3.3.6   Normal approximation to the binomial distribution

The normal distribution can be used to approximate other distributions, such as the binomial distribution. The binomial formula is cumbersome when sample size is large, particularly when calculating probabilities for a large number of observations. Under certain conditions, the normal distribution can be used to approximate binomial probabilities. This method was widely used when calculating binomial probabilities by hand was the only option. Nowadays, modern statistical software is capable of calculating exact binomial probabilities even for very large n. The normal approximation to the binomial is discussed here since it is an important result that will be revisited in later chapters.

Consider the binomial model when probability of success is $p = 0.10$. Figure 3.12 shows four hollow histograms for simulated samples from the binomial distribution using four different sample sizes: $n = 10, 30, 100, 300$. As the sample size increases from $n = 10$ to $n = 300$, the distribution is transformed from a blocky and skewed distribution into one resembling the normal curve.

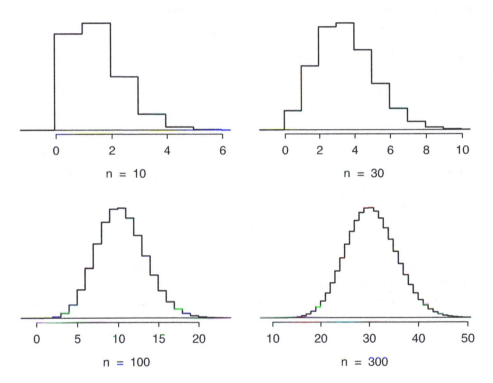

Figure 3.12: Hollow histograms of samples from the binomial model when $p = 0.10$. The sample sizes for the four plots are $n = 10, 30, 100,$ and 300, respectively.

NORMAL APPROXIMATION OF THE BINOMIAL DISTRIBUTION

The binomial distribution with probability of success p is nearly normal when the sample size n is sufficiently large such that np and $n(1 - p)$ are both at least 10. The approximate normal distribution has parameters corresponding to the mean and standard deviation of the binomial distribution:

$$\mu = np \qquad\qquad \sigma = \sqrt{np(1 - p)}$$

EXAMPLE 3.29

Approximately 20% of the US population smokes cigarettes. A local government commissioned a survey of 400 randomly selected individuals to investigate whether their community might have a lower smoker rate than 20%. The survey found that 59 of the 400 participants smoke cigarettes. If the true proportion of smokers in the community is 20%, what is the probability of observing 59 or fewer smokers in a sample of 400 people?

The desired probability is equivalent to the sum of the individual probabilities of observing $k = 0$, $1, ..., 58$, or 59 smokers in a sample of $n = 400$: $P(X \le 59)$. Confirm that the normal approximation is valid: $np = 400 \times 0.20 = 80$, $n(1-p) = 400 \times 0.8 = 320$. To use the normal approximation, calculate the mean and standard deviation from the binomial model:

$$\mu = np = 80 \qquad\qquad\qquad \sigma = \sqrt{np(1-p)} = 8.$$

Convert 59 to a Z-score: $Z = \dfrac{59 - 80}{8} = -2.63$. Use the normal probability table to identify the left tail area, which is 0.0043.

This estimate is very close to the answer derived from the exact binomial calculation:

$$P(k = 0 \text{ or } k = 1 \text{ or } \cdots \text{ or } k = 59) = P(k = 0) + P(k = 1) + \cdots + P(k = 59) = 0.0041.$$

However, even when the conditions for using the approximation are met, the normal approximation to the binomial tends to perform poorly when estimating the probability of a small range of counts. Suppose the normal approximation is used to compute the probability of observing 69, 70, or 71 smokers in 400 when $p = 0.20$. In this setting, the exact binomial and normal approximation result in notably different answers: the approximation gives 0.0476, while the binomial returns 0.0703.

The cause of this discrepancy is illustrated in Figure 3.13, which shows the areas representing the binomial probability (outlined) and normal approximation (shaded). Notice that the width of the area under the normal distribution is 0.5 units too slim on both sides of the interval.

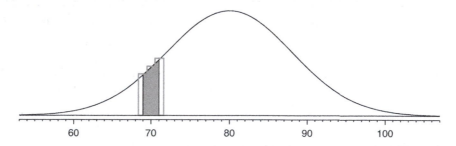

Figure 3.13: A normal curve with the area between 69 and 71 shaded. The outlined area represents the exact binomial probability.

The normal approximation can be improved if the cutoff values for the range of observations is modified slightly: the lower value should be reduced by 0.5 and the upper value increased by 0.5. The normal approximation with continuity correction gives 0.0687 for the probability of observing 69, 70, or 71 smokers in 400 when $p = 0.20$, which is closer to the exact binomial result of 0.0703.

This adjustment method is known as a continuity correction, which allows for increased accuracy when a continuous distribution is used to approximate a discrete one. The modification is typically not necessary when computing a tail area, since the total interval in that case tends to be quite wide.

3.3.7 Evaluating the normal approximation

The normal model can also be used to approximate data distributions. While using a normal model can be convenient, it is important to remember that normality is always an approximation. Testing the appropriateness of the normal assumption is a key step in many data analyses.

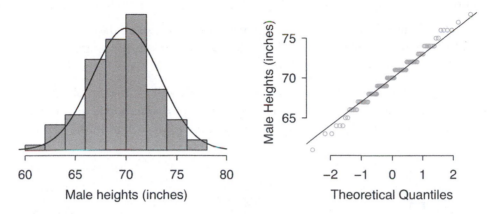

Figure 3.14: A sample of 100 male heights. Since the observations are rounded to the nearest whole inch, the points in the normal probability plot appear to jump in increments.

Example 3.27 suggests the distribution of heights of US males is well approximated by the normal model. There are two visual methods used to assess the assumption of normality. The first is a simple histogram with the best fitting normal curve overlaid on the plot, as shown in the left panel of Figure 3.14. The sample mean $\bar{x}$ and standard deviation s are used as the parameters of the best fitting normal curve. The closer this curve fits the histogram, the more reasonable the normal model assumption. More commonly, a **normal probability plot** is used, such as the one shown in the right panel of Figure 3.14.[19] If the points fall on or near the line, the data closely follow the normal model.

---

[19] Also called a **quantile-quantile plot**, or Q-Q plot.

EXAMPLE 3.30

Three datasets were simulated from a normal distribution, with sample sizes $n = 40$, $n = 100$, and $n = 400$; the histograms and normal probability plots of the datasets are shown in Figure 3.15. What happens as sample size increases?

As sample size increases, the data more closely follows the normal distribution; the histograms become more smooth, and the points on the Q-Q plots show fewer deviations from the line.

It is important to remember that when evaluating normality in a small dataset, apparent deviations from normality may simply be due to small sample size. Remember that all three of these simulated datasets are drawn from a normal distribution.

When assessing the normal approximation in real data, it will be rare to observe a Q-Q plot as clean as the one shown for $n = 400$. Typically, the normal approximation is reasonable even if there are some small observed departures from normality in the tails, such as in the plot for $n = 100$.

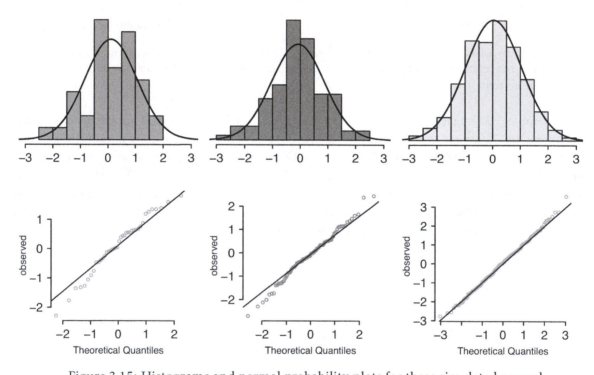

Figure 3.15: Histograms and normal probability plots for three simulated normal data sets; $n = 40$ (left), $n = 100$ (middle), $n = 400$ (right).

EXAMPLE 3.31

Would it be reasonable to use the normal distribution to accurately calculate percentiles of heights of NBA players? Consider all 435 NBA players from the 2008-9 season presented in Figure 3.16.[20]

The histogram in the left panel is slightly left skewed, and the points in the normal probability plot do not closely follow a straight line, particularly in the upper quantiles. The normal model is not an accurate approximation of NBA player heights.

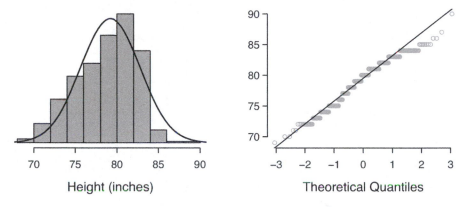

Figure 3.16: Histogram and normal probability plot for the NBA heights from the 2008-9 season.

EXAMPLE 3.32

Consider the poker winnings of an individual over 50 days. A histogram and normal probability plot of these data are shown in Figure 3.17 Evaluate whether a normal approximation is appropriate.

The data are very strongly right skewed in the histogram, which corresponds to the very strong deviations on the upper right component of the normal probability plot. These data show very strong deviations from the normal model; the normal approximation should not be applied to these data.

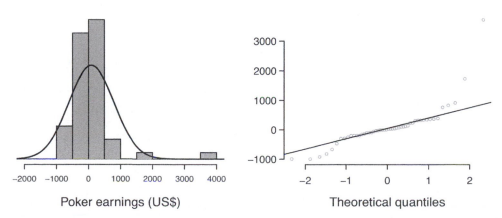

Figure 3.17: A histogram of poker data with the best fitting normal plot and a normal probability plot.

[20]These data were collected from www.nba.com.

GUIDED PRACTICE 3.33

Determine which data sets represented in Figure 3.18 plausibly come from a nearly normal distribution.[21]

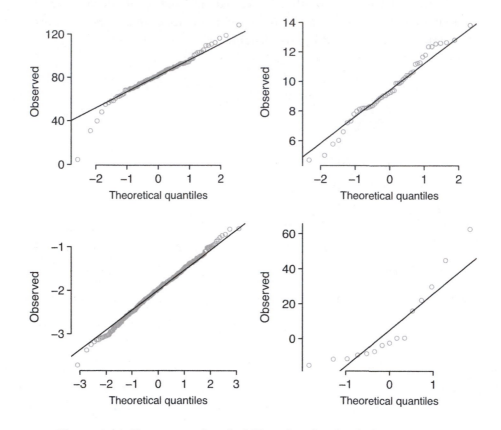

Figure 3.18: Four normal probability plots for Guided Practice 3.33.

---

[21]Answers may vary. The top-left plot shows some deviations in the smallest values in the dataset; specifically, the left tail shows some large outliers. The top-right and bottom-left plots do not show any obvious or extreme deviations from the lines for their respective sample sizes, so a normal model would be reasonable. The bottom-right plot has a consistent curvature that suggests it is not from the normal distribution. From examining the vertical coordinates of the observations, most of the data are between -20 and 0, then there are about five observations scattered between 0 and 70; this distribution has strong right skew.

When observations spike downwards on the left side of a normal probability plot, this indicates that the data have more outliers in the left tail expected under a normal distribution. When observations spike upwards on the right side, the data have more outliers in the right tail than expected under the normal distribution.

GUIDED PRACTICE 3.34

Figure 3.19 shows normal probability plots for two distributions that are skewed. One distribution is skewed to the low end (left skewed) and the other to the high end (right skewed). Which is which?[22]

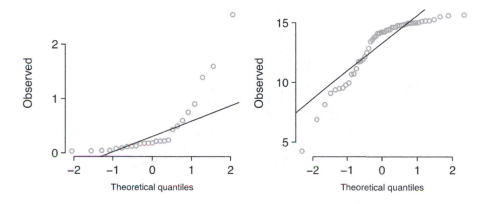

Figure 3.19: Normal probability plots for Guided Practice 3.34.

---

[22]Examine where the points fall along the vertical axis. In the first plot, most points are near the low end with fewer observations scattered along the high end; this describes a distribution that is skewed to the high end. The second plot shows the opposite features, and this distribution is skewed to the low end.

3.4 Poisson distribution

The **Poisson distribution** is a discrete distribution used to calculate probabilities for the number of occurrences of a rare event. In technical terms, it is used as a model for count data. For example, historical records of hospitalizations in New York City indicate that among a population of approximately 8 million people, 4.4 people are hospitalized each day for an acute myocardial infarction (AMI), on average. A histogram of showing the distribution of the number of AMIs per day on 365 days for NYC is shown in Figure 3.20.[23]

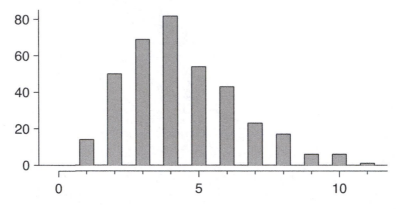

Figure 3.20: A histogram of the number of people hospitalized for an AMI on 365 days for NYC, as simulated from a Poisson distribution with mean 4.4.

POISSON DISTRIBUTION

$Pois(\lambda)$

Poisson dist. with rate λ

The Poisson distribution is a probability model for the number of events that occur in a population. The probability that exactly k events occur is given by

$$P(X = k) = \frac{e^{-\lambda}(\lambda)^k}{k!},$$

where k may take a value 0, 1, 2, ... The mean and standard deviation of this distribution are λ and $\sqrt{\lambda}$, respectively. A Poisson random variable X can be expressed as $X \sim Pois(\lambda)$.

λ

Rate for the Poisson dist.

When events accumulate over time in such a way that the probability an event occurs in an interval is proportional to the length of an interval and that the number of events in non-overlapping intervals are independent, the parameter λ (the Greek letter *lambda*) represents the average number of events per unit time; i.e., the rate per unit time.

In this setting, the number of events in t units of time has probability

$$P(X = k) = \frac{e^{-\lambda t}(\lambda t)^k}{k!},$$

where k takes on values 0, 1, 2, ... When used this way, the mean and standard deviation are λt and $\sqrt{\lambda t}$, respectively. The rate parameter λ represents the expected number of events per unit time, while the quantity λt represents the expected number events over a time period of t units.

The histogram in Figure 3.20 approximates a Poisson distribution with rate equal to 4.4 events per day, for a population of 8 million.

---

[23]These data are simulated. In practice, it would be important to check for an association between successive days.

EXAMPLE 3.35

In New York City, what is the probability that 2 individuals are hospitalized for AMI in seven days, if the rate is known to be 4.4 deaths per day?

From the given information, $\lambda = 4.4$, $k = 2$, and $t = 7$.

$$P(X = k) = \frac{e^{-\lambda t}(\lambda t)^k}{k!}$$

$$P(X = 2) = \frac{e^{-4.4 \times 7}(4.4 \times 7)^2}{2!} = 1.99 \times 10^{-11}.$$

GUIDED PRACTICE 3.36

In New York City, what is the probability that (a) at most 2 individuals are hospitalized for AMI in seven days, (b) at least 3 individuals are hospitalized for AMI in seven days?[24]

A rigorous set of conditions for the Poisson distribution is not discussed here. Generally, the Poisson distribution is used to calculate probabilities for rare events that accumulate over time, such as the occurrence of a disease in a population.

EXAMPLE 3.37

For children ages 0 - 14, the incidence rate of acute lymphocytic leukemia (ALL) was approximately 30 diagnosed cases per million children per year in 2010. Approximately 20% of the US population of 319,055,000 are in this age range. What is the expected number of cases of ALL in the US over five years?

The incidence rate for one year can be expressed as $30/1,000,000 = 0.00003$; for five years, the rate is $(5)(0.00003) = 0.00015$. The number of children age 0-14 in the population is $(0.20)(319,055,000) \approx 63,811,000$.

$$\lambda = \text{(relevant population size)(rate per child)}$$
$$= 63,811,000 \times 0.00015$$
$$= 9,571.5$$

The expected number of cases over five years is 9,571.5 cases.

---

[24](a) $P(X \leq 2) = P(X = 0) + P(X = 1) + P(X = 2) = \frac{e^{-4.4 \times 7}(4.4 \times 7)^0}{0!} + \frac{e^{-4.4 \times 7}(4.4 \times 7)^1}{1!} + \frac{e^{-4.4 \times 7}(4.4 \times 7)^2}{2!} = 2.12 \times 10^{-11}$ (b) $P(X \geq 3) = 1 - P(X < 3) = 1 - P(X \leq 2) = 1 - 2.12 \times 10^{-11} \approx 1$

3.5   Distributions related to Bernoulli trials

The binomial distribution is not the only distribution that can be built from a series of repeated Bernoulli trials. This section discusses the geometric, negative binomial, and hypergeometric distributions.

3.5.1   Geometric distribution

The geometric distribution describes the waiting time until one success for a series of independent Bernoulli random variables, in which the probability of success p remains constant.

EXAMPLE 3.38

Recall that in the Milgram shock experiments, the probability of a person refusing to give the most severe shock is $p = 0.35$. Suppose that participants are tested one at a time until one person refuses; i.e., until the first occurrence of a successful trial. What are the chances that the first occurrence happens with the first trial? The second trial? The third?

The probability that the first trial is successful is simply $p = 0.35$.

If the second trial is the first successful one, then the first one must have been unsuccessful. Thus, the probability is given by $(0.65)(0.35) = 0.228$.

Similarly, the probability that the first success is the third trial: $(0.65)(0.65)(0.35) = 0.148$.

This can be stated generally. If the first success is on the nth trial, then there are $n - 1$ failures and finally 1 success, which corresponds to the probability $(0.65)^{n-1}(0.35)$.

The geometric distribution from Example 3.38 is shown in Figure 3.21. In general, the probabilities for a geometric distribution decrease **exponentially**.

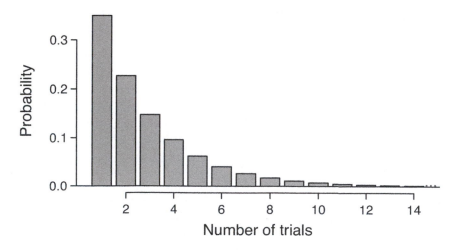

Figure 3.21: The geometric distribution when the probability of success is $p = 0.35$.

GEOMETRIC DISTRIBUTION

If the probability of a success in one trial is p and the probability of a failure is $1-p$, then the probability of finding the first success in the kth trial is given by

$$P(X = k) = (1-p)^{k-1}p.$$

Geom(p)

Geometric dist. with p prob. of success

The mean (i.e. expected value), variance, and standard deviation of this wait time are given by

$$\mu = \frac{1}{p} \qquad\qquad \sigma^2 = \frac{1-p}{p^2} \qquad\qquad \sigma = \sqrt{\frac{1-p}{p^2}}$$

A geometric random variable X can be expressed as $X \sim \text{Geom}(p)$.

GUIDED PRACTICE 3.39

If individuals were examined until one did not administer the most severe shock, how many might need to be tested before the first success? [25]

EXAMPLE 3.40

What is the probability of the first success occurring within the first 4 people?

This is the probability it is the first ($k = 1$), second ($k = 2$), third ($k = 3$), or fourth ($k = 4$) trial that is the first success, which represent four disjoint outcomes. Compute the probability of each case and add the separate results:

$$P(X = 1, 2, 3, \text{ or } 4)$$
$$= P(X = 1) + P(X = 2) + P(X = 3) + P(X = 4)$$
$$= (0.65)^{1-1}(0.35) + (0.65)^{2-1}(0.35) + (0.65)^{3-1}(0.35) + (0.65)^{4-1}(0.35)$$
$$= 0.82.$$

Alternatively, find the complement of P(X = 0), since the described event is the complement of no success in 4 trials: $1 - (0.65)^4(0.35)^0 = 0.82$.

There is a 0.82 probability that the first success occurs within 4 trials.

Note that there are differing conventions for defining the geometric distribution; while this text uses the definition that the distribution describes the total number of trials *including* the success, others define the distribution as the number of trials required before the success is obtained. In R, the latter definition is used.

---

[25]About $1/p = 1/0.35 = 2.86$ individuals.

3.5.2 Negative binomial distribution

The geometric distribution describes the probability of observing the first success on the kth trial. The **negative binomial distribution** is more general: it describes the probability of observing the rth success on the kth trial.

Suppose a research assistant needs to successfully extract RNA from four plant samples before leaving the lab for the day. Yesterday, it took 6 attempts to attain the fourth successful extraction. The last extraction must have been a success; that leaves three successful extractions and two unsuccessful ones that make up the first five attempts. There are ten possible sequences, which are shown in 3.22.

Extraction Attempt

	1	2	3	4	5	6
1	F	F	S^1	S^2	S^3	S^4
2	F	S^1	F	S^2	S^3	S^4
3	F	S^1	S^2	F	S^3	S^4
4	F	S^1	S^2	S^3	F	S^4
5	S^1	F	F	S^2	S^3	S^4
6	S^1	F	S^2	F	S^3	S^4
7	S^1	F	S^2	S^3	F	S^4
8	S^1	S^2	F	F	S^3	S^4
9	S^1	S^2	F	S^3	F	S^4
10	S^1	S^2	S^3	F	F	S^4

Figure 3.22: The ten possible sequences when the fourth successful extraction is on the sixth attempt.

GUIDED PRACTICE 3.41

Each sequence in Figure 3.22 has exactly two failures and four successes with the last attempt always being a success. If the probability of a success is $p = 0.8$, find the probability of the first sequence.[26]

If the probability of a successful extraction is $p = 0.8$, what is the probability that it takes exactly six attempts to reach the fourth successful extraction? As expressed by 3.41, there are 10 different ways that this event can occur. The probability of the first sequence was identified in Guided Practice 3.41 as 0.0164, and each of the other sequences have the same probability. Thus, the total probability is $(10)(0.0164) = 0.164$.

---

[26]The first sequence: $0.2 \times 0.2 \times 0.8 \times 0.8 \times 0.8 \times 0.8 = 0.0164$.

A general formula for computing a negative binomial probability can be generated using similar logic as for binomial probability. The probability is comprised of two pieces: the probability of a single sequence of events, and then the number of possible sequences. The probability of observing r successes out of k attempts can be expressed as $(1-p)^{k-r}p^r$. Next, identify the number of possible sequences. In the above example, 10 sequences were identified by fixing the last observation as a success and looking for ways to arrange the other observations. In other words, the goal is to arrange $r-1$ successes in $k-1$ trials. This can be expressed as:

$$\binom{k-1}{r-1} = \frac{(k-1)!}{(r-1)!\,((k-1)-(r-1))!} = \frac{(k-1)!}{(r-1)!\,(k-r)!}.$$

NEGATIVE BINOMIAL DISTRIBUTION

The negative binomial distribution describes the probability of observing the rth success on the kth trial, for independent trials:

$$P(X=k) = \binom{k-1}{r-1}p^r(1-p)^{k-r}, \qquad (3.42)$$

where p is the probability an individual trial is a success.

The mean and variance are given by

$$\mu = \frac{r}{p} \qquad\qquad \sigma^2 = \frac{r(1-p)}{p^2}$$

A negative binomial random variable X can be expressed as $X \sim NB(r,p)$.

$NB(r,p)$

Neg. Bin. dist. with k successes & p prob. of success

IS IT NEGATIVE BINOMIAL? FOUR CONDITIONS TO CHECK.

(1) The trials are independent.
(2) Each trial outcome can be classified as a success or failure.
(3) The probability of a success (p) is the same for each trial.
(4) The last trial must be a success.

EXAMPLE 3.43

Calculate the probability of a fourth successful extraction on the fifth attempt.

The probability of a single success is $p = 0.8$, the number of successes is $r = 4$, and the number of necessary attempts under this scenario is $k = 5$.

$$\binom{k-1}{r-1}p^r(1-p)^{k-r} = \frac{4!}{3!1!}(0.8)^4(0.2) = 4 \times 0.08192 = 0.328.$$

GUIDED PRACTICE 3.44

Assume that each extraction attempt is independent. What is the probability that the fourth success occurs within 5 attempts?[27]

BINOMIAL VERSUS NEGATIVE BINOMIAL

The binomial distribution is used when considering the number of successes for a fixed number of trials. For negative binomial problems, there is a fixed number of successes and the goal is to identify the number of trials necessary for a certain number of successes (note that the last observation must be a success).

GUIDED PRACTICE 3.45

On 70% of days, a hospital admits at least one heart attack patient. On 30% of the days, no heart attack patients are admitted. Identify each case below as a binomial or negative binomial case, and compute the probability. (a) What is the probability the hospital will admit a heart attack patient on exactly three days this week? (b) What is the probability the second day with a heart attack patient will be the fourth day of the week? (c) What is the probability the fifth day of next month will be the first day with a heart attack patient?[28]

In R, the negative binomial distribution is defined as the number of failures that occur before a target number of successes is reached; i.e., $k - r$. In this text, the distribution is defined in terms of the total number of trials required to observe r successes, where the last trial is necessarily a success.

---

[27]If the fourth success ($r = 4$) is within five attempts, it either took four or five tries ($k = 4$ or $k = 5$):

$$P(k = 4 \text{ OR } k = 5) = P(k = 4) + P(k = 5)$$

$$= \binom{4-1}{4-1}0.8^4 + \binom{5-1}{4-1}(0.8)^4(1-0.8) = 1 \times 0.41 + 4 \times 0.082 = 0.41 + 0.33 = 0.74.$$

[28]In each part, $p = 0.7$. (a) The number of days is fixed, so this is binomial. The parameters are $k = 3$ and $n = 7$: 0.097. (b) The last "success" (admitting a patient) is fixed to the last day, so apply the negative binomial distribution. The parameters are $r = 2$, $k = 4$: 0.132. (c) This problem is negative binomial with $r = 1$ and $k = 5$: 0.006. Note that the negative binomial case when $r = 1$ is the same as using the geometric distribution.

3.5.3 Hypergeometric distribution

Suppose that a large number of deer live in a forest. Researchers are interested in using the capture-recapture method to estimate total population size. A number of deer are captured in an initial sample and marked, then released; at a later time, another sample of deer are captured, and the number of marked and unmarked deer are recorded.[29] An estimate of the total population can be calculated based on the assumption that the proportion of marked deer in the second sample should equal the proportion of marked deer in the entire population. For example, if 50 deer were initially captured and marked, and then 5 out of 40 deer (12.5%) in a second sample are found to be marked, then the population estimate is 400 deer, since 50 out of 400 is 12.5%.

The capture-recapture method sets up an interesting scenario that requires a new probability distribution. Let N represent the total number of deer in the forest, m the number of marked deer captured in the original sample, and n the number of deer in the second sample. What are the probabilities of obtaining $0, 1, ..., m$ marked deer in the second sample, if N and m are known?

It is helpful to think in terms of a series of Bernoulli trials, where each capture in the second sample represents a trial; consider the trial a success if a marked deer is captured, and a failure if an unmarked deer is captured. If the deer were sampled *with replacement*, such that one deer was sampled, checked if it were marked versus unmarked, then released before another deer was sampled, then the probability of obtaining some number of marked deer in the second sample would be binomially distributed with probability of success m/N (out of n trials). The trials are independent, and the probability of success remains constant across trials.

However, in capture-recapture, the goal is to collect a representative sample such that the proportion of marked deer in the sample can be used to estimate the total population—the sampling is done *without replacement*. Once a trial occurs and a deer is sampled, it is not returned to the population before the next trial. The probability of success is not constant from trial to trial; i.e., these trials are dependent. For example, if a marked deer has just been sampled, then the probability of sampling a marked deer in the next trial decreases, since there is one fewer marked deer available.

Suppose that out of 9 deer, 4 are marked. What is the probability of observing 1 marked deer in a sample of size 3, if the deer are sampled without replacement? First, consider the total number of ways to draw 3 deer from the population; As shown in Figure 3.23, samples may consist of 3, 2, 1, or 0 marked deer. There are $\binom{4}{3}$ ways to obtain a sample consisting of 3 marked deer out of the 4 total marked deer. By independence, there are $\binom{4}{2}\binom{5}{1}$ ways to obtain a sample consisting of exactly 2 marked deer and 1 unmarked deer. In total, there are 84 possible combinations; this quantity is equivalent to $\binom{9}{3}$. Only $\binom{4}{1}\binom{5}{2} = 40$ of those combinations represent the desired event of exactly 1 marked deer. Thus, the probability of observing 1 marked deer in a sample of size 3, under sampling without replacement, equals $40/84 = 0.476$.

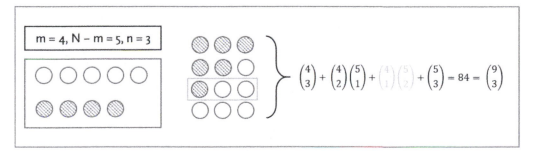

Figure 3.23: Possible samples of marked and unmarked deer in a sample $n = 3$, where $m = 4$ and $N - m = 5$. Striped circles represent marked deer, and empty circles represent unmarked deer.

---

[29]It is assumed that enough time has passed so that the marked deer redistribute themselves in the population, and that marked and unmarked deer have equal probability of being captured in the second sample.

GUIDED PRACTICE 3.46

Suppose that out of 9 deer, 4 are marked. What is the probability of observing 1 marked deer in a sample of size 3, if the deer are sampled with replacement?[30]

HYPERGEOMETRIC DISTRIBUTION

The hypergeometric distribution describes the probability of observing k successes in a sample of size n, from a population of size N, where there are m successes, and individuals are sampled without replacement:

$$P(X = k) = \frac{\binom{m}{k}\binom{N-m}{n-k}}{\binom{N}{n}}.$$

Let $p = m/N$, the probability of success. The mean and variance are given by

$$\mu = np \qquad\qquad\qquad \sigma^2 = np(1-p)\frac{N-n}{N-1}$$

A hypergeometric random variable X can be written as $X \sim \text{HGeom}(m, N - m, n)$.

IS IT HYPERGEOMETRIC? THREE CONDITIONS TO CHECK.

(1) The trials are dependent.
(2) Each trial outcome can be classified as a success or failure.
(3) The probability of a success is different for each trial.

GUIDED PRACTICE 3.47

A small clinic would like to draw a random sample of 10 individuals from their patient list of 120, of which 30 patients are smokers. (a) What is the probability of 6 individuals in the sample being smokers? (b) What is the probability that at least 2 individuals in the sample smoke?[31]

---

[30]Let X represent the number of marked deer in the sample of size 3. If the deer are sampled with replacement, $X \sim$ Bin$(3, 4/9)$, and $P(X = 1) = \binom{3}{1}(4/9)^1(5/9)^2 = 0.412$.

[31](a) Let X represent the number of smokers in the sample. $P(X = 6) = \frac{\binom{30}{6}\binom{90}{4}}{\binom{120}{10}} = 0.013$. (b) $P(X \geq 2) = 1 - P(X \leq 1) =$

$1 - P(X = 0) - P(X = 1) = 1 - \frac{\binom{30}{0}\binom{90}{10}}{\binom{120}{10}} - \frac{\binom{30}{1}\binom{90}{9}}{\binom{120}{10}} = 0.768$.

3.6 Distributions for pairs of random variables

Example 3.8 calculated the variability in health care costs for an employee and her partner relying on the assumption that the number of health episodes between the two are not related. It could be reasonable to assume that the health status of one person gives no information about the other's health, given that the two are not physically related and were not previously living together. However, associations between random variables can be subtle. For example, couples are often attracted to each other because of common interests or lifestyles, which suggests that health status may actually be related.

The relationship between a pair of discrete random variables is a feature of the **joint distribution** of the pair. In this example the joint distribution of annual costs is a table of all possible combinations of costs for the employee and her partner, using the probabilities and costs from the last 10 years (these costs were previously calculated in Example 3.6 and Guided Practice 3.7). Entries in the table are probabilities of pairs of annual costs. For example, the entry 0.25 in the second row and second column of Figure 3.24 indicates that in approximately 25% of the last 10 years, the employee paid $1,008 in costs while her partner paid $988.

	Partner costs, Y	
Employee costs, X	$968	$988
$968	0.18	0.12
$1,008	0.15	0.25
$1,028	0.04	0.16
$1,108	0.03	0.07

Figure 3.24: Joint distribution of health care costs.

More generally, the definition of a joint distribution for a pair of random variables X and Y uses the notion of joint probabilities discussed in Section 2.2.1.

JOINT DISTRIBUTION

The **joint distribution** $p_{X,Y}(x,y)$ for a pair of random variables (X,Y) is the collection of probabilities

$$p(x_i, y_j) = P(X = x_i \text{ and } Y = y_j)$$

for all pairs of values (x_i, y_j) that the random variables X and Y take on.

Joint distributions are often displayed in tabular form as in Figure 3.24. If X and Y have k_1 and k_2 possible values respectively, there will be $(k_1)(k_2)$ possible (x, y) pairs. This is unlike pairs of values (x, y) observed in a dataset, where each observed value of x is usually paired with only one value of y. A joint distribution is often best displayed as a table of probabilities, with $(k_1)(k_2)$ entries. Figure 3.25 shows the general form of the table for the joint distribution of two discrete distributions.

	Values of Y			
Values of X	y_1	y_2	$\cdots$	y_{k_2}
x_1	$p(x_1, y_1)$	$p(x_1, y_2)$	$\cdots$	$p(x_1, y_{k_2})$
x_2	$p(x_2, y_1)$	$p(x_2, y_2)$	$\cdots$	$p(x_2, y_{k_2})$
$\vdots$	$\cdots$	$\cdots$	$\cdots$	$\cdots$
x_{k_1}	$p(x_{k_1}, y_1)$	$p(x_{k_1}, y_2)$	$\cdots$	$p(x_{k_1}, y_{k_2})$

Figure 3.25: Table for a joint distribution. Entries are probabilities for pairs (x_i, y_j). These probabilities can be written as $p(x_i, y_j)$ or more specifically, $p_{X,Y}(x_i, y_j)$.

When two variables X and Y have a joint distribution, the **marginal distribution** of X is the collection of probabilities for X when Y is ignored.[32] If X represents employee costs and Y represents partner costs, the event $(X = \$968)$ consists of the two disjoint events $(X = \$968, Y = \$968)$ and $(X = \$968, Y = \$988)$, so $P(X = \$968) = 0.18 + 0.12 = 0.30$, the sum of the first row of the table. The row sums are the values of the marginal distribution of X, while the column sums are the values of the marginal distributions of Y. The marginal distributions of X and Y are shown in Figure 3.26, along with the joint distribution of X and Y. The term marginal distribution is apt in this setting—the marginal probabilities appear in the table margins.

	Partner Costs, Y		
Employee costs, X	**\$968**	**\$988**	**Marg. Dist., X**
\$968	0.18	0.12	0.30
\$1,008	0.15	0.25	0.40
\$1,028	0.04	0.16	0.20
\$1,108	0.03	0.07	0.10
Marg. Dist., Y	0.40	0.60	1.00

Figure 3.26: Joint and marginal distributions of health care costs

For a pair of random variables X and Y, the **conditional distribution** of Y given a value x of the variable X is the probability distribution of Y when its values are restricted to the value x for X. Just as marginal and joint probabilities are used to calculate conditional probabilities, joint and marginal distributions can be used to obtain conditional distributions. If information is observed about the value of one of the correlated random variables, such as X, then this information can be used to obtain an updated distribution for Y; unlike the marginal distribution of Y, the conditional distribution of Y given X accounts for information from X.

---

[32]The marginal distribution of X can be written as $p_X(x)$, and a specific value in the marginal distribution written as $p_X(x_i)$.

CONDITIONAL DISTRIBUTION

The **conditional distribution** $p_{Y|X}(y|x)$ for a pair of random variables (X, Y) is the collection of probabilities

$$P(Y = y_j | X = x_i) = \frac{P(Y = y_j \text{ and } X = x_i)}{P(X = x_j)}$$

for all pairs of values (x_i, y_j) that the random variables X and Y take on.

EXAMPLE 3.48

If it is known that the employee's annual health care cost is \$968, what is the conditional distribution of the partner's annual health care cost?

Note that there is a different conditional distribution of Y for every possible value of X; this problem specifically asks for the conditional distribution of Y given that $X = \$968$.

$$p_{Y|X}(\$968|\$968) = P(Y = \$968 | X = \$968) = \frac{P(Y = \$968 \text{ and } X = \$968)}{P(X = \$968)} = \frac{0.18}{0.30} = 0.60$$

$$p_{Y|X}(\$988|\$968) = P(Y = \$988 | X = \$968) = \frac{P(Y = \$988 \text{ and } X = \$968)}{P(X = \$968)} = \frac{0.12}{0.30} = 0.40$$

With the knowledge that the employee's annual health care cost is \$968, there is a probability of 0.60 that the partner's cost is \$968 and 0.40 that the partner's cost is \$988.

GUIDED PRACTICE 3.49

Consider two random variables, X and Y, with the joint distribution shown in Figure 3.27.
(a) Compute the marginal distributions of X and Y.
(b) Identify the joint probability $p_{X,Y}(1, 2)$.
(c) What is the value of $p_{X,Y}(2, 1)$?
(d) Compute the conditional distribution of X given that $Y = 2$.[33]

	$Y = 1$	$Y = 2$
$X = 1$	0.20	0.40
$X = 4$	0.30	0.10

Figure 3.27: Joint distribution of X and Y

The variance calculation in Example 3.8 relied on the assumption that the patterns of health care expenses for the two partners were unrelated. In Example 3.48, 0.40 is the conditional probability that the partner's health care costs will be \$988, given that the employee's cost is \$968. The marginal probability that the partner's health care cost is \$988 is 0.60, which is different from 0.40. The patterns of health care costs are related in that knowing the value of the employee's costs changes the probabilities associated with partner's costs. The marginal and conditional distributions of the partner's costs are not the same.

The notion of independence of two events discussed in Chapter 2 can be applied to the setting of random variables. Recall that two events A and B are independent if the conditional proba-

---

[33](a) The marginal distribution of X: $p_X(1) = 0.60$, $p_X(4) = 0.40$. The marginal distribution of Y: $p_Y(1) = 0.50$, $p_Y(2) = 0.50$ (b) $p_{X,Y}(1, 2) = P(X = 1, Y = 2) = 0.40$ (c) Since X cannot take on value 2, $p_{X,Y}(2, 1) = 0$. (d) The conditional distribution of X given that $Y = 2$: $p_{X|Y}(1|2) = \frac{p_{X,Y}(1,2)}{p_Y(2)} = \frac{0.40}{0.50} = 0.80$, $p_{X|Y}(4|2) = \frac{p_{X,Y}(4,2)}{p_Y(2)} = \frac{0.10}{0.50} = 0.20$.

bility $P(A|B)$ equals the marginal probability $P(A)$ or equivalently, if the product of the marginal probabilities $P(A)$ and $P(B)$ equals the joint probability $P(A \text{ and } B)$.

A pair (X, Y) of random variables are called **independent random variables** if the conditional distribution for Y, given any value of X, is the same as the marginal distribution of Y. Additionally, if all joint probabilities $P(X = x_i, Y = y_j)$ that comprise the joint distribution of X and Y can be computed from the product of the marginal probabilities, $P(X = x_i)P(Y = y_j)$, X and Y are independent.

INDEPENDENT RANDOM VARIABLES

Two random variables X and Y are independent if the probabilities

$$P(Y = y_j | X = x_i) = P(Y = y_j)$$

for all pairs of values (x_i, y_j).
Equivalently, X and Y are independent if the probabilities

$$P(Y = y_j \text{ and } X = x_i) = P(Y = y_j)P(X = x_i)$$

for all pairs of values (x_i, y_j).

EXAMPLE 3.50

Demonstrate that the employee's health care costs and the partner's health care costs are not independent random variables.

As shown in Example 3.48, the conditional distribution of the partner's annual health care cost given that the employee's annual cost is \$968 is $P(Y = \$968 | X = \$968) = 0.60$, $P(Y = \$988 | X = \$968) = 0.40$. However, the marginal distribution of the partner's annual health care cost is $P(Y = \$968) = 0.40$, $P(Y = \$968) = 0.60$. Thus, X and Y are not independent.

This can also be demonstrated from examining the joint distribution, as shown in Figure 3.26. The probability that the employee's cost and partner's cost are both \$968 is 0.18. The marginal probabilities $P(X = \$968)$ and $P(Y = \$968)$, respectively, are 0.30 and 0.40. Since $(0.40)(0.30) \neq 0.18$, X and Y are dependent random variables.

Note that demonstrating $P(Y = y_j | X = x_i) = P(Y = y_j)$ or $P(Y = y_j \text{ and } X = x_i) = P(Y = y_j)P(X = x_i)$ does not hold for any one (x_i, y_j) pair is sufficient to prove that X and Y are not independent, since independence requires these conditions to hold over *all* pairs of values (x_i, y_j).

GUIDED PRACTICE 3.51

Based on Figure 3.27, check whether X and Y are independent.[34]

Two random variables that are not independent are called **correlated random variables**. The correlation between two random variables is a measure of the strength of the relationship between them, just as it was for pairs of data points explored in Section 1.6.1. There are many examples of correlated random variables, such as height and weight in a population of individuals, or the gestational age and birth weight of newborns.

When two random variables are positively correlated, they tend to increase or decrease together. If one of the variables increases while the other decreases (or vice versa) they are negatively

---

[34]X and Y are not independent. One way to demonstrate this is to compare $p_X(1)$ with $p_{X|Y}(1|2)$. If X were independent of Y, then conditioning on $Y = 2$ should not provide any information about X, and $p_X(1)$ should equal $p_{X|Y}(1|2)$. However, $p_X(1) = 0.60$ and $p_{X|Y}(1|2) = 0.80$. Thus, X and Y are not independent.

correlated. Correlation is easy to identify in a scatterplot, but is more difficult to identify in a table of a joint distribution. Fortunately, there is a formula to calculate correlation for a joint distribution specified in a table.

Correlation between random variables is similar to correlation between pairs of observations in a dataset, with some important differences. Calculating a correlation r in a dataset was introduced in Section 1.6.1 and uses the formula:

$$r = \frac{1}{n-1} \sum_{i=1}^{n} \left(\frac{x_i - \bar{x}}{s_x} \right) \left(\frac{y_i - \bar{y}}{s_y} \right). \tag{3.52}$$

The correlation coefficient r is an average of products, with each term in the product measuring the distance between x and its mean $\bar{x}$ and the distance between y and its mean $\bar{y}$, after the distances have been scaled by respective standard deviations.

The compact formula for the correlation between two random variables X and Y uses the same idea:

$$\rho_{X,Y} = E\left(\frac{X - \mu_X}{\sigma_X} \right) \left(\frac{Y - \mu_Y}{\sigma_Y} \right), \tag{3.53}$$

where $\rho_{X,Y}$ is the correlation between the two variables, and $\mu_X, \mu_Y, \sigma_X, \sigma_Y$ are the respective means and standard deviations for X and Y. Just as with the mean of a random variable, the expectation in the formula for correlation is a weighted sum of products, with each term weighted by the probability of values for the pair (X, Y). Equation 3.53 is useful for understanding the analogy between correlation of random variables and correlation of observations in a dataset, but it cannot be used to calculate $\rho_{X,Y}$ without the probability weights. The weights come from the **joint distribution** of the pair of variables (X, Y).

Equation 3.54 is an expansion of Equation 3.53. The double summation adds up terms over all combinations of the indices i and j.

$$\rho_{X,Y} = \sum_i \sum_j p(i,j) \frac{(x_i - \mu_X)}{sd(X)} \frac{(y_j - \mu_Y)}{sd(Y)}. \tag{3.54}$$

EXAMPLE 3.55

Compute the correlation between annual health care costs for the employee and her partner.

As calculated previously, $E(X) = \$1010$, $Var(X) = 1556$, $E(Y) = \$980$, and $Var(Y) = 96$. Thus, $SD(X) = \$39.45$ and $SD(Y) = \$9.80$.

$$\rho_{X,Y} = p(x_1, y_1) \frac{(x_1 - \mu_X)}{sd(X)} \frac{(y_1 - \mu_Y)}{sd(Y)} + p(x_1, y_2) \frac{(x_1 - \mu_X)}{sd(X)} \frac{(y_2 - \mu_Y)}{sd(Y)}$$

$$+ \cdots + p(x_4, y_1) \frac{(x_4 - \mu_X)}{sd(X)} \frac{(y_1 - \mu_Y)}{sd(Y)} + p(x_4, y_2) \frac{(x_4 - \mu_X)}{sd(X)} \frac{(y_2 - \mu_Y)}{sd(Y)}$$

$$= (0.18) \frac{(968 - 1010)}{39.45} \frac{(968 - 980)}{9.8} + (0.12) \frac{(968 - 1010)}{39.45} \frac{(988 - 980)}{9.8}$$

$$+ \cdots + (0.03) \frac{(1108 - 1010)}{39.45} \frac{(968 - 980)}{9.8} + (0.07) \frac{(1108 - 1010)}{39.45} \frac{(988 - 980)}{9.8}$$

$$= 0.22.$$

The correlation between annual health care costs for these two individuals is positive. It is reasonable to expect that there might be a positive correlation in health care costs for two individuals in a relationship; for example, if one person contracts the flu, then it is likely the other person will also contract the flu, and both may need to see a doctor.

GUIDED PRACTICE 3.56

Based on Figure 3.27, compute the correlation between X and Y. For your convenience, the following values are provided: $E(X) = 2.2$, $Var(X) = 2.16$, $E(Y) = 1.5$, $Var(Y) = 0.25$.[35]

When two random variables X and Y are correlated:

$$\text{Variance}(X + Y) = \text{Variance}(X) + \text{Variance}(Y) + 2\sigma_X\sigma_Y\text{Correlation}(X, Y) \qquad (3.57)$$

$$\text{Variance}(X - Y) = \text{Variance}(X) + \text{Variance}(Y) - 2\sigma_X\sigma_Y\text{Correlation}(X, Y). \qquad (3.58)$$

When random variables are positively correlated the variance of the sum or the difference of two variables will be larger than the sum of the two variances. When they are negatively correlated the variance of the sum or difference will be smaller than the sum of the two variances.

The standard deviation for the sum or difference will always be the square root of the variance.

EXAMPLE 3.59

Calculate the standard deviation of the sum of the health care costs for the couple.

This calculation uses Equation 3.57 to calculate the variance of the sum. The standard deviation will be the square root of the variance.

$$\begin{aligned}
\text{Var}(X + Y) &= \text{Var}(X) + \text{Var}(Y) + 2\sigma_X\sigma_Y\rho_{X,Y} \\
&= (1556 + 96) + (2)(39.45)(9.80)(0.22) \\
&= 1822.10.
\end{aligned}$$

The standard deviation is $\sqrt{1822.10} = \$42.69$. Because the health care costs are correlated, the standard deviation of the total cost is larger than the value calculated in Example 3.8 under the assumption that the annual costs were independent.

GUIDED PRACTICE 3.60

Compute the standard deviation of $X - Y$ for the pair of random variables shown in Figure 3.27.[36]

---

[35]The correlation between X and Y is $\rho_{X,Y} = (0.20)\frac{(1-2.2)}{\sqrt{2.16}}\frac{(1-1.5)}{\sqrt{0.25}} + \cdots + (0.10)\frac{(4-2.2)}{\sqrt{2.16}}\frac{(2-1.5)}{\sqrt{0.25}} = -0.0208$.

[36]$\text{Var}(X - Y) = \text{Var}(X) + \text{Var}(Y) - 2\sigma_X\sigma_Y\rho_{X,Y} = 2.16 + 0.25 - 2(\sqrt{2.16})(\sqrt{0.25})(-0.0208) = 2.44$. Thus, $SD(X - Y) = \sqrt{2.44} = 1.56$.

EXAMPLE 3.61

The Association of American Medical Colleges (AAMC) introduced a new version of the Medical College Admission Test (MCAT) in the spring of 2015. Data from the scores were recently released by AAMC.[37] The test consists of 4 components: chemical and physical foundations of biological systems; critical analysis and reasoning skills; biological and biochemical foundations of living systems; psychological, social and biological foundations of behavior. The overall score is the sum of the individual component scores. The grading for each of the four components is scaled so that the mean score is 125. The means and standard deviations for the four components and the total scores for the population taking the exam in May 2015 exam are shown in Figure 3.28.

Show that the standard deviation in the table for the total score does not agree with that obtained under the assumption of independence.

The variance of each component of the score is the square of each standard deviation. Under the assumption of independence, the variance of the total score would be

$$\text{Var(Total Score)} = 3.0^2 + 3.0^2 + 3.0^2 + 3.1^2$$
$$= 36.61,$$

so the standard deviation is 6.05, which is less than 10.6.

Since the observed standard deviation is larger than that calculated under independence, this suggests the component scores are positively correlated.

It would not be reasonable to expect that the component scores are independent. Think about a student taking the MCAT exam: someone who scores well on one component of the exam is likely to score well on the other parts.

Component	Mean	Standard Deviation
Chem. Phys. Found.	125	3.0
Crit. Analysis	125	3.0
Living Systems	125	3.0
Found. Behavior	125	3.1
Total Score	500	10.6

Figure 3.28: Means and Standard Deviations for MCAT Scores

---

[37] https://www.aamc.org/students/download/434504/data/percentilenewmcat.pdf

3.7 Notes

Thinking in terms of random variables and distributions of probabilities makes it easier to describe all possible outcomes of an experiment or process of interest, versus only considering probabilities on the scale of individual outcomes or sets of outcomes. Several of the fundamental concepts of probability can naturally be extended to probability distributions. For example, the process of obtaining a conditional distribution is analogous to the one for calculating a conditional probability.

Many processes can be modeled using a specific named distribution. The statistical techniques discussed in later chapters, such as hypothesis testing and regression, are often based on particular distributional assumptions. In particular, many methods rely on the assumption that data are normally distributed.

The discussion of random variables and their distribution provided in this chapter only represents an introduction to the topic. In this text, properties of random variables such as expected value or correlation are presented in the context of discrete random variables; these concepts are also applicable to continuous random variables. A course in probability theory will cover additional named distributions as well as more advanced methods for working with distributions.

Lab 1 introduces the general notion of a random variable and its distribution using a simulation, then discusses the binomial distribution. Lab 2 discusses the normal distribution and working with normal probabilities, as well as the Poisson distribution. Lab 3 covers the geometric, negative binomial, and hypergeometric distributions. All three labs include practice problems that illustrate the use of R functions for probability distributions and introduce additional features of the R programming language. Lab 4 discusses distributions for pairs of random variables and some R functions useful for matrix calculations.

3.8 Exercises

3.8.1 Random variables

3.1 College smokers. At a university, 13% of students smoke.

(a) Calculate the expected number of smokers in a random sample of 100 students from this university.

(b) The university gym opens at 9 am on Saturday mornings. One Saturday morning at 8:55 am there are 27 students outside the gym waiting for it to open. Should you use the same approach from part (a) to calculate the expected number of smokers among these 27 students?

3.2 Ace of clubs wins. Consider the following card game with a well-shuffled deck of cards. If you draw a red card, you win nothing. If you get a spade, you win $5. For any club, you win $10 plus an extra $20 for the ace of clubs.

(a) Create a probability model for the amount you win at this game. Also, find the expected winnings for a single game and the standard deviation of the winnings.

(b) What is the maximum amount you would be willing to pay to play this game? Explain your reasoning.

3.3 Hearts win. In a new card game, you start with a well-shuffled full deck and draw 3 cards without replacement. If you draw 3 hearts, you win $50. If you draw 3 black cards, you win $25. For any other draws, you win nothing.

(a) Create a probability model for the amount you win at this game, and find the expected winnings. Also compute the standard deviation of this distribution.

(b) If the game costs $5 to play, what would be the expected value and standard deviation of the net profit (or loss)?

(c) If the game costs $5 to play, should you play this game? Explain.

3.4 Baggage fees. An airline charges the following baggage fees: $25 for the first bag and $35 for the second. Suppose 54% of passengers have no checked luggage, 34% have one piece of checked luggage and 12% have two pieces. We suppose a negligible portion of people check more than two bags.

(a) Build a probability model, compute the average revenue per passenger, and compute the corresponding standard deviation.

(b) About how much revenue should the airline expect for a flight of 120 passengers? With what standard deviation? Note any assumptions you make and if you think they are justified.

3.5 Gull clutch size. Large black-tailed gulls usually lay one to three eggs, and rarely have a fourth egg clutch. It is thought that clutch sizes are effectively limited by how effectively parents can incubate their eggs. Suppose that on average, gulls have a 25% of laying 1 egg, 40% of laying 2 eggs, 30% chance of laying 3 eggs, and 5% chance of laying 4 eggs.

(a) Calculate the expected number of eggs laid by a random sample of 100 gulls.

(b) Calculate the standard deviation of the number of eggs laid by a random sample of 100 gulls.

3.6 Scooping ice cream. Ice cream usually comes in 1.5 quart boxes (48 fluid ounces), and ice cream scoops hold about 2 ounces. However, there is some variability in the amount of ice cream in a box as well as the amount of ice cream scooped out. We represent the amount of ice cream in the box as X and the amount scooped out as Y. Suppose these random variables have the following means, standard deviations, and variances:

	mean	SD	variance
X	48	1	1
Y	2	0.25	0.0625

(a) An entire box of ice cream, plus 3 scoops from a second box is served at a party. How much ice cream do you expect to have been served at this party? What is the standard deviation of the amount of ice cream served?

(b) How much ice cream would you expect to be left in the box after scooping out one scoop of ice cream? That is, find the expected value of $X - Y$. What is the standard deviation of the amount left in the box?

(c) Using the context of this exercise, explain why we add variances when we subtract one random variable from another.

3.8.2 Binomial distribution

3.7 Underage drinking, Part I. Data collected by the Substance Abuse and Mental Health Services Administration (SAMHSA) suggests that 69.7% of 18-20 year olds consumed alcoholic beverages in any given year.[38]

(a) Suppose a random sample of ten 18-20 year olds is taken. Is the use of the binomial distribution appropriate for calculating the probability that exactly six consumed alcoholic beverages? Explain.

(b) Calculate the probability that exactly 6 out of 10 randomly sampled 18- 20 year olds consumed an alcoholic drink.

(c) What is the probability that exactly four out of ten 18-20 year olds have *not* consumed an alcoholic beverage?

(d) What is the probability that at most 2 out of 5 randomly sampled 18-20 year olds have consumed alcoholic beverages?

(e) What is the probability that at least 1 out of 5 randomly sampled 18-20 year olds have consumed alcoholic beverages?

3.8 Chickenpox, Part I. The US CDC estimates that 90% of Americans have had chickenpox by the time they reach adulthood.

(a) Suppose we take a random sample of 100 American adults. Is the use of the binomial distribution appropriate for calculating the probability that exactly 97 out of 100 randomly sampled American adults had chickenpox during childhood? Explain.

(b) Calculate the probability that exactly 97 out of 100 randomly sampled American adults had chickenpox during childhood.

(c) What is the probability that exactly 3 out of a new sample of 100 American adults have *not* had chickenpox in their childhood?

(d) What is the probability that at least 1 out of 10 randomly sampled American adults have had chickenpox?

(e) What is the probability that at most 3 out of 10 randomly sampled American adults have *not* had chickenpox?

---

[38]SAMHSA, Office of Applied Studies, National Survey on Drug Use and Health, 2007 and 2008.

3.9 Underage drinking, Part II. We learned in Exercise 3.7 that about 70% of 18-20 year olds consumed alcoholic beverages in any given year. We now consider a random sample of fifty 18-20 year olds.

(a) How many people would you expect to have consumed alcoholic beverages? And with what standard deviation?

(b) Would you be surprised if there were 45 or more people who have consumed alcoholic beverages?

(c) What is the probability that 45 or more people in this sample have consumed alcoholic beverages? How does this probability relate to your answer to part (b)?

3.10 Chickenpox, Part II. We learned in Exercise 3.8 that about 90% of American adults had chickenpox before adulthood. We now consider a random sample of 120 American adults.

(a) How many people in this sample would you expect to have had chickenpox in their childhood? And with what standard deviation?

(b) Would you be surprised if there were 105 people who have had chickenpox in their childhood?

(c) What is the probability that 105 or fewer people in this sample have had chickenpox in their childhood? How does this probability relate to your answer to part (b)?

3.11 Donating blood. When patients receive blood transfusions, it is critical that the blood type of the donor is compatible with the patients, or else an immune system response will be triggered. For example, a patient with Type O- blood can only receive Type O- blood, but a patient with Type O+ blood can receive either Type O+ or Type O-. Furthermore, if a blood donor and recipient are of the same ethnic background, the chance of an adverse reaction may be reduced. According to a 10-year donor database, 0.37 of white, non-Hispanic donors are O+ and 0.08 are O-.

(a) Consider a random sample of 15 white, non-Hispanic donors. Calculate the expected value of individuals who could be a donor to a patient with Type O+ blood. With what standard deviation?

(b) What is the probability that 3 or more of the people in this sample could donate blood to a patient with Type O- blood?

3.12 Sickle cell anemia. Sickle cell anemia is a genetic blood disorder where red blood cells lose their flexibility and assume an abnormal, rigid, "sickle" shape, which results in a risk of various complications. If both parents are carriers of the disease, then a child has a 25% chance of having the disease, 50% chance of being a carrier, and 25% chance of neither having the disease nor being a carrier. If two parents who are carriers of the disease have 3 children, what is the probability that

(a) two will have the disease?

(b) none will have the disease?

(c) at least one will neither have the disease nor be a carrier?

(d) the first child with the disease will the be 3rd child?

3.13 Hepatitis C. Hepatitis C is spread primarily through contact with the blood of an infected person, and is nearly always transmitted through needle sharing among intravenous drug users. Suppose that in a month's time, an IV drug user has a 30% chance of contracting hepatitis C through needle sharing. What is the probability that 3 out of 5 IV drug users contract hepatitis C in a month? Assume that the drug users live in different parts of the country.

3.14 Arachnophobia. A Gallup Poll found that 7% of teenagers (ages 13 to 17) suffer from arachnophobia and are extremely afraid of spiders. At a summer camp there are 10 teenagers sleeping in each tent. Assume that these 10 teenagers are independent of each other.[39]

(a) Calculate the probability that at least one of them suffers from arachnophobia.

(b) Calculate the probability that exactly 2 of them suffer from arachnophobia.

(c) Calculate the probability that at most 1 of them suffers from arachnophobia.

(d) If the camp counselor wants to make sure no more than 1 teenager in each tent is afraid of spiders, does it seem reasonable for him to randomly assign teenagers to tents?

---

[39]Gallup Poll, What Frightens America's Youth?, March 29, 2005.

3.15 Wolbachia infection. Approximately 12,500 stocks of *Drosophila melanogaster* flies are kept at The Bloomington *Drosophila* Stock Center for research purposes. A 2006 study examined how many stocks were infected with Wolbachia, an intracellular microbe that can manipulate host reproduction for its own benefit. About 30% of stocks were identified as infected. Researchers working with infected stocks should be cautious of the potential confounding effects that Wolbachia infection may have on experiments. Consider a random sample of 250 stocks.

(a) Calculate the probability that exactly 60 stocks are infected.

(b) Calculate the probability that at most 60 stocks are infected.

(c) Calculate the probability that at least 80 stocks are infected.

(d) If a researcher wants to make sure that no more than 40% of the stocks used for an experiment are infected, does it seem reasonable to take a random sample of 250?

3.16 Male children. While it is often assumed that the probabilities of having a boy or a girl are the same, the actual probability of having a boy is slightly higher at 0.51. Suppose a couple plans to have 3 kids.

(a) Use the binomial model to calculate the probability that two of them will be boys.

(b) Write out all possible orderings of 3 children, 2 of whom are boys. Use these scenarios to calculate the same probability from part (a) but using the addition rule for disjoint outcomes. Confirm that your answers from parts (a) and (b) match.

(c) If we wanted to calculate the probability that a couple who plans to have 8 kids will have 3 boys, briefly describe why the approach from part (b) would be more tedious than the approach from part (a).

3.17 Hyponatremia. Hyponatremia (low sodium levels) occurs in a certain proportion of marathon runners during a race. Suppose that historically, the proportion of runners who develop hyponatremia is 0.12. In a certain marathon, there are 200 runners participating.

(a) How many cases of hyponatremia are expected during the marathon?

(b) What is the probability of more than 30 cases of hyponatremia occurring?

3.18 Sleep deprivation. Consider a senior Statistics concentrator with a packed extracurricular schedule, taking five classes, and writing a thesis. Each time she takes an exam, she either scores very well (at least two standard deviations above the mean) or does not. Her performance on any given exam depends on whether she is operating on a reasonable amount of sleep the night before (more than 7 hours), relatively little sleep (between 4 - 7 hours, inclusive), or practically no sleep (less than 4 hours).

When she has had practically no sleep, she scores very well about 30% of the time. When she has had relatively little sleep, she scores very well 40% of the time. When she has had a reasonable amount of sleep, she scores very well 42% of the time. Over the course of a semester, she has a reasonable amount of sleep 50% of nights, and practically no sleep 30% of nights.

(a) What is her overall probability of scoring very well on an exam?

(b) What is the probability she had practically no sleep the night before an exam where she scored very well?

(c) Suppose that one day she has three exams scheduled. What is the probability that she scores very well on exactly two of the exams, under the assumption that her performance on each exam is independent of her performance on another exam?

(d) What is the probability that she had practically no sleep the night prior to a day when she scored very well on exactly two out of three exams?

3.8.3 Normal distribution

3.19 Area under the curve, Part I. What percent of a standard normal distribution $N(\mu = 0, \sigma = 1)$ is found in each region? Be sure to draw a graph.

(a) $Z < -1.35$ (b) $Z > 1.48$ (c) $-0.4 < Z < 1.5$ (d) $|Z| > 2$

3.20 Area under the curve, Part II. What percent of a standard normal distribution $N(\mu = 0, \sigma = 1)$ is found in each region? Be sure to draw a graph.

(a) $Z > -1.13$ (b) $Z < 0.18$ (c) $Z > 8$ (d) $|Z| < 0.5$

3.21 The standard normal distribution. Consider the standard normal distribution with mean $\mu = 0$ and standard deviation $\sigma = 1$.

(a) What is the probability that an outcome Z is greater than 2.60?

(b) What is the probability that Z is less than 1.35?

(c) What is the probability that Z is between -1.70 and 3.10?

(d) What value of Z cuts off the upper 15% of the distribution?

(e) What value of Z marks off the lower 20% of the distribution?

3.22 Triathlon times. In triathlons, it is common for racers to be placed into age and gender groups. The finishing times of men ages 30-34 has mean of 4,313 seconds with a standard deviation of 583 seconds. The finishing times of the women ages 25-29 has a mean of 5,261 seconds with a standard deviation of 807 seconds. The distribution of finishing times for both groups is approximately normal. Note that a better performance corresponds to a faster finish.

(a) If a man of the 30-34 age group finishes the race in 4,948 seconds, what percent of the triathletes in the group did he finish faster than?

(b) If a woman of the 25-29 age group finishes the race in 5,513 seconds, what percent of the triathletes in the group did she finish faster than?

(c) Calculate the cutoff time for the fastest 5% of athletes in the men's group.

(d) Calculate the cutoff time for the slowest 10% of athletes in the women's group.

3.23 GRE scores. The Graduate Record Examination (GRE) is a standardized test commonly taken by graduate school applicants in the United States. The total score is comprised of three components: Quantitative Reasoning, Verbal Reasoning, and Analytical Writing. The first two components are scored from 130 - 170. The mean score for Verbal Reasoning section for all test takers was 151 with a standard deviation of 7, and the mean score for the Quantitative Reasoning was 153 with a standard deviation of 7.67. Suppose that both distributions are nearly normal.

(a) A student scores 160 on the Verbal Reasoning section and 157 on the Quantitative Reasoning section. Relative to the scores of other students, which section did the student perform better on?

(b) Calculate the student's percentile scores for the two sections. What percent of test takers performed better on the Verbal Reasoning section?

(c) Compute the score of a student who scored in the 80th percentile on the Quantitative Reasoning section.

(d) Compute the score of a student who scored worse than 70% of the test takers on the Verbal Reasoning section.

3.24 Osteoporosis. The World Health Organization defines osteoporosis in young adults as a measured bone mineral density 2.5 or more standard deviations below the mean for young adults. Assume that bone mineral density follows a normal distribution in young adults. What percentage of young adults suffer from osteoporosis according to this criterion?

3.25 LA weather. The average daily high temperature in June in LA is 77°F with a standard deviation of 5°F. Suppose that the temperatures in June closely follow a normal distribution.

(a) What is the probability of observing an 83°F temperature or higher in LA during a randomly chosen day in June?

(b) How cold are the coldest 10% of the days during June in LA?

3.26 Clutch volume. A study investigating maternal investment in a frog species found on the Tibetan Plateau reported data on the volume of egg clutches measured across 11 study sites. The distribution is roughly normal, with approximate distribution $N(882.5, 380)$ mm^3.

(a) What is the probability of observing an egg clutch between volume 700-800 mm^3?

(b) How large are the largest 5% of egg clutches?

3.27 Glucose levels. Fasting blood glucose levels for normal non-diabetic individuals are normally distributed in the population, with mean $\mu = 85$ mg/dL and standard deviation $\sigma = 7.5$ mg/dL.

(a) What is the probability that a randomly chosen member of the population has a fasting glucose level higher than 100 mg/dL?

(b) What value of fasting glucose level defines the lower 5th percentile of the distribution?

3.28 Arsenic poisoning. Arsenic blood concentration is normally distributed with mean $\mu = 3.2$ μg/dl and standard deviation $\sigma = 1.5$ μg/dl. What range of arsenic blood concentration defines the middle 95% of this distribution?

3.29 Age at childbirth. In the last decade, the average age of a mother at childbirth is 26.4 years, with standard deviation 5.8 years. The distribution of age at childbirth is approximately normal.

(a) What proportion of women who give birth are 21 years of age or older?

(b) Giving birth at what age puts a woman in the upper 2.5% of the age distribution?

3.30 Find the SD. Find the standard deviation of the distribution in the following situations.

(a) MENSA is an organization whose members have IQs in the top 2% of the population. IQs are normally distributed with mean 100, and the minimum IQ score required for admission to MENSA is 132.

(b) Cholesterol levels for women aged 20 to 34 follow an approximately normal distribution with mean 185 milligrams per deciliter (mg/dl). Women with cholesterol levels above 220 mg/dl are considered to have high cholesterol and about 18.5% of women fall into this category.

3.31 Underage drinking, Part III. As first referenced in Exercise 3.7, about 70% of 18-20 year olds consumed alcoholic beverages in 2008. Consider a random sample of fifty 18-20 year olds.

(a) Of these fifty people, how many would be expected to have consumed alcoholic beverages? With what standard deviation?

(b) Evaluate the conditions for using the normal approximation to the binomial. What is the probability that 45 or more people in this sample have consumed alcoholic beverages?

3.32 Chickenpox, Part III. As first referenced in Exercise 3.8, about 90% of American adults had chickenpox before adulthood. Consider a random sample of 120 American adults.

(a) How many people in this sample would be expected to have had chickenpox in their childhood? With what standard deviation?

(b) Evaluate the conditions for using the normal approximation to the binomial. What is the probability that 105 or fewer people in this sample have had chickenpox in their childhood?

3.33   University admissions. Suppose a university announced that it admitted 2,500 students for the following year's freshman class. However, the university has dorm room spots for only 1,786 freshman students. If there is a 70% chance that an admitted student will decide to accept the offer and attend this university, what is the approximate probability that the university will not have enough dormitory room spots for the freshman class?

3.34   SAT scores. SAT scores (out of 2400) are distributed normally with a mean of 1500 and a standard deviation of 300. Suppose a school council awards a certificate of excellence to all students who score at least 1900 on the SAT, and suppose we pick one of the recognized students at random. What is the probability this student's score will be at least 2100? (The material covered in Section 2.2 would be useful for this question.)

3.35   Scores on stats final. The final exam scores of 20 introductory statistics students are plotted below. Do these data appear to follow a normal distribution? Explain your reasoning.

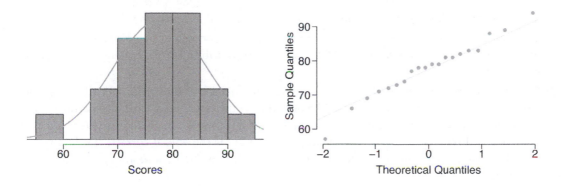

3.36   Heights of female college students. The heights of 25 female college students are plotted below. Do these data appear to follow a normal distribution? Explain your reasoning.

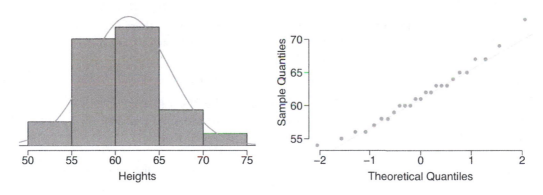

3.8.4   Poisson distribution

3.37   Computing Poisson probabilities. This is a simple exercise in computing probabilities for a Poisson random variable. Suppose that X is a Poisson random variable with rate parameter $\lambda = 2$. Calculate $P(X = 2)$, $P(X \leq 2)$, and $P(X \geq 3)$.

3.38   Stenographer's typos. A very skilled court stenographer makes one typographical error (typo) per hour on average.

(a) What are the mean and the standard deviation of the number of typos this stenographer makes in an hour?

(b) Calculate the probability that this stenographer makes at most 3 typos in a given hour.

(c) Calculate the probability that this stenographer makes at least 5 typos over 3 hours.

3.39 Customers at a coffee shop. A coffee shop serves an average of 75 customers per hour during the morning rush.

(a) What are the mean and the standard deviation of the number of customers this coffee shop serves in one hour during this time of day?

(b) Would it be considered unusually low if only 60 customers showed up to this coffee shop in one hour during this time of day?

(c) Calculate the probability that this coffee shop serves 70 customers in one hour during this time of day.

3.40 Osteosarcoma in NYC. Osteosarcoma is a relatively rare type of bone cancer. It occurs most often in young adults, age 10 - 19; it is diagnosed in approximately 8 per 1,000,000 individuals per year in that age group. In New York City (including all five boroughs), the number of young adults in this age range is approximately 1,400,000.

(a) What is the expected number of cases of osteosarcoma in NYC in a given year?

(b) What is the probability that 15 or more cases will be diagnosed in a given year?

(c) The largest concentration of young adults in NYC is in the borough of Brooklyn, where the population in that age range is approximately 450,000. What is the probability of 10 or more cases in Brooklyn in a given year?

(d) Suppose that in a given year, 10 cases of osteosarcoma were observed in NYC, with all 10 cases occurring among young adults living in Brooklyn. An official from the NYC Public Health Department claims that the probability of this event (that is, the probability of 10 or more cases being observed, and all of them occurring in Brooklyn) is what was calculated in part c). Is the official correct? Explain your answer. You may assume that your answer to part c) is correct. This question can be answered without doing any calculations.

(e) Suppose that over five years, there was one year in which 10 or more cases of osteosarcoma were observed in Brooklyn. Is the probability of this event equal to the probability calculated in part c)? Explain your answer.

3.41 How many cars show up? For Monday through Thursday when there isn't a holiday, the average number of vehicles that visit a particular retailer between 2pm and 3pm each afternoon is 6.5, and the number of cars that show up on any given day follows a Poisson distribution.

(a) What is the probability that exactly 5 cars will show up next Monday?

(b) What is the probability that 0, 1, or 2 cars will show up next Monday between 2pm and 3pm?

(c) There is an average of 11.7 people who visit during those same hours from vehicles. Is it likely that the number of people visiting by car during this hour is also Poisson? Explain.

3.42 Lost baggage. Occasionally an airline will lose a bag. Suppose a small airline has found it can reasonably model the number of bags lost each weekday using a Poisson model with a mean of 2.2 bags.

(a) What is the probability that the airline will lose no bags next Monday?

(b) What is the probability that the airline will lose 0, 1, or 2 bags on next Monday?

(c) Suppose the airline expands over the course of the next 3 years, doubling the number of flights it makes, and the CEO asks you if it's reasonable for them to continue using the Poisson model with a mean of 2.2. What is an appropriate recommendation? Explain.

3.43 Hemophilia. Hemophilia is a sex-linked bleeding disorder that slows the blood clotting process. In severe cases of hemophilia, continued bleeding occurs after minor trauma or even in the absence of injury. Hemophilia affects 1 in 5,000 male births. In the United States, about 400 males are born with hemophilia each year; there are approximately 4,000,000 births per year. *Note: this problem is best done using statistical software.*

(a) What is the probability that at most 380 newborns in a year are born with hemophilia?

(b) What is the probability that 450 or more newborns in a year are born with hemophilia?

(c) Consider a hypothetical country in which there are approximately 1.5 million births per year. If the incidence rate of hemophilia is equal to that in the US, how many newborns are expected to have hemophilia in a year, with what standard deviation?

3.44 Opioid overdose. The US Centers for Disease Control (CDC) has been monitoring the rate of deaths from opioid overdoses for at least the last 15 years. In 2013, the rate of opioid-related deaths has risen to 6.8 deaths per year per 100,000 non-Hispanic white members. In 2014-2015, the population of Essex County, MA, was approximately 769,000, of whom 73% are non-Hispanic white. Assume that incidence rate of opioid deaths in Essex County is the same as the 2013 national rate. *Note: this problem is best done using statistical software.*

(a) In 2014, Essex County reported 146 overdose fatalities from opioids. Assume that all of these deaths occurred in the non-Hispanic white members of the population. What is the probability of 146 or more such events a year?

(b) What was the observed rate of opioid-related deaths in Essex County in 2014, stated in terms of deaths per 100,000 non-Hispanic white members of the population?

(c) In 2015, Essex County reported 165 opioid-related deaths in its non-Hispanic white population. Using the rate from part (b), calculate the probability of 165 or more such events.

3.8.5 Distributions related to Bernoulli trials

3.45 Married women. The 2010 American Community Survey estimates that 47.1% of women ages 15 years and over are married. Suppose that a random sample of women in this age group are selected for a research study.[40]

(a) On average, how many women would need to be sampled in order to select a married woman? What is the standard deviation?

(b) If the proportion of married women were actually 30%, what would be the new mean and standard deviation?

(c) Based on the answers to parts (a) and (b), how does decreasing the probability of an event affect the mean and standard deviation of the wait time until success?

---

[40] U.S. Census Bureau, 2010 American Community Survey, Marital Status.

3.46 Donating blood, Part II. Recall from Problem 3.11 that a patient with Type O+ blood can receive either Type O+ or Type O- blood, while a patient with Type O- blood can only receive Type O- blood. According to data collected from blood donors, 0.37 of white, non-Hispanic donors are Type O+ and 0.08 are Type O-. For the following questions, assume that only white, non-Hispanic donors are being tested.

(a) On average, how many donors would need to be randomly sampled for a Type O+ donor to be identified? With what standard deviation?

(b) What is the probability that 4 donors must be sampled to identify a Type O+ donor?

(c) What is the probability that more than 4 donors must be sampled to identify a Type O+ donor?

(d) What is the probability of the first Type O- donor being found within the first 4 people?

(e) On average, how many donors would need to be randomly sampled for a Type O- donor to be identified? With what standard deviation?

(f) What is the probability that fewer than 4 donors must be tested before a Type O- donor is found?

3.47 Wolbachia infection, Part II. Recall from Problem 3.15 that 30% of the *Drosophila* stocks at the BDSC are infected with Wolbachia. Suppose a research assistant randomly samples a stock one at a time until identifying an infected stock.

(a) Calculate the probability that an infected stock is found within the first 5 stocks sampled.

(b) What is the probability that no more than 5 stocks must be tested before an infected one is found?

(c) Calculate the probability that at least 3 stocks must be tested for an infected one to be found.

3.48 With and without replacement. In the following situations assume that half of the specified population is male and the other half is female.

(a) Suppose you're sampling from a room with 10 people. What is the probability of sampling two females in a row when sampling with replacement? What is the probability when sampling without replacement?

(b) Now suppose you're sampling from a stadium with 10,000 people. What is the probability of sampling two females in a row when sampling with replacement? What is the probability when sampling without replacement?

(c) We often treat individuals who are sampled from a large population as independent. Using your findings from parts (a) and (b), explain whether or not this assumption is reasonable.

3.49 Eye color. A husband and wife both have brown eyes but carry genes that make it possible for their children to have brown eyes (probability 0.75), blue eyes (0.125), or green eyes (0.125).

(a) What is the probability the first blue-eyed child they have is their third child? Assume that the eye colors of the children are independent of each other.

(b) On average, how many children would such a pair of parents have before having a blue-eyed child? What is the standard deviation of the number of children they would expect to have until the first blue-eyed child?

3.50 Defective rate. A machine that produces a special type of transistor (a component of computers) has a 2% defective rate. The production is considered a random process where each transistor is independent of the others.

(a) What is the probability that the 10th transistor produced is the first with a defect?

(b) What is the probability that the machine produces no defective transistors in a batch of 100?

(c) On average, how many transistors would you expect to be produced before the first with a defect? What is the standard deviation?

(d) Another machine that also produces transistors has a 5% defective rate where each transistor is produced independent of the others. On average how many transistors would you expect to be produced with this machine before the first with a defect? What is the standard deviation?

(e) Based on your answers to parts (c) and (d), how does increasing the probability of an event affect the mean and standard deviation of the wait time until success?

3.51 Rolling a die. Calculate the following probabilities and indicate which probability distribution model is appropriate in each case. You roll a fair die 5 times. What is the probability of rolling

(a) the first 6 on the fifth roll?

(b) exactly three 6s?

(c) the third 6 on the fifth roll?

3.52 Playing darts. Calculate the following probabilities and indicate which probability distribution model is appropriate in each case. A very good darts player can hit the direct center of the board 65% of the time. What is the probability that a player:

(a) hits the bullseye for the 10th time on the 15th try?

(b) hits the bullseye 10 times in 15 tries?

(c) hits the first bullseye on the third try?

3.53 Cilantro preference. Cilantro leaves are widely used in many world cuisines. While some people enjoy it, others claim that it has a soapy, pungent aroma. A recent study conducted on participants of European ancestry identified a genetic variant that is associated with soapy-taste detection. In the initial questionnaire, 1,994 respondents out of 14,604 reported that they thought cilantro tasted like soap. Suppose that participants are randomly selected one by one.

(a) What is the probability that the first soapy-taste detector is the third person selected?

(b) What is the probability that in a sample of ten people, no more than two are soapy-taste detectors?

(c) What is the probability that three soapy-taste detectors are identified from sampling ten people?

(d) What is the mean and standard deviation of the number of people that must be sampled if the goal is to identify four soapy-taste detectors?

3.54 Serving in volleyball. A not-so-skilled volleyball player has a 15% chance of making the serve, which involves hitting the ball so it passes over the net on a trajectory such that it will land in the opposing team's court. Suppose that serves are independent of each other.

(a) What is the probability that on the 10th try, the player makes their 3rd successful serve?

(b) Suppose that the player has made two successful serves in nine attempts. What is the probability that their 10th serve will be successful?

(c) Even though parts (a) and (b) discuss the same scenario, explain the reason for the discrepancy in probabilities.

3.55 Cilantro preference, Part II. Recall from Problem 3.53 that in a questionnaire, 1,994 respondents out of 14,604 reported that they thought cilantro tasted like soap. Suppose that a random sample of 15 individuals are selected for further study.

(a) What is the mean and variance of the number of people sampled that are soapy-taste detectors?

(b) What is the probability that 4 of the people sampled are soapy-taste detectors?

(c) What is the probability that at most 2 of the people sampled are soapy-taste detectors?

(d) Suppose that the 15 individuals were sampled with replacement. What is the probability of selecting 4 soapy-taste detectors?

(e) Compare the answers from parts (b) and (d). Explain why the answers are essentially the same.

3.56 Dental caries. A study to examine oral health of schoolchildren in Belgium found that of the 4,351 children examined, 44% were caries free (i.e., free of decay, restorations, and missing teeth). Suppose that children are sampled one by one.

(a) What is the probability that at least three caries free children are identified from sampling seven children?

(b) What is the probability that the first caries free child is the second one selected?

(c) Suppose that in a single school of 350 children, the incidence rate of caries equals the national rate. If 10 schoolchildren are selected at random, what is the probability that at most 2 have caries?

(d) What is the probability that in a sample of 50 children, no more than 15 are caries free?

3.8.6 Distributions for pairs of random variables

3.57 Joint distributions, Part I. Suppose X and Y have the following joint distribution.

	Y = -1	Y = 1
X = 0	0.20	0.40
X = 1	0.30	0.10

(a) Calculate the marginal distributions of X and Y.

(b) Calculate the mean, variance, and standard deviation of X.

(c) What are the standardized values of X?

(d) The mean and standard deviation of Y are 0 and 1, respectively, and the two standardized values for Y are -1 and 1. Calculate $\rho_{X,Y}$, the correlation coefficient of X and Y.

(e) Are X and Y independent? Explain your answer.

3.58 Joint distributions, Part II. Suppose X and Y have the following joint distribution.

	Y = -1	Y = 1
X = 0	0.25	0.25
X = 1	0.25	0.25

(a) Are X and Y independent?

(b) Calculate the correlation between X and Y.

(c) Are your answers to parts (a) and (b) consistent? Explain your answer.

3.59 Joint distributions, Part III. Consider the following joint probability distribution:

	Y		
X	-1	0	1
-1	0.10	0	0.35
0	0	0.10	0.10
1	0.15	0.10	0.10

(a) Calculate the marginal distributions.

(b) Compute μ_X.

(c) Compute σ_Y^2.

(d) Calculate the conditional distribution of X, given $Y = 0$.

3.60 Dice rolls and coin tosses. Let X represent the outcome from a roll of a fair six-sided die. Then, toss a fair coin X times and let Y denote the number of tails observed.

(a) Consider the joint probability table of X and Y. How many entries are in the table for the joint distribution of X and Y? How many entries equal 0?

(b) Compute the joint probability $P(X = 1, Y = 0)$.

(c) Compute the joint probability $P(X = 1, Y = 2)$.

(d) Compute the joint probability $P(X = 6, Y = 3)$.

3.61 Health insurance claims. In the health insurance example introduced in Example 3.6, the largest annual expense for the annual employee ($1,108) was caused by 8 visits to a provider for a knee injury requiring physical therapy. The couple has confidence that this or a similar injury will not happen again in the coming year, and wonders about the effect of reduced visits on expected total health care costs and its variability.

A new joint distribution of health care cost for the couple is shown in the following table:

	Partner costs, Y	
Employee costs, X	$968	$988
$968	0.18	0.12
$1,008	0.15	0.25
$1,028	0.07	0.23

(a) For the partner, will there be a change to the expected cost and its standard deviation?

(b) Calculated the expected value and standard deviation for the employee's costs.

(c) Calculate the expected total cost for the couple.

(d) Calculate the new correlation for employee and partner costs.

(e) Calculate the standard deviation of the total cost.

Chapter 4

Foundations for inference

Not surprisingly, many studies are now demonstrating the adverse effect of obesity on health outcomes. A 2017 study conducted by the consortium studying the global burden of disease estimates that high body mass index (a measure of body fat that adjusts for height and weight) may account for as many as 4.0 million deaths globally.[1] In addition to the physiologic effects of being overweight, other studies have shown that perceived weight status (feeling that one is overweight or underweight) may have a significant effect on self-esteem.[2,3]

As stated in its mission statement, the United States Centers for Disease Control and Prevention (US CDC) "serves as the national focus for developing and applying disease prevention and control, environmental health, and health promotion and health education activities designed to improve the health of the people of the United States".[4] Since it is not feasible to measure the health status and outcome of every single US resident, the CDC estimates features of health from samples taken from the population, via large surveys that are repeated periodically. These surveys include the National Health Interview Survey (NHIS), the National Health and Nutrition Examination Survey (NHANES), the Youth Risk Behavior Surveillance System (YRBSS) and the Behavior Risk Factor Surveillance System (BRFSS). In the language of statistics, the average weight of all US adults is a **population parameter**; the mean weight in a sample or survey is an **estimate** of population average weight. The principles of statistical inference provide not only estimates of population parameters, but also measures of uncertainty that account for the fact that different random samples will produce different estimates because of the variability of random sampling; i.e., two different random samples will not include exactly the same people.

This chapter introduces the important ideas in drawing estimates from samples by discussing methods of inference for a population mean, μ, including three widely used tools: point estimates for a population mean, interval estimates that include both a point estimate and a margin of error, and a method for testing scientific hypotheses about μ. The concepts used in this chapter will appear throughout the rest of the book, which discusses inference for other settings. While particular equations or formulas may change to reflect the details of a problem at hand, the fundamental ideas will not.

[1] DOI: 10.1056/NEJMoa1614362

[2] J Ment Health Policy Econ. 2010 Jun;13(2):53-63

[3] DOI: 10.1186/1471-2458-7-80

[4] https://www.cdc.gov/maso/pdf/cdcmiss.pdf

The BRFSS was established in 1984 in 15 states to collect data using telephone interviews about health-related risk behaviors, chronic health conditions, and the use of preventive services. It now collects data in all 50 states and the District of Columbia from more than 400,000 interviews conducted each year. The data set cdc contains a small number of variables from a random sample of 20,000 responses from the 264,684 interviews from the BRFSS conducted in the year 2000. Part of this dataset is shown in Figure 4.1, with the variables described in Figure 4.2.[5]

	case	age	gender	weight	wtdesire	height	genhlth
1	1	77	m	175	175	70	good
2	2	33	f	125	115	64	good
3	3	49	f	105	105	60	good
20000	20000	83	m	170	165	69	good

Figure 4.1: Four cases from the cdc dataset.

Variable	Variable definition.
case	Case number in the dataset, ranging from 1 to 20,000.
age	Age in years.
gender	A factor variable, with levels m for male, f for female.
weight	Weight in pounds.
wtdesire	Weight that the respondent wishes to be, in pounds.
height	Height in inches.
genhlth	A factor variable describing general health status, with levels excellent, very good, good, fair, poor.

Figure 4.2: Some variables and their descriptions for the cdc dataset.

Few studies are as large as the original BRFSS dataset (more than 250,000 cases); in fact, few are as large as the 20,000 cases in the dataset cdc. The dataset cdc is large enough that estimates calculated from cdc can be thought of as essentially equivalent to the population characteristics of the entire US adult population. This chapter uses a random sample of 60 cases from cdc, stored as cdc.samp, to illustrate the effect of sampling variability and the ideas behind inference. In other words, suppose that cdc represents the population, and that cdc.samp is a sample from the population; the goal is to estimate characteristics of the population of 20,000 using only the data from the 60 individuals in the sample.

For labs, slides, and other resources, please visit
www.openintro.org/book/biostat

---

[5]With small modifications (character strings re-coded as factors), the data appears in this text as it does in an *OpenIntro* lab. https://www.openintro.org/go?id=statlab_r_core_intro_to_data

4.1 Variability in estimates

A natural way to estimate features of the population, such as the population mean weight, is to use the corresponding summary statistic calculated from the sample.[6] The mean weight in the sample of 60 adults in cdc.samp is $\bar{x}_{weight} = 173.3$ lbs; this sample mean is a **point estimate** of the population mean, μ_{weight}. If a different random sample of 60 individuals were taken from cdc, the new sample mean would likely be different as a result of **sampling variation**. While estimates generally vary from one sample to another, the population mean is a fixed value.

GUIDED PRACTICE 4.1

How would one estimate the difference in average weight between men and women? Given that $\bar{x}_{men} = 185.1$ lbs and $\bar{x}_{women} = 162.3$ lbs, what is a good point estimate for the population difference?[7]

Point estimates become more accurate with increasing sample size. Figure 4.3 shows the sample mean weight calculated for random samples drawn from cdc, where sample size increases by 1 for each draw until sample size equals 500. The red dashed horizontal line in the figure is drawn at the average weight of all adults in cdc, 169.7 lbs, which represents the population mean weight.[8]

Figure 4.3: The mean weight computed for a random sample from cdc, increasing sample size one at a time until $n = 500$. The sample mean approaches the population mean (i.e., mean weight in cdc) as sample size increases.

Note how a sample size around 50 may produce a sample mean that is as much as 10 lbs higher or lower than the population mean. As sample size increases, the fluctuations around the population mean decrease; in other words, as sample size increases, the sample mean becomes less variable and provides a more reliable estimate of the population mean.

---

[6]Other population parameters, such as population median or population standard deviation, can also be estimated using sample versions.

[7]Given that $\bar{x}_{men} = 185.1$ lbs and $\bar{x}_{women} = 162.3$ lbs, the difference of the two sample means, $185.1 - 162.3 = 22.8$lbs, is a point estimate of the difference. The data in the random sample suggests that adult males are, on average, about 23 lbs heavier than adult females.

[8]It is not exactly the mean weight of all US adults, but will be very close since cdc is so large.

4.1.1 The sampling distribution for the mean

The sample mean weight calculated from cdc.samp is 173.3 lbs. Another random sample of 60 participants might produce a different value of $\overline{x}$, such as 169.5 lbs; repeated random sampling could result in additional different values, perhaps 172.1 lbs, 168.5 lbs, and so on. Each sample mean $\overline{x}$ can be thought of as a single observation from a random variable $\overline{X}$. The distribution of $\overline{X}$ is called the **sampling distribution of the sample mean**, and has its own mean and standard deviation like the random variables discussed in Chapter 3. The concept of a sampling distribution can be illustrated by taking repeated random samples from cdc. Figure 4.4 shows a histogram of sample means from 1,000 random samples of size 60 from cdc. The histogram provides an approximation of the theoretical sampling distribution of $\overline{X}$ for samples of size 60.

Figure 4.4: A histogram of 1000 sample means for weight among US adults, where the samples are of size $n = 60$.

SAMPLING DISTRIBUTION

The sampling distribution is the distribution of the point estimates based on samples of a fixed size from a certain population. It is useful to think of a particular point estimate as being drawn from a sampling distribution.

Since the complete sampling distribution consists of means for all possible samples of size 60, drawing a much larger number of samples provides a more accurate view of the distribution; the left panel of Figure 4.5 shows the distribution calculated from 100,000 sample means.

A normal probability plot of these sample means is shown in the right panel of Figure 4.5. All of the points closely fall around a straight line, implying that the distribution of sample means is nearly normal (see Section 3.3). This result follows from the Central Limit Theorem.

CENTRAL LIMIT THEOREM, INFORMAL DESCRIPTION

If a sample consists of at least 30 independent observations and the data are not strongly skewed, then the distribution of the sample mean is well approximated by a normal model.

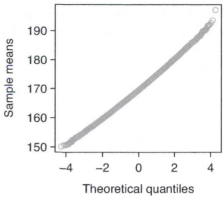

Figure 4.5: The left panel shows a histogram of the sample means for 100,000 random samples. The right panel shows a normal probability plot of those sample means.

The sampling distribution for the mean is unimodal and symmetric around the mean of the random variable $\overline{X}$. Statistical theory can be used to show that the mean of the sampling distribution for $\overline{X}$ is exactly equal to the population mean μ.

However, in almost any study, conclusions about a population parameter must be drawn from the data collected from a single sample. The sampling distribution of $\overline{X}$ is a theoretical concept, since obtaining repeated samples by conducting a study many times is not possible. In other words, it is not feasible to calculate the population mean μ by finding the mean of the sampling distribution for $\overline{X}$.

4.1.2 Standard error of the mean

The **standard error (SE)** of the sample mean measures the sample-to-sample variability of $\overline{X}$, the extent to which values of the repeated sample means oscillate around the population mean. The theoretical standard error of the sample mean is calculated by dividing the population standard deviation (σ_x) by the square root of the sample size n. Since the population standard deviation σ is typically unknown, the sample standard deviation s is often used in the definition of a standard error; s is a reasonably good estimate of σ. If $\overline{X}$ represents the sample mean weight, its standard error (denoted by SE) is

$$\text{SE}_{\overline{X}} = \frac{s_x}{\sqrt{n}} = \frac{49.04}{\sqrt{60}} = 6.33.$$

SE
standard
error

This estimate tends to be sufficiently good when the sample size is at least 30 and the population distribution is not strongly skewed. In the case of skewed distributions, a larger sample size is necessary.

The probability tools of Section 3.1 can be used to derive the formula $\sigma_{\overline{X}} = \sigma_x/\sqrt{n}$, but the derivation is not shown here. Larger sample sizes produce sampling distributions that have lower variability. Increasing the sample size causes the distribution of $\overline{X}$ to be clustered more tightly around the population mean μ, allowing for more accurate estimates of μ from a single sample, as shown in Figure 4.6. When sample size is large, it is more likely that any particular sample will have a mean close to the population mean.

(a)   (b)

Figure 4.6: (a) Reproduced from Figure 4.4, an approximation of the sampling distribution of $\overline{X}$ with $n = 60$. (b) An approximation of the sampling distribution of $\overline{X}$ with $n = 200$.

THE STANDARD ERROR (SE) OF THE SAMPLE MEAN

Given n independent observations from a population with standard deviation σ, the standard error of the sample mean is equal to

$$SE_{\overline{X}} = \frac{s_x}{\sqrt{n}}.$$

This is an accurate estimate of the theoretical standard deviation of $\overline{X}$ when the sample size is at least 30 and the population distribution is not strongly skewed.

SUMMARY: POINT ESTIMATE TERMINOLOGY

- The population mean and standard deviation are denoted by μ and σ.

- The sample mean and standard deviation are denoted by $\overline{x}$ and s.

- The distribution of the random variable $\overline{X}$ refers to the collection of sample means if multiple samples of the same size were repeatedly drawn from a population.

- The mean of the random variable $\overline{X}$ equals the population mean μ. In the notation of Chapter 3, $\mu_{\overline{X}} = E(\overline{X}) = \mu$.

- The standard deviation of $\overline{X}$ ($\sigma_{\overline{X}}$) is called the standard error (SE) of the sample mean.

- The theoretical standard error of the sample mean, as calculated from a single sample of size n, is equal to $\frac{\sigma}{\sqrt{n}}$. The standard error is abbreviated by SE and is usually estimated by using s, the sample standard deviation, such that $SE = \frac{s}{\sqrt{n}}$.

4.2 Confidence intervals

4.2.1 Interval estimates for a population parameter

While a point estimate consists of a single value, an interval estimate provides a plausible range of values for a parameter. When estimating a population mean μ, a **confidence interval** for μ has the general form

$$(\overline{x} - m, \ \overline{x} + m) = \overline{x} \pm m,$$

where m is the **margin of error**. Intervals that have this form are called **two-sided confidence intervals** because they provide both lower and upper bounds, $\overline{x} - m$ and $\overline{x} + m$, respectively. One-sided sided intervals are discussed in Section 4.2.3.

The standard error of the sample mean is the standard deviation of its distribution; additionally, the distribution of sample means is nearly normal and centered at μ. Under the normal model, the sample mean $\overline{x}$ will be within 1.96 standard errors (i.e., standard deviations) of the population mean μ approximately 95% of the time.[9] Thus, if an interval is constructed that spans 1.96 standard errors from the point estimate in either direction, a data analyst can be 95% **confident** that the interval

$$\overline{x} \pm 1.96 \times \text{SE} \tag{4.2}$$

contains the population mean. The value 95% is an approximation, accurate when the sampling distribution for the sample mean is close to a normal distribution. This assumption holds when the sample size is sufficiently large (guidelines for 'sufficiently large' are given in Section 4.4).

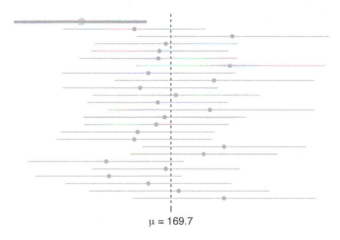

$\mu = 169.7$

Figure 4.7: Twenty-five samples of size $n = 60$ were taken from cdc. For each sample, a 95% confidence interval was calculated for the population average adult weight. Only 1 of these 25 intervals did not contain the population mean, $\mu = 169.7$ lbs.

The phrase "95% confident" has a subtle interpretation: if many samples were drawn from a population, and a confidence interval is calculated from each one using Equation 4.2, about 95% of those intervals would contain the population mean μ. Figure 4.7 illustrates this process with 25 samples taken from cdc. Of the 25 samples, 24 contain the mean weight in cdc of 169.7 lbs, while one does not.

---

[9]In other words, the Z-score of 1.96 is associated with 2.5% area to the right (and $Z = -1.96$ has 2.5% area to the left); this can be found on normal probability tables or from using statistical software.

Just as with the sampling distribution of the sample mean, the interpretation of a confidence interval relies on the abstract construct of repeated sampling. A data analyst, who can only observe one sample, does not know whether the population mean lies within the single interval calculated. The uncertainty is due to random sampling—by chance, it is possible to select a sample from the population that has unusually high (or low) values, resulting in a sample mean $\bar{x}$ that is relatively far from μ, and by extension, a confidence interval that does not contain μ.

EXAMPLE 4.3

The sample mean adult weight from the 60 observations in cdc.samp is $\bar{x}_{\text{weight}} = 173.3$ lbs, and the standard deviation is $s_{\text{weight}} = 49.04$ lbs. Use Equation 4.2 to calculate an approximate 95% confidence interval for the average adult weight in the US population.

The standard error for the sample mean is $\text{SE}_{\bar{x}} = \frac{49.04}{\sqrt{60}} = 6.33$ lbs. The 95% confidence interval is

$$\bar{x}_{\text{weight}} \pm 1.96\text{SE}_{\bar{x}} = 173.3 \pm (1.96)(6.33) = (160.89, 185.71) \text{ lbs.}$$

The data support the conclusion that, with 95% confidence, the average weight of US adults is between approximately 161 and 186 lbs.

Figure 4.5 visually shows that the sampling distribution is nearly normal. To assess normality of the sampling distribution without repeated sampling, it is necessary to check whether the data are skewed. Although Figure 4.8 shows some skewing, the sample size is large enough that the confidence interval should be reasonably accurate.

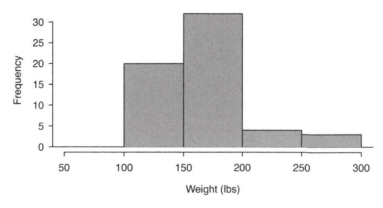

Figure 4.8: Histogram of weight in cdc.samp

GUIDED PRACTICE 4.4

There are 31 females in the sample of 60 US adults, and the average and standard deviation of weight for these individuals are 162.3 lbs and 57.74 lbs, respectively. A histogram of weight for the 31 females is shown in Figure 4.9. Calculate an approximate 95% confidence interval for the average weight of US females. Is the interval likely to be accurate?[10]

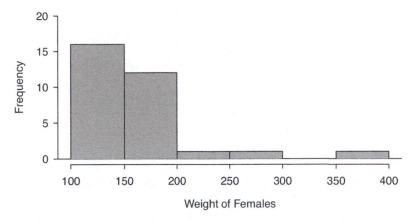

Figure 4.9: Histogram of weight for the 31 females in cdc.samp.

4.2.2 Changing the confidence level

Ninety-five percent confidence intervals are the most commonly used interval estimates, but intervals with confidence levels other than 95% can also be constructed. The general formula for a confidence interval (for the population mean μ) is given by

$$\bar{x} \pm z^\star \times \text{SE}, \qquad (4.5)$$

where $z^\star$ is chosen according to the confidence level. When calculating a 95% confidence level, $z^\star$ is 1.96, since the area within 1.96 standard deviations of the mean captures 95% of the distribution.

To construct a 99% confidence interval, $z^\star$ must be chosen such that 99% of the normal curve is captured between $-z^\star$ and $z^\star$.

EXAMPLE 4.6

Let Y be a normally distributed random variable. Ninety-nine percent of the time, Y will be within how many standard deviations of the mean?

This is equivalent to the z-score with 0.005 area to the right of z and 0.005 to the left of $-z$. In the normal probability table, this is the z-value that with 0.005 area to its right and 0.995 area to its left. The closest two values are 2.57 and 2.58; for convenience, round up to 2.58. The unobserved random variable Y will be within 2.58 standard deviations of μ 99% of the time, as shown in Figure 4.10.

---

[10]Applying Equation 4.2: $162.3 \pm (1.96)(57.73/\sqrt{31}) \rightarrow (149.85, 174.67)$. The usual interpretation would be that a data analyst can be about 95% confident the average weight of US females is between approximately 150 and 175 lbs. However, the histogram of female weights shows substantial right skewing, and several females with recorded weights larger than 200 lbs. The confidence interval is probably not accurate; a larger sample should be collected in order for the sampling distribution of the mean to be approximately normal. Chapter 5 will introduce the t-distribution, which is more reliable with small sample sizes than the z-distribution.

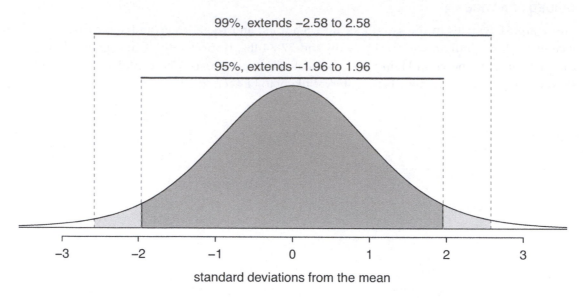

Figure 4.10: The area between $-z^\star$ and $z^\star$ increases as $|z^\star|$ becomes larger. If the confidence level is 99%, $z^\star$ is chosen such that 99% of the normal curve is between $-z^\star$ and $z^\star$, which corresponds to 0.5% in the lower tail and 0.5% in the upper tail: $z^\star = 2.58$.

A 99% confidence interval will have the form

$$\bar{x} \pm 2.58 \times SE, \tag{4.7}$$

and will consequently be wider than a 95% interval for μ calculated from the same data, since the margin of error m is larger.

EXAMPLE 4.8

Create a 99% confidence interval for the average adult weight in the US population using the data in cdc.samp. The point estimate is $\bar{x}_{weight} = 173.3$ and the standard error is $SE_{\bar{x}} = 6.33$.

Apply the 99% confidence interval formula: $\bar{x}_{weight} \pm 2.58 \times SE_{\bar{x}} \rightarrow (156.97, 189.63)$. A data analyst can be 99% confident that the average adult weight is between 156.97 and 189.63 lbs.

The 95% confidence interval for the average adult weight is (160.89, 185.71) lbs. Increasing the confidence level to 99% results in the interval (156.97, 189.63) lbs; this wider interval is more likely to contain the population mean μ. However, increasing the confidence level comes at a cost: a wider interval is less informative in providing a precise estimate of the population mean. Consider the extreme: to be "100% confident" that an interval contains μ, the interval must span all possible values of μ. For example, with 100% confidence the average weight is between 0 and 1000 lbs; while this interval necessarily contains μ, it has no interpretive value and is completely uninformative.[11]

Decreasing the confidence level produces a narrower interval; the estimate is more precise, but also more prone to inaccuracy. For example, consider a 50% confidence interval for average adult weight using cdc.samp: the $z^\star$ value is 0.67, and the confidence interval is (169.06, 177.54) lbs. This interval provides a more precise estimate of the population average weight μ than the 99% or 95% confidence intervals, but the increased precision comes with less confidence about whether the

---

[11] Strictly speaking, to be 100% confident requires an interval spanning all positive numbers; 1000 lbs has been arbitrarily chosen as an upper limit for human weight.

interval contains μ. In a theoretical setting of repeated sampling, if 100 50% confidence intervals were computed, only half could be expected to contain μ.

The choice of confidence level is a trade-off between obtaining a precise estimate and calculating an interval that can be reasonably expected to contain the population parameter. In published literature, the most used confidence intervals are the 90%, 95%, and 99%.

4.2.3 One-sided confidence intervals

One-sided confidence intervals for a population mean provide either a lower bound or an upper bound, but not both. One-sided confidence intervals have the form

$$(\bar{x} - m, \infty) \text{ or } (-\infty, \bar{x} + m).$$

While the margin of error m for a one-sided interval is still calculated from the standard error of $\bar{x}$ and a $z^\star$ value, the choice of $z^\star$ is a different than for a two-sided interval. For example, the intent of a 95% one-sided upper confidence interval is to provide an upper bound m such that a data analyst can be 95% confident that a population mean μ is less than $\bar{x} + m$. The $z^\star$ value must correspond to the point on the normal distribution that has 0.05 area in the right tail, $z^\star = 1.645$.[12] A one-sided upper 95% confidence interval will have the form

$$(-\infty, \bar{x} + 1.645 \times \text{SE}).$$

EXAMPLE 4.9

Calculate a lower 95% confidence interval for the population average adult weight in the United States. In the sample of 60 adults in cdc.samp, the mean and standard error are $\bar{x} = 173.3$ and $SE = 6.33$ days.

The lower bound is $173.3 - (1.645 \times 6.33) = 163.89$. The lower 95% interval $(163.89, \infty)$ suggests that one can be 95% confident that the population average adult weight is at least 163.9 lbs.

GUIDED PRACTICE 4.10

Calculate an upper 99% confidence interval for the population average adult weight in the United States. The mean and standard error for weight in cdc.samp are $\bar{x} = 173.3$ and $SE = 6.33$ days.[13]

4.2.4 Interpreting confidence intervals

The correct interpretation of an XX% confidence interval is, "We are XX% confident that the population parameter is between ..." While it may be tempting to say that a confidence interval captures the population parameter with a certain probability, this is a common error. The confidence level only quantifies how plausible it is that the parameter is within the interval; there is no probability associated with whether a parameter is contained in a specific confidence interval. The confidence coefficient reflects the nature of a procedure that is correct XX% of the time, given that the assumptions behind the calculations are true.

---

[12]Previously, with a two-sided interval, 1.96 was chosen in order to have a total area of 0.05 from both the right and left tails.

[13]For a one-sided 99% confidence interval, the $z^\star$ value corresponds to the point with 0.01 area in the right tail, $z^\star = 2.326$. Thus, the upper bound for the interval is $173.3 + (2.326 \times 6.33) = 188.024$. The upper 99% interval $(-\infty, 188.024)$ suggests that one can be 99% confident that the population average adult weight is at most 188.0 lbs.

The conditions regarding the validity of the normal approximation can be checked using the numerical and graphical summaries discussed in Chapter 1. However, the condition that data should be from a random sample is sometimes overlooked. If the data are not from a random sample, then the confidence interval no longer has interpretive value, since there is no population mean to which the confidence interval applies. For example, while only simple arithmetic is needed to calculate a confidence interval for BMI from the famuss dataset in Chapter 1, the participants in the study are almost certainly not a random sample from some population; thus, a confidence interval should not be calculated in this setting.

EXAMPLE 4.11

Body mass index (BMI) is one measure of body weight that adjusts for height. The National Health and Nutrition Examination Survey (NHANES) consists of a set of surveys and measurements conducted by the US CDC to assess the health and nutritional status of adults and children in the United States. The dataset nhanes.samp contains 76 variables and is a random sample of 200 individuals from the measurements collected in the years 2009-2010 and 2012-2013.[14] Use nhanes.samp to calculate a 95% confidence interval for adult BMI in the US population, and assess whether the data suggest Americans tend to be overweight.

In the random sample of 200 participants, BMI is available for all 135 of the participants that are 21 years of age or older. As shown in the histogram (Figure 4.11), the data are right-skewed, with one large outlier. The outlier corresponds to an implausibly extreme BMI value of 69.0; since it seems likely that the value represents an error from when the data was recorded, this data point is excluded from the following analysis.

The mean and standard deviation in this sample of 134 are 28.8 and 6.7 kg/meter^2, respectively. The sample size is large enough to justify using the normal approximation when computing the confidence interval. The standard error of the mean is SE $= 6.7/\sqrt{134} = 0.58$, so the 95% confidence interval is given by

$$\bar{x}_{BMI} \pm (1.96)(SE) = 28.8 \pm (1.96)(0.58)$$
$$= (27.7, 29.9).$$

Based on this sample, a data analyst can be 95% confident that the average BMI of US adults is between 27.7 and 29.9 kg/m^2.

The World Health Organization (WHO) and other agencies use BMI to set normative guidelines for body weight. The current guidelines are shown in Figure 4.12.

The confidence interval (27.7, 29.9) kg/m^2 certainly suggests that the average BMI in the US population is higher than 21.7, the middle of the range for normal BMIs, and even higher than 24.99, the upper limit of the normal weight category. These data indicate that Americans tend to be overweight.

[14]The sample was drawn from a larger sample of 20,293 participants in the **NHANES** package, available from The Comprehensive R Archive Network (CRAN). The CDC uses a complex sampling design that samples some demographic subgroups with larger probabilities, but nhanes.samp has been adjusted so that it can be viewed as a random sample of the US population.

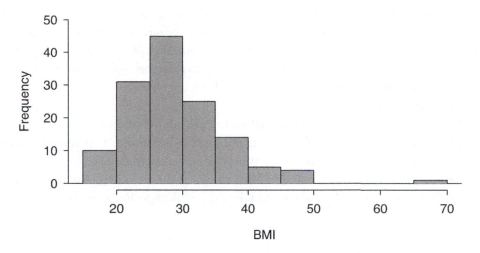

Figure 4.11: The distribution of BMI for the 135 adults in nhanes.samp.

Category	BMI range
Underweight	< 18.50
Normal (healthy weight)	18.5-24.99
Overweight	≥ 25
Obese	≥ 30

Figure 4.12: WHO body weight categories based on BMI.

4.3   Hypothesis testing

Important decisions in science, such as whether a new treatment for a disease should be approved for the market, are primarily data-driven. For example, does a clinical study of a new cholesterol-lowering drug provide robust evidence of a beneficial effect in patients at risk for heart disease? A confidence interval can be calculated from the study data to provide a plausible range of values for a population parameter, such as the population average decrease in cholesterol levels. A drug is considered to have a beneficial effect on a population of patients if the population average effect is large enough to be clinically important. It is also necessary to evaluate the strength of the evidence that a drug is effective; in other words, is the observed effect larger than would be expected from chance variation alone?

Hypothesis testing is a method for calculating the probability of making a specific observation under a working hypothesis, called the null hypothesis. By assuming that the data come from a distribution specified by the null hypothesis, it is possible to calculate the likelihood of observing a value as extreme as the one represented by the sample. If the chances of such an extreme observation are small, there is enough evidence to reject the null hypothesis in favor of an alternative hypothesis.

NULL AND ALTERNATIVE HYPOTHESES

The **null hypothesis** (H_0) often represents either a skeptical perspective or a claim to be tested. The **alternative hypothesis** (H_A) is an alternative claim and is often represented by a range of possible parameter values.

Generally, an investigator suspects that the null hypothesis is not true and performs a hypothesis test in order to evaluate the strength of the evidence against the null hypothesis. The logic behind rejecting or failing to reject the null hypothesis is similar to the principle of presumption of innocence in many legal systems. In the United States, a defendant is assumed innocent until proven guilty; a verdict of guilty is only returned if it has been established beyond a reasonable doubt that the defendant is not innocent. In the formal approach to hypothesis testing, the null hypothesis (H_0) is not rejected unless the evidence contradicting it is so strong that the only reasonable conclusion is to reject H_0 in favor of H_A.

The next section presents the steps in formal hypothesis testing, which is applied when data are analyzed to support a decision or make a scientific claim.

4.3.1   The Formal Approach to Hypothesis Testing

In this section, hypothesis testing will be used to address the question of whether Americans generally wish to be heavier or lighter than their current weight. In the cdc data, the two variables weight and wtdesire are, respectively, the recorded actual and desired weights for each respondent, measured in pounds.

Suppose that μ is the population average of the difference weight − wtdesire. Using the observations from cdc.samp, assess the strength of the claim that, on average, there is no systematic preference to be heavier or lighter.

Step 1: Formulating null and alternative hypotheses

The claim to be tested is that the population average of the difference between actual and desired weight for US adults is equal to 0.

$$H_0 : \mu = 0.$$

In the absence of prior evidence that people typically wish to be lighter (or heavier), it is reasonable to begin with an alternative hypothesis that allows for differences in either direction.

$$H_A : \mu \neq 0.$$

The alternative hypothesis $H_A : \mu \neq 0$ is called a **two-sided alternative**. A one-sided alternative could be used if, for example, an investigator felt there was prior evidence that people typically wish to weigh less than they currently do: $H_A : \mu > 0$.

More generally, when testing a hypothesis about a population mean μ, the null and alternative hypotheses are written as follows

– For a two-sided alternative:

$$H_0 : \mu = \mu_0, \ H_A : \mu \neq \mu_0.$$

– For a one-sided alternative:

$$H_0 : \mu = \mu_0, \ H_A : \mu < \mu_0 \quad \text{or} \quad H_0 : \mu = \mu_0, \ H_A : \mu > \mu_0.$$

The symbol μ denotes a population mean, while μ_0 refers to the numeric value specified by the null hypothesis; in this example, $\mu_0 = 0$. Note that null and alternative hypotheses are statements about the underlying population, not the observed values from a sample.

Step 2: Specifying a significance level, α

It is important to specify how rare or unlikely an event must be in order to represent sufficient evidence against the null hypothesis. This should be done during the design phase of a study, to prevent any bias that could result from defining 'rare' only after analyzing the results.

When testing a statistical hypothesis, an investigator specifies a **significance level**, α, that defines a 'rare' event. Typically, α is chosen to be 0.05, though it may be larger or smaller, depending on context; this is discussed in more detail in Section 4.3.4. An α level of 0.05 implies that an event occurring with probability lower than 5% will be considered sufficient evidence against H_0.

Step 3: Calculating the test statistic

Calculating the test statistic t is analogous to standardizing observations with Z-scores as discussed in Chapter 3. The test statistic quantifies the number of standard deviations between the sample mean $\overline{x}$ and the population mean μ:

$$t = \frac{\overline{x} - \mu_0}{s/\sqrt{n}},$$

where s is the sample standard deviation and n is the number of observations in the sample. If $x =$ weight $-$ wtdesire, then for the 60 recorded differences in cdc.samp, $\overline{x} = 18.2$ and $s = 33.46$. In this sample, respondents weigh on average about 18 lbs more than they wish. The test statistic is

$$t = \frac{18.2 - 0}{33.46/\sqrt{60}} = 4.22.$$

The observed sample mean is 4.22 standard deviations to the right of $\mu_0 = 0$.

Step 4: Calculating the *p*-value

The ***p*-value** is the probability of observing a sample mean as or more extreme than the observed value, under the assumption that the null hypothesis is true. In samples of size 40 or more, the *t*-statistic will have a standard normal distribution unless the data are strongly skewed or extreme outliers are present. Recall that a standard normal distribution has mean 0 and standard deviation 1.

For two-sided tests, with $H_A : \mu \neq \mu_0$, the *p*-value is the sum of the area of the two tails defined by the *t*-statistic: $2P(Z \geq |t|) = P(Z \leq -|t|) + P(Z \geq |t|)$ (Figure 4.13).

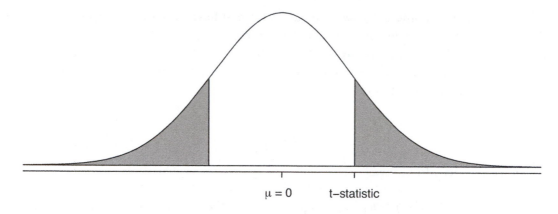

Figure 4.13: A two-sided *p*-value for $H_A : \mu \neq \mu_0$ on a standard normal distribution. The shaded regions represent observations as or more extreme than $\bar{x}$ in either direction.

For one-sided tests with $H_A : \mu > \mu_0$, the *p*-value is given by $P(Z \geq t)$, as shown in Figure 4.14. If $H_A : \mu < \mu_0$, the *p*-value is the area to the left of the *t*-statistic, $P(Z \leq t)$.

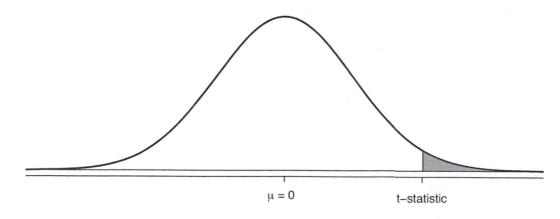

Figure 4.14: A one-sided *p*-value for $H_A : \mu > \mu_0$ on a standard normal distribution is represented by the shaded area to the right of the *t*-statistic. This area equals the probability of making an observation as or more extreme than $\bar{x}$, if the null hypothesis is true.

The *p*-value can either be calculated from software or from the normal probability tables. For the weight-difference example, the *p*-value is vanishingly small: $p = P(Z \leq -4.22) + P(Z > 4.22) < 0.001$.

Step 5: Drawing a conclusion

To reach a conclusion about the null hypothesis, directly compare p and α. Note that for a conclusion to be informative, it must be presented in the context of the original question; it is not useful to only state whether or not H_0 is rejected.

If $p > \alpha$, the observed sample mean is not extreme enough to warrant rejecting H_0; more formally stated, there is insufficient evidence to reject H_0. A high p-value suggests that the difference between the observed sample mean and μ_0 can reasonably be attributed to random chance.

If $p \leq \alpha$, there is sufficient evidence to reject H_0 and accept H_A. In the cdc.samp weight-difference data, the p-value is very small, with the t-statistic lying to the right of the population mean. The chance of drawing a sample with mean as large or larger than 18.2 if the distribution were centered at 0 is less than 0.001. Thus, the data support the conclusion that on average, the difference between actual and desired weight is not 0 and is positive; people generally seem to feel they are overweight.

GUIDED PRACTICE 4.12

Suppose that the mean weight difference in the sampled group of 60 adults had been 7 pounds instead of 18.2 pounds, but with the same standard deviation of 33.46 pounds. Would there still be enough evidence at the $\alpha = 0.05$ level to reject $H_0 : \mu = 0$ in favor of $H_A : \mu \neq 0$?[15]

---

[15] Re-calculate the t-statistic: $(7 - 0)/(33.46/\sqrt{60}) = 1.62$. The p-value $P(Z \leq -1.62) + P(Z \geq 1.62) = 0.105$. Since $p > \alpha$, there is insufficient evidence to reject H_0. In this case, a sample average difference of 7 is not large enough to discount the possibility that the observed difference is due to sampling variation, and that the observations are from a distribution centered at 0.

4.3.2 Two examples

EXAMPLE 4.13

While fish and other types of seafood are important for a healthy diet, nearly all fish and shellfish contain traces of mercury. Dietary exposure to mercury can be particularly dangerous for young children and unborn babies. Regulatory organizations such as the US Food and Drug Administration (FDA) provide guidelines as to which types of fish have particularly high levels of mercury and should be completely avoided by pregnant women and young children; additionally, certain species known to have low mercury levels are recommended for consumption. While there is no international standard that defines excessive mercury levels in saltwater fish species, general consensus is that fish with levels above 0.50 parts per million (ppm) should not be consumed. A study conducted to assess mercury levels for saltwater fish caught off the coast of New Jersey found that a sample of 23 bluefin tuna had mean mercury level of 0.52 ppm, with standard deviation 0.16 ppm.[16] Based on these data, should the FDA add bluefin tuna from New Jersey to the list of species recommended for consumption, or should a warning be issued about their mercury levels?

Let μ be the population average mercury content for bluefin tuna caught off the coast of New Jersey. Conduct a two-sided test of the hypothesis $\mu = 0.50$ ppm in order to assess the evidence for either definitive safety or potential danger.

Formulate the null and alternative hypotheses. $H_0 : \mu = 0.50$ ppm vs. $H_A : \mu \neq 0.50$ ppm

Specify the significance level, α. A significance level of $\alpha = 0.05$ seems reasonable.

Calculate the test statistic. The t-statistic has value

$$t = \frac{\overline{x} - \mu_0}{s/\sqrt{n}} = \frac{0.52 - 0.50}{0.16/\sqrt{23}} = 0.599.$$

Calculate the p-value. For this two-sided alternative $H_A : \mu \neq 0.50$, the p-value is

$$P(Z \leq -|t|) + P(Z \geq |t|) = 2 \times P(Z \geq 0.599) = 0.549.$$

Draw a conclusion. The p-value is larger than the specified significance level α, as shown in Figure 4.15.[17] The data do not show that the mercury content of bluefin tuna caught off the coast of New Jersey differs significantly from 0.50 ppm. Since $p > \alpha$, there is insufficient evidence to reject the null hypothesis that the mean mercury level for the New Jersey coastal population of bluefin tuna is 0.50 ppm.

Note that "failure to reject" is not equivalent to "accepting" the null hypothesis. Recall the earlier analogy related to the principle of "innocent until proven guilty". If there is not enough evidence to prove that the defendant is guilty, the official decision must be "not guilty", since the defendant may not necessarily be innocent. Similarly, while there is not enough evidence to suggest that μ is not equal to 0.5 ppm, it would be incorrect to claim that the evidence states that μ *is* 0.5 ppm.

From these data, there is not statistically significant evidence to either recommend these fish as clearly safe for consumption or to warn consumers against eating them. Based on these data, the Food and Drug Administration might decide to monitor this species more closely and conduct further studies.

---

[16]J. Burger, M. Gochfeld, Science of the Total Environment 409 (2011) 1418–1429

[17]The grey shaded regions are bounded by -1.96 and 1.96, since the area within 1.96 standard deviations of the mean captures 95% of the distribution.

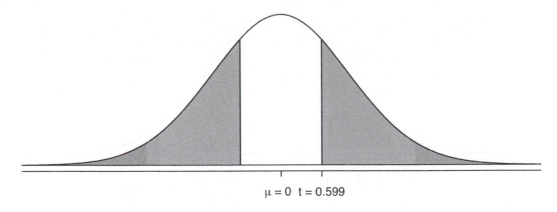

$$\mu = 0 \quad t = 0.599$$

Figure 4.15: The large blue shaded regions represent the *p*-value, the area to the right of $t = 0.599$ and to the left of $-t = -0.599$. The smaller grey shaded regions represents the **rejection region** as defined by α; in this case, an area of 0.025 in each tail. The *t*-statistic calculated from $\overline{x}$ would have to lie within either of the extreme tail areas to constitute sufficient evidence against the null hypothesis.

EXAMPLE 4.14

In 2015, the National Sleep Foundation published new guidelines for the amount of sleep recommended for adults: 7-9 hours of sleep per night.[18] The NHANES survey includes a question asking respondents about how many hours per night they sleep; the responses are available in nhanes.samp. In the sample of 134 adults used in the BMI example, the average reported hours of sleep is 6.90, with standard deviation 1.39. Is there evidence that American adults sleep less than 7 hours per night?

Let μ be the population average of hours of sleep per night for US adults. Conduct a one-sided test, since the question asks whether the average amount of sleep per night might be less than 7 hours.

Formulate the null and alternative hypotheses. $H_0 : \mu = 7$ hours vs. $H_A : \mu < 7$ hours.

Specify the significance level, α. Let $\alpha = 0.05$, since the question does not reference a different value.

Calculate the test statistic. The *t*-statistic has value

$$t = \frac{\overline{x} - \mu_0}{s/\sqrt{n}} = \frac{6.90 - 7.00}{1.33/\sqrt{134}} = -0.864.$$

Calculate the p-value.

For this one-sided alternative $H_A : \mu < 7$, the *p*-value is

$$P(Z \le t) = P(Z < -0.864) = 0.19.$$

Since the alternative states that μ_0 is less than 7, the *p*-value is represented by the area to the left of $t = -0.864$, as shown in Figure 4.16.

Draw a conclusion. The *p*-value is larger than the specified significance level α. The null hypothesis is not rejected since the data do not represent sufficient evidence to support the claim that American adults sleep less than 7 hours per night.

---

[18]Sleep Health: Journal of the National Sleep Foundation, Vol. 1, Issue 1, pp. 40 - 43

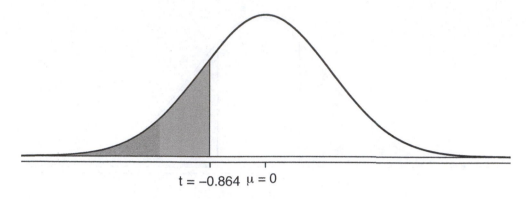

$t = -0.864 \quad \mu = 0$

Figure 4.16: The large blue shaded region represents the p-value, the area to the left of $t = -0.864$. The smaller grey shaded region represents the rejection region of area 0.05 in the left tail.

GUIDED PRACTICE 4.15

From these data, is there sufficient evidence at the $\alpha = 0.10$ significance level to support the claim that American adults sleep more than 7 hours per night?[19]

4.3.3 Hypothesis testing and confidence intervals

The relationship between a hypothesis test and the corresponding confidence interval is defined by the significance level α; the two approaches are based on the same inferential logic, and differ only in perspective. The hypothesis testing approach asks whether $\bar{x}$ is far enough away from μ_0 to be considered extreme, while the confidence interval approach asks whether μ_0 is close enough to $\bar{x}$ to be plausible. In both cases, "far enough" and "close enough" are defined by α, which determines the $z^\star$ used to calculate the margin of error $m = z^\star(s/\sqrt{n})$.

Hypothesis Test. For a two-sided test, $\bar{x}$ needs to be at least m units away from μ_0 in either direction to be considered extreme. The t-points marking off the rejection region are equal to the $z^\star$ value used in the confidence interval, with the positive and negative t-points accounting for the ± structure in the confidence interval.

Confidence Interval. The plausible range of values for μ_0 around $\bar{x}$ is defined as $(\bar{x} - m, \bar{x} + m)$. If μ_0 is plausible, it can at most be m units away in either direction from $\bar{x}$. If the interval does not contain μ_0, then μ_0 is implausible according to α and there is sufficient evidence to reject H_0.

---

[19]The t-statistic does not change from 1.65. Re-calculate the p-value since the alternative hypothesis is now $H_A : \mu > 7$: $P(Z \geq -0.864) = 0.81$. Since $p > \alpha$, there is insufficient evidence to reject H_0 at $\alpha = 0.10$. A common error when conducting one-sided tests is to assume that the p-value will always be the area in the smaller of the two tails to the right or left of the observed value. It is important to remember that the area corresponding to the p-value is in the direction specified by the alternative hypothesis.

$t = -0.864$

Suppose that a two-sided test is conducted at significance level α; the confidence level of the matching interval is $(1-\alpha)\%$. For example, a two-sided hypothesis test with $\alpha = 0.05$ can be compared to a 95% confidence interval. A hypothesis test will reject at $\alpha = 0.05$ if the 95% confidence interval does not contain the null hypothesis value of the population mean (μ_0).

THE RELATIONSHIP BETWEEN TWO-SIDED HYPOTHESIS TESTS AND CONFIDENCE INTERVALS

When testing the null hypothesis $H_0 : \mu = \mu_0$ against the two-sided alternative $H_A : \mu \neq \mu_0$, H_0 will be rejected at significance level α when the $100(1-\alpha)\%$ confidence interval for μ does not contain μ_0.

EXAMPLE 4.16

Calculate the confidence interval for the average mercury level for bluefin tuna caught off the coast of New Jersey. The summary statistics for the sample of 21 fish are $\overline{x} = 0.53$ ppm and $s = 0.16$ ppm. Does the interval agree with the results of Example 4.13?

The 95% confidence interval is:

$$\overline{x} \pm 1.96 \frac{s}{\sqrt{n}} = 0.53 \pm 1.96 \frac{0.16}{\sqrt{21}} = (0.462, 0.598) \text{ ppm.}$$

The confidence interval is relatively wide, containing values below 0.50 ppm that might be regarded as safe, in addition to values that might be regarded as potentially dangerous. This interval supports the conclusion reached from hypothesis testing; the sample data does not suggest that the mercury level differs significantly from 0.50 ppm in either direction.

The same relationship applies for one-sided hypothesis tests. For example, a one-sided hypothesis test with $\alpha = 0.05$ and $H_A : \mu > \mu_0$ corresponds to a one-sided 95% confidence interval that has a lower bound, but no upper bound (i.e., $(\overline{x} - m, \infty)$).

THE RELATIONSHIP BETWEEN ONE-SIDED HYPOTHESIS TESTS AND CONFIDENCE INTERVALS

- When testing the null hypothesis $H_0 : \mu = \mu_0$ against the one-sided alternative $H_A : \mu > \mu_0$, H_0 will be rejected at significance level α when μ_0 is smaller than the lower bound of the $100(1-\alpha)\%$ confidence interval for μ. This is equivalent to μ_0 having a value outside the lower one-sided confidence interval $(\overline{x} - m, \infty)$.

- When testing the null hypothesis $H_0 : \mu = \mu_0$ against the one-sided alternative $H_A : \mu < \mu_0$, H_0 will be rejected at significance level α whenever μ_0 is larger than the upper bound of the $100(1-\alpha)\%$ confidence interval for μ. This is equivalent to μ_0 having a value outside the upper one-sided confidence interval $(-\infty, \overline{x} + m)$.

EXAMPLE 4.17

Previously, a hypothesis test was conducted at $\alpha = 0.05$ to test the null hypothesis $H_0 : \mu = 7$ hours against the alternative $H_A : \mu < 7$ hours, for the average sleep per night US adults. Calculate the corresponding one-sided confidence interval and compare the information obtained from a confidence interval versus a hypothesis test. The summary statistics for the sample of 134 adults are $\overline{x} = 6.9$ and $s = 1.39$.

In theory, a one-sided upper confidence interval extends to ∞ on the left side, but since it is impossible to get negative sleep, it is more sensible to bound this confidence interval by 0. The upper one-sided 95% confidence interval is

$$(0, \overline{x} + 1.645 \frac{s}{\sqrt{n}}) = (0, 6.9 + 1.645 \frac{1.39}{\sqrt{134}}) = (0,\ 7.1)\ \text{hours}.$$

From these data, we can be 95% confident that the average sleep per night among US adults is at most 7.1 hours per night. The μ_0 value of 7 hours is inside the one-sided interval; thus, there is not sufficient evidence to reject the null hypothesis $H_0 : \mu = 7$ against the one-sided alternative $H_0 : \mu < 7$ hours at $\alpha = 0.05$.

The interval provides a range of plausible values for a parameter based on the observed sample; in this case, the data suggest that the population average sleep per night for US adults is no larger than 7.1 hours. The p-value from a hypothesis test represents a measure of the strength of the evidence against the null hypothesis, indicating how unusual the observed sample would be under H_0; the hypothesis test indicated that the data do not seem extreme enough ($p = 0.19$) to contradict the hypothesis that the population average sleep hours per night is 7.

In practice, both a p-value and a confidence interval are computed when using a sample to make inferences about a population parameter.

4.3.4 Decision errors

Hypothesis tests can potentially result in incorrect decisions, such as rejecting the null hypothesis when the null is actually true. Figure 4.17 shows the four possible ways that the conclusion of a test can be right or wrong.

		Test conclusion	
		Fail to reject H_0	Reject H_0 in favor of H_A
	H_0 True	Correct Decision	Type 1 Error
Reality	H_A True	Type 2 Error	Correct Decision

Figure 4.17: Four different scenarios for hypothesis tests.

Rejecting the null hypothesis when the null is true represents a **Type I error**, while a **Type II error** refers to failing to reject the null hypothesis when the alternative is true.

EXAMPLE 4.18

In a trial, the defendant is either innocent (H_0) or guilty (H_A). After hearing evidence from both the prosecution and the defense, the court must reach a verdict. What does a Type I Error represent in this context? What does a Type II Error represent?

If the court makes a Type I error, this means the defendant is innocent, but wrongly convicted (rejecting H_0 when H_0 is true). A Type II error means the court failed to convict a defendant that was guilty (failing to reject H_0 when H_0 is false).

The probability of making a Type I error is the same as the significance level α, since α determines the cutoff point for rejecting the null hypothesis. For example, if α is chosen to be 0.05, then there is a 5% chance of incorrectly rejecting H_0.

The rate of Type I error can be reduced by lowering α (e.g., to 0.01 instead of 0.05); doing so requires an observation to be more extreme to qualify as sufficient evidence against the null hypothesis. However, this inevitably raises the rate of Type II errors, since the test will now have a higher chance of failing to reject the null hypothesis when the alternative is true.

EXAMPLE 4.19

In a courtroom setting, how might the rate of Type I errors be reduced? What effect would this have on the rate of Type II errors?

Lowering the rate of Type I error is equivalent to raising the standards for conviction such that fewer people are wrongly convicted. This increases Type II error, since higher standards for conviction leads to fewer convictions for people who are actually guilty.

GUIDED PRACTICE 4.20

In a courtroom setting, how might the rate of Type II errors be reduced? What effect would this have on the rate of Type I errors?[20]

Choosing a significance level

Reducing the error probability of one type of error increases the chance of making the other type. As a result, the significance level is often adjusted based on the consequences of any decisions that might follow from the result of a significance test.

By convention, most scientific studies use a significance level of $\alpha = 0.05$; small enough such that the chance of a Type I error is relatively rare (occurring on average 5 out of 100 times), but also large enough to prevent the null hypothesis from almost never being rejected. If a Type I error is especially dangerous or costly, a smaller value of α is chosen (e.g., 0.01). Under this scenario, it is better to be cautious about rejecting the null hypothesis, so very strong evidence against H_0 is required in order to reject the null and accept the alternative. Conversely, if a Type II error is relatively dangerous, then a larger value of α is chosen (e.g., 0.10). Hypothesis tests with larger values of α will reject H_0 more often.

For example, in the early stages of assessing a drug therapy, it may be important to continue further testing even if there is not very strong initial evidence for a beneficial effect. If the scientists conducting the research know that any initial positive results will eventually be more rigorously tested in a larger study, they might choose to use $\alpha = 0.10$ to reduce the chances of making a Type II error: prematurely ending research on what might turn out to be a promising drug.

---

[20]To lower the rate of Type II error, the court could lower the standards for conviction, or in other words, lower the bar for what constitutes sufficient evidence of guilt (increase α, e.g. to 0.10 instead of 0.05). This will result in more guilty people being convicted, but also increase the rate of wrongful convictions, increasing the Type I error.

A government agency responsible for approving drugs to be marketed to the general population, however, would likely be biased towards minimizing the chances of making a Type I error—approving a drug that turns out to be unsafe or ineffective. As a result, they might conduct tests at significance level 0.01 in order to reduce the chances of concluding that a drug works when it is in fact ineffective. The US FDA and the European Medical Agency (EMA) customarily require that two independent studies show the efficacy of a new drug or regimen using $\alpha = 0.05$, though other values are sometimes used.

4.3.5   Choosing between one-sided and two-sided tests

In some cases, the choice of a one-sided or two-sided test can influence whether the null hypothesis is rejected. For example, consider a sample for which the t-statistic is 1.80. If a two-sided test is conducted at $\alpha = 0.05$, the p-value is

$$P(Z \leq -|t|) + P(Z \geq |t|) = 2P(Z \geq 1.80) = 0.072.$$

There is insufficient evidence to reject H_0, since $p > \alpha$. However, what if a one-sided test is conducted at $\alpha = 0.05$, with $H_A : \mu > \mu_0$? In this case, the p-value is

$$P(Z \geq t) = P(Z \geq 1.80) = 0.036.$$

The conclusion of the test is different: since $p < \alpha$, there is sufficient evidence to reject H_0 in favor of the alternative hypothesis. Figure 4.18 illustrates the different outcomes from the tests.

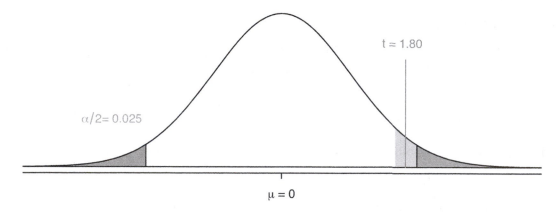

Figure 4.18: Under a one-sided test at significance level $\alpha = 0.05$, a t-statistic of 1.80 is within the rejection region (shaded light blue). However, it would not be within the rejection region under a two-sided test with $\alpha = 0.05$ (darker blue).

Two-sided tests are more "conservative" than one-sided tests; it is more difficult to reject the null hypothesis with a two-sided test. The p-value for a one-sided test is exactly half the p-value for a two-sided test conducted at the same significance level; as a result, it is easier for the p-value from a one-sided test to be smaller than α. Additionally, since the rejection region for a two-sided test is divided between two tails, a test statistic needs to be more extreme in order to fall within a rejection region. While the t-statistic of 1.80 is not within the two-sided rejection region, it is within the one-sided rejection region.[21]

---

[21] The two-sided rejection regions are bounded by -1.96 and 1.96, while the one-sided rejection region begins at 1.65.

For a fixed sample size, a one-tailed test will have a smaller probability of Type II error in comparison to a two-tailed test conducted at the same α level. In other words, with a one-sided test, it is easier to reject the null hypothesis if the alternative is actually true.

The choice of test should be driven by context, although it is not always clear which test is appropriate. Since it is easier to reject H_0 with the one-tailed test, it might be tempting to always use a one-tailed test when a significant result in a particular direction would be interesting or desirable.

However, it is important to consider the potential consequences of missing a significant difference in the untested direction. Generally, a two-sided test is the safest option, since it does not incorporate any existing biases about the direction of the results and can detect a difference at either the upper or lower tail. In the 1980s, researchers were interested in assessing a new set of drugs expected to be more effective at reducing heart arrhythmias than previously available therapies. They designed a one-sided clinical trial, convinced that the newer therapy would reduce mortality. The trial was quickly terminated due to an unanticipated effect of the drug; an independent review board found that the newer therapy was almost 4 times as likely to kill patients as a placebo! In a clinical research setting, it can be dangerous and even unethical to conduct a one-sided test under the belief that there is no possibility of patient harm from the drug intervention being tested.

One-sided tests are appropriate if the consequences of missing an effect in the untested direction are negligible, or if a large observed difference in the untested direction and a conclusion of "no difference" lead to the same decision. For example, suppose that a company has developed a drug to reduce blood pressure that is cheaper to produce than current options available on the market. If the drug is shown to be equally effective or more effective than an existing drug, the company will continue investing in it. Thus, they are only interested in testing the alternative hypothesis that the new drug is less effective than the existing drug, in which case, they will stop the project. It is acceptable to conduct a one-sided test in this situation since missing an effect in the other direction causes no harm.

The decision as to whether to use a one-sided or two-sided test must be made before data analysis begins, in order to avoid biasing conclusions based on the results of a hypothesis test. In particular, changing to a one-sided test after discovering that the results are "almost" significant for the two-sided test is unacceptable. Manipulating analyses in order to achieve low p-values leads to invalid results that are often not replicable. Unfortunately, this kind of "significance-chasing" has become widespread in published science, leading to concern that most current published research findings are false.

4.3.6   The informal use of p-values

Formal hypothesis tests are designed for settings where a decision or a claim about a hypothesis follows a test, such as in scientific publications where an investigator wishes to claim that an intervention changes an outcome. However, progress in science is usually based on a collection of studies or experiments, and it is often the case that the results of one study are used as a guide for the next study or experiment.

Sir Ronald Fisher was the first to propose using p-values as one of the statistical tools for evaluating an experiment. In his view, an outcome from an experiment that would only happen 1 in 20 times ($p = 0.05$) was worth investigating further. The use of p-values for formal decision making came later. While valuable, formal hypothesis testing can often be overused; not all significant results should lead to a definitive claim, but instead prompt further analysis.

The formal use of p-values is emphasized here because of its prominence in the scientific literature, and because the steps outlined are fundamental to the scientific method for empirical research: specify hypotheses, state in advance how strong the evidence should be to constitute sufficient evidence against the null, specify the method of analysis and compute the test statistic, draw a conclusion. These steps are designed to avoid the pitfall of choosing a hypothesis or method of analysis that is biased by the data and hence reaches a conclusion that may not be reproducible.

4.4 Notes

Confidence intervals and hypothesis testing are two of the central concepts in inference for a population based on a sample. The confidence interval shows a range of population parameter values consistent with the observed sample, and is often used to design additional studies. Hypothesis testing is a useful tool for evaluating the strength of the evidence against a working hypothesis according to a pre-specified standard for accepting or rejecting hypotheses.

The calculation of p-values and confidence intervals is relatively straightforward; given the necessary summary statistics, α, and confidence coefficients, finding any p-value or confidence interval simply involves a set of formulaic steps. However, the more difficult parts of any inference problem are the steps that do not involve any calculations. Specifying appropriate null and alternative hypotheses for a test relies on an understanding of the problem context and the scientific setting of the investigation. Similarly, a choice about a confidence coefficient for an interval relies on judgment as to balancing precision against the chance of possible error. It is also not necessarily obvious when a significance level other than $\alpha = 0.05$ should be applied. These choices represent the largest distinction between a true statistics problem as compared to a purely mathematical exercise.

Furthermore, in order to rely on the conclusions drawn from making inferences, it is necessary to consider factors such as study design, measurement quality, and the validity of any assumptions made. For example, is it valid to use the normal approximation to calculate p-values? In small to moderate sample sizes ($30 \leq n \leq 50$), it may not be clear that the normal model is accurate. It is even necessary to be cautious about the use and interpretation of the p-value. For example, an article published in *Nature* about the mis-use of p-values references a published study that showed people who meet their spouses online are more likely to have marital satisfaction, with p-value less than 0.001. However, statistical significance does not measure the importance or practical relevance of a result; in this case, the change in happiness moved from 5.48 to 5.64 on a 7-point scale. A p-value reported without context or other evidence is uninformative and potentially deceptive.

These nuanced issues cannot be adequately covered in any introduction to statistics. It is unrealistic to encourage students to use their own judgment with aspects of inference that even experienced investigators find challenging. At the same time, it would also be misleading to suggest that the choices are always clear-cut in practice. It seems best to offer some practical guidance for getting started:

- The default choice of α is 0.05; similarly, the default confidence coefficient for a confidence interval is 95%.

- Unless it is clear from the context of a problem that change in only one direction from the null hypothesis is of interest, the alternative hypothesis should be two-sided.

- The use of a standard normal distribution to calculate p-values is reasonable for sample sizes of 30 or more if the distribution of data are not strongly skewed and there are no large outliers. If there is skew or a few large outliers, sample sizes of 50 or more are usually sufficient.

- Pay attention to the context of a problem, particularly when formulating hypotheses and drawing conclusions.

The next chapters will discuss methods of inference in specific settings, such as comparing two groups. These settings expand on the concepts discussed in this chapter and offer additional opportunities to practice calculating tests and intervals, reading problems for context, and checking underlying assumptions behind methods of inference.

The labs for the chapter reinforce conceptual understanding of confidence intervals and hypothesis tests, and their link to sampling variability using the data from the YRBSS and NHANES. Both datasets are large enough to be viewed in an instructional setting as populations from which repeated samples can be drawn. They are useful platforms for illustrating the conceptual role of hypothetical repeated sampling in the properties of tests and intervals, a topic which many students find difficult. Students may find the last lab for this chapter (Lab 4) particularly helpful for understanding conceptual details of inference, such as the distinction between the significance level α and the p-value, and the definition of α as the Type I error rate.

4.5 Exercises

4.5.1 Variability in estimates

4.1 Egg coloration. The evolutionary role of variation in bird egg coloration remains mysterious to biologists. One hypothesis suggests that egg color may play a role in sexual selection. For example, perhaps healthier females are able to deposit more blue-green pigment into eggshells instead of using it themselves as an antioxidant. Researchers measured the blue-green chroma (BGC) of 70 different collared flycatcher nests in an area of the Czech Republic.

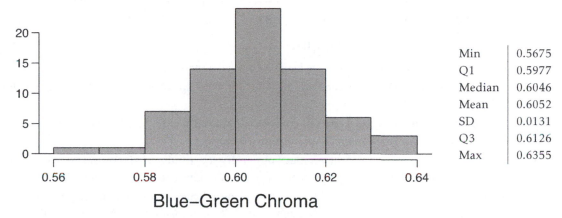

Min	0.5675
Q1	0.5977
Median	0.6046
Mean	0.6052
SD	0.0131
Q3	0.6126
Max	0.6355

(a) What is the point estimate for the average BGC of nests?

(b) What is the point estimate for the standard deviation of the BGC of eggs across nests?

(c) Would a nest with average BGC of 0.63 be considered unusually high? Explain your reasoning.

(d) Compute the standard error of the sample mean using the summary statistics.

4.2 Heights of adults. Researchers studying anthropometry collected body girth measurements and skeletal diameter measurements, as well as age, weight, height and gender, for 507 physically active individuals. The histogram below shows the sample distribution of heights in centimeters.[22]

Min	147.2
Q1	163.8
Median	170.3
Mean	171.1
SD	9.4
Q3	177.8
Max	198.1

(a) What is the point estimate for the average height of active individuals?

(b) What is the point estimate for the standard deviation of the heights of active individuals? What about the IQR?

(c) Is a person who is 1m 80cm (180 cm) tall considered unusually tall? And is a person who is 1m 55cm (155cm) considered unusually short? Explain your reasoning.

(d) The researchers take another random sample of physically active individuals. Would you expect the mean and the standard deviation of this new sample to be the ones given above? Explain your reasoning.

(e) The sample means obtained are point estimates for the mean height of all active individuals, if the sample of individuals is equivalent to a simple random sample. What measure is used to quantify the variability of such an estimate? Compute this quantity using the data from the original sample under the condition that the data are a simple random sample.

4.3 Hen eggs. The distribution of the number of eggs laid by a certain species of hen during their breeding period is on average, 35 eggs, with a standard deviation of 18.2. Suppose a group of researchers randomly samples 45 hens of this species, counts the number of eggs laid during their breeding period, and records the sample mean. They repeat this 1,000 times, and build a distribution of sample means.

(a) What is this distribution called?

(b) Would you expect the shape of this distribution to be symmetric, right skewed, or left skewed? Explain your reasoning.

(c) Calculate the variability of this distribution and state the appropriate term used to refer to this value.

(d) Suppose the researchers' budget is reduced and they are only able to collect random samples of 10 hens. The sample mean of the number of eggs is recorded, and we repeat this 1,000 times, and build a new distribution of sample means. How will the variability of this new distribution compare to the variability of the original distribution?

---

[22]G. Heinz et al. "Exploring relationships in body dimensions". In: *Journal of Statistics Education* 11.2 (2003).

4.5.2 Confidence intervals

4.4 Mental health, Part I. The 2010 General Social Survey asked the question: "For how many days during the past 30 days was your mental health, which includes stress, depression, and problems with emotions, not good?" Based on responses from 1,151 US residents, the survey reported a 95% confidence interval of 3.40 to 4.24 days in 2010.

(a) Interpret this interval in context of the data.

(b) What does "95% confident" mean? Explain in the context of the application.

(c) If a new survey were to be done with 500 Americans, would the standard error of the estimate be larger, smaller, or about the same? Assume the standard deviation has remained constant since 2010.

4.5 Relaxing after work, Part I. The 2010 General Social Survey asked the question: "After an average work day, about how many hours do you have to relax or pursue activities that you enjoy?" to a random sample of 1,155 Americans.[23] A 95% confidence interval for the mean number of hours spent relaxing or pursuing activities they enjoy is (1.38, 1.92).

(a) Interpret this interval in context of the data.

(b) Suppose another set of researchers reported a confidence interval with a larger margin of error based on the same sample of 1,155 Americans. How does their confidence level compare to the confidence level of the interval stated above?

(c) Suppose next year a new survey asking the same question is conducted, and this time the sample size is 2,500. Assuming that the population characteristics, with respect to how much time people spend relaxing after work, have not changed much within a year. How will the margin of error of the new 95% confidence interval compare to the margin of error of the interval stated above?

(d) Suppose the researchers think that 90% confidence interval would be more appropriate. Will this new interval be smaller or larger than the original 95% confidence interval? Justify your answer. (Assume that the standard deviation remains constant).

---

[23]National Opinion Research Center, General Social Survey, 2010.

4.6 Thanksgiving spending, Part I. The 2009 holiday retail season, which kicked off on November 27, 2009 (the day after Thanksgiving), had been marked by somewhat lower self-reported consumer spending than was seen during the comparable period in 2008. To get an estimate of consumer spending, 436 randomly sampled American adults were surveyed. Daily consumer spending for the six-day period after Thanksgiving, spanning the Black Friday weekend and Cyber Monday, averaged $84.71. A 95% confidence interval based on this sample is ($80.31, $89.11). Determine whether the following statements are true or false, and explain your reasoning.

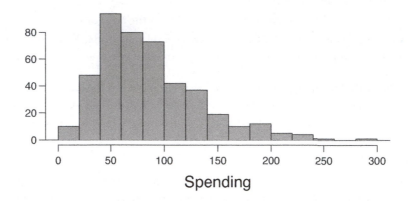

(a) We are 95% confident that the average spending of these 436 American adults is between $80.31 and $89.11.

(b) This confidence interval is not valid since the distribution of spending in the sample is right skewed.

(c) 95% of random samples have a sample mean between $80.31 and $89.11.

(d) We are 95% confident that the average spending of all American adults is between $80.31 and $89.11.

(e) A 90% confidence interval would be narrower than the 95% confidence interval.

(f) The margin of error is 4.4.

4.7 Waiting at an ER, Part I. A hospital administrator hoping to improve wait times decides to estimate the average emergency room waiting time at her hospital. She collects a simple random sample of 64 patients and determines the time (in minutes) between when they checked in to the ER until they were first seen by a doctor. A 95% confidence interval based on this sample is (128 minutes, 147 minutes), which is based on the normal model for the mean. Determine whether the following statements are true or false, and explain your reasoning.

(a) This confidence interval is not valid since we do not know if the population distribution of the ER wait times is nearly Normal.

(b) We are 95% confident that the average waiting time of these 64 emergency room patients is between 128 and 147 minutes.

(c) We are 95% confident that the average waiting time of all patients at this hospital's emergency room is between 128 and 147 minutes.

(d) 95% of random samples have a sample mean between 128 and 147 minutes.

(e) A 99% confidence interval would be narrower than the 95% confidence interval since we need to be more sure of our estimate.

(f) The margin of error is 9.5 and the sample mean is 137.5.

(g) Halving the margin of error of a 95% confidence interval requires doubling the sample size.

4.8 Age at first marriage, Part I. The National Survey of Family Growth conducted by the Centers for Disease Control gathers information on family life, marriage and divorce, pregnancy, infertility, use of contraception, and men's and women's health. One of the variables collected on this survey is the age at first marriage. The histogram below shows the distribution of ages at first marriage of 5,534 randomly sampled women between 2006 and 2010. The average age at first marriage among these women is 23.44 with a standard deviation of 4.72.[24]

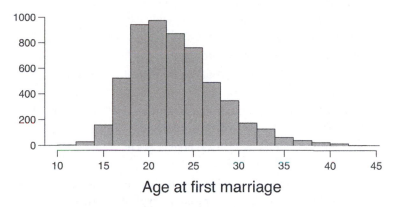

Estimate the average age at first marriage of women using a 95% confidence interval, and interpret this interval in context. Discuss any relevant assumptions.

4.9 Mental health, Part II. The General Social Survey (GSS) is a sociological survey used to collect data on demographic characteristics and attitudes of residents of the United States. The 2010 General Social Survey asked the question, "For how many days during the past 30 days was your mental health not good?" Based on responses from 1,151 US adults, the survey reported a 95% confidence interval of (3.40, 4.24) days. Assume that the sampled US adults are representative of all US adults.

(a) Identify each of the following statements as true or false. Justify your answers.

    i. The confidence interval of (3.40, 4.24) contains the mean days out of the past 30 days that U.S. adults experienced poor mental health.

    ii. There is a 95% chance that the mean days out of the past 30 days that U.S. adults experienced poor mental health is within the confidence interval (3.40, 4.24).

    iii. If we repeated this survey 1,000 times and constructed a 95% confidence interval each time, then approximately 950 of those intervals would contain the true mean days out of the past 30 days that U.S. adults experienced poor mental health.

    iv. The survey provides statistically significant evidence at the $\alpha = 0.05$ significance level that the mean days out of the past 30 days that U.S. adults experienced poor mental health is not 4.5 days.

    v. We can be 95% confident that the mean days out of the past 30 days that U.S. adults experienced poor mental health is 3.82 days.

    vi. We can be 95% confident that the interval (3.40, 4.24) days contains the mean days out of the past 30 days that the sampled adults experienced poor mental health.

(b) Would you expect the 90% confidence interval to be larger or smaller than the 95% confidence interval? Explain your reasoning.

(c) Calculate the 90% confidence interval.

---

[24]Centers for Disease Control and Prevention, National Survey of Family Growth, 2010.

4.10 Leisure time, Part III. In 2010, the General Social Survey collected responses from 1,154 US residents. The survey is conducted face-to-face with an in-person interview of a randomly selected sample of adults. One of the questions on the survey is "After an average workday, about how many hours do you have to relax or pursue activities that you enjoy?" A 95% confidence interval from the 2010 GSS survey for the collected answers is 3.53 to 3.83 hours. Identify each of the following statements as true or false. Explain your answers.

(a) If the researchers wanted to report a confidence interval with a smaller margin of error based on the same sample of 1,154 Americans, the confidence interval would be larger.

(b) We can be 95% confident that the interval (3.53, 3.83) hours contains the mean hours that the sampled adults have for leisure time after an average workday.

(c) The confidence interval of (3.53, 3.83) hours contains the mean hours that U.S. adults have for leisure time after an average workday.

(d) The survey provides statistically significant evidence at the $\alpha = 0.05$ significance level that the mean hours U.S. adults have for leisure time after the average workday is 3.6 hours.

(e) There is a 5% chance that the interval (3.53, 3.83) hours does not contain the mean hours that U.S. adults have for leisure time after an average workday.

(f) The interval (3.53, 3.83) hours provides evidence at the $\alpha = 0.05$ significance level that U.S. adults, on average, have fewer than 3.9 hours of leisure time after a typical workday.

4.5.3 Hypothesis testing

4.11 Identify hypotheses, Part I. Write the null and alternative hypotheses in words and then symbols for each of the following situations.

(a) New York is known as "the city that never sleeps". A random sample of 25 New Yorkers were asked how much sleep they get per night. Do these data provide convincing evidence that New Yorkers on average sleep less than 8 hours a night?

(b) Employers at a firm are worried about the effect of March Madness, a basketball championship held each spring in the US, on employee productivity. They estimate that on a regular business day employees spend on average 15 minutes of company time checking personal email, making personal phone calls, etc. They also collect data on how much company time employees spend on such non- business activities during March Madness. They want to determine if these data provide convincing evidence that employee productivity decreases during March Madness.

4.12 Identify hypotheses, Part II. Write the null and alternative hypotheses in words and using symbols for each of the following situations.

(a) Since 2008, chain restaurants in California have been required to display calorie counts of each menu item. Prior to menus displaying calorie counts, the average calorie intake of diners at a restaurant was 1100 calories. After calorie counts started to be displayed on menus, a nutritionist collected data on the number of calories consumed at this restaurant from a random sample of diners. Do these data provide convincing evidence of a difference in the average calorie intake of a diners at this restaurant?

(b) Based on the performance of those who took the GRE exam between July 1, 2004 and June 30, 2007, the average Verbal Reasoning score was calculated to be 462. In 2011 the average verbal score was slightly higher. Do these data provide convincing evidence that the average GRE Verbal Reasoning score has changed since 2004?

4.13 Online communication. A study suggests that the average college student spends 10 hours per week communicating with others online. You believe that this is an underestimate and decide to collect your own sample for a hypothesis test. You randomly sample 60 students from your dorm and find that on average they spent 13.5 hours a week communicating with others online. A friend of yours, who offers to help you with the hypothesis test, comes up with the following set of hypotheses. Indicate any errors you see.

$$H_0 : \bar{x} < 10 \ hours$$
$$H_A : \bar{x} > 13.5 \ hours$$

4.14 Age at first marriage, Part II. Exercise 4.8 presents the results of a 2006 - 2010 survey showing that the average age of women at first marriage is 23.44. Suppose a social scientist believes that this value has increased in 2012, but she would also be interested if she found a decrease. Below is how she set up her hypotheses. Indicate any errors you see.

$$H_0 : \bar{x} = 23.44 \ years$$
$$H_A : \bar{x} > 23.44 \ years$$

4.15 Waiting at an ER, Part II. Exercise 4.7 provides a 95% confidence interval for the mean waiting time at an emergency room (ER) of (128 minutes, 147 minutes). Answer the following questions based on this interval.

(a) A local newspaper claims that the average waiting time at this ER exceeds 3 hours. Is this claim supported by the confidence interval? Explain your reasoning.

(b) The Dean of Medicine at this hospital claims the average wait time is 2.2 hours. Is this claim supported by the confidence interval? Explain your reasoning.

(c) Without actually calculating the interval, determine if the claim of the Dean from part (b) would be supported based on a 99% confidence interval?

4.16 Gifted children, Part I. Researchers investigating characteristics of gifted children collected data from schools in a large city on a random sample of thirty-six children who were identified as gifted children soon after they reached the age of four. The following histogram shows the distribution of the ages (in months) at which these children first counted to 10 successfully. Also provided are some sample statistics.[25]

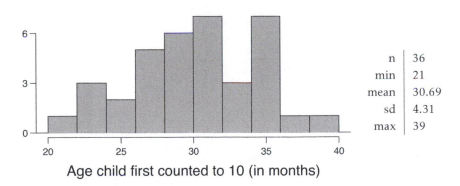

n	36
min	21
mean	30.69
sd	4.31
max	39

Age child first counted to 10 (in months)

(a) Are conditions for inference satisfied?

(b) Suppose an online survey reports that children first count to 10 successfully when they are 32 months old, on average. Perform a hypothesis test to evaluate if these data provide convincing evidence that the average age at which gifted children first count to 10 successfully is less than the general average of 32 months. Use a significance level of 0.10.

(c) Interpret the p-value in context of the hypothesis test and the data.

(d) Calculate a 90% confidence interval for the average age at which gifted children first count to 10 successfully.

(e) Do your results from the hypothesis test and the confidence interval agree? Explain.

---

[25]F.A. Graybill and H.K. Iyer. *Regression Analysis: Concepts and Applications.* Duxbury Press, 1994, pp. 511–516.

4.17 Nutrition labels. The nutrition label on a bag of potato chips says that a one ounce (28 gram) serving of potato chips has 130 calories and contains ten grams of fat, with three grams of saturated fat. A random sample of 35 bags yielded a sample mean of 134 calories with a standard deviation of 17 calories. Is there evidence that the nutrition label does not provide an accurate measure of calories in the bags of potato chips? We have verified the independence, sample size, and skew conditions are satisfied.

4.18 Waiting at an ER, Part III. The hospital administrator mentioned in Exercise 4.7 randomly selected 64 patients and measured the time (in minutes) between when they checked in to the ER and the time they were first seen by a doctor. The average time is 137.5 minutes and the standard deviation is 39 minutes. She is getting grief from her supervisor on the basis that the wait times in the ER has increased greatly from last year's average of 127 minutes. However, she claims that the increase is probably just due to chance.

(a) Calculate a 95% confidence interval. Is the change in wait times statistically significant at the $\alpha = 0.05$ level?

(b) Would the conclusion in part (a) change if the significance level were changed to $\alpha = 0.01$?

(c) Is the supervisor justified in criticizing the hospital administrator regarding the change in ER wait times? How might you present an argument in favor of the administrator?

4.19 Birth weights. Suppose an investigator takes a random sample of $n = 50$ birth weights from several teaching hospitals located in an inner-city neighborhood. In her random sample, the sample mean $\bar{x}$ is 3,150 grams and the standard deviation is 250 grams.

(a) Calculate a 95% confidence interval for the population mean birth weight in these hospitals.

(b) The typical weight of a baby at birth for the US population is 3,250 grams. The investigator suspects that the birth weights of babies in these teaching hospitals is different than 3,250 grams, but she is not sure if it is smaller (from malnutrition) or larger (because of obesity prevalence in mothers giving birth at these hospitals). Carry out the hypothesis test that she would conduct.

4.20 Gifted children, Part II. Exercise 4.16 describes a study on gifted children. In this study, along with variables on the children, the researchers also collected data on the mother's and father's IQ of the 36 randomly sampled gifted children. The histogram below shows the distribution of mother's IQ. Also provided are some sample statistics.

n	36
min	101
mean	118.2
sd	6.5
max	131

(a) Perform a hypothesis test to evaluate if these data provide convincing evidence that the average IQ of mothers of gifted children is different than the average IQ for the population at large, which is 100. Use a significance level of 0.10.

(b) Calculate a 90% confidence interval for the average IQ of mothers of gifted children.

(c) Do your results from the hypothesis test and the confidence interval agree? Explain.

4.21 Testing for fibromyalgia. A patient named Diana was diagnosed with fibromyalgia, a long-term syndrome of body pain, and was prescribed anti-depressants. Being the skeptic that she is, Diana didn't initially believe that anti-depressants would help her symptoms. However after a couple months of being on the medication she decides that the anti-depressants are working, because she feels like her symptoms are in fact getting better.

(a) Write the hypotheses in words for Diana's skeptical position when she started taking the anti-depressants.

(b) What is a Type 1 Error in this context?

(c) What is a Type 2 Error in this context?

4.22 Testing for food safety. A food safety inspector is called upon to investigate a restaurant with a few customer reports of poor sanitation practices. The food safety inspector uses a hypothesis testing framework to evaluate whether regulations are not being met. If he decides the restaurant is in gross violation, its license to serve food will be revoked.

(a) Write the hypotheses in words.

(b) What is a Type 1 Error in this context?

(c) What is a Type 2 Error in this context?

(d) Which error is more problematic for the restaurant owner? Why?

(e) Which error is more problematic for the diners? Why?

(f) As a diner, would you prefer that the food safety inspector requires strong evidence or very strong evidence of health concerns before revoking a restaurant's license? Explain your reasoning.

4.23 Which is higher? In each part below, there is a value of interest and two scenarios (I and II). For each part, report if the value of interest is larger under scenario I, scenario II, or whether the value is equal under the scenarios.

(a) The standard error of $\bar{x}$ when $s = 120$ and (I) $n = 25$ or (II) $n = 125$.

(b) The margin of error of a confidence interval when the confidence level is (I) 90% or (II) 80%.

(c) The p-value for a Z-statistic of 2.5 when (I) $n = 500$ or (II) $n = 1000$.

(d) The probability of making a Type 2 Error when the alternative hypothesis is true and the significance level is (I) 0.05 or (II) 0.10.

4.24 True or false. Determine if the following statements are true or false, and explain your reasoning. If false, state how it could be corrected.

(a) If a given value (for example, the null hypothesized value of a parameter) is within a 95% confidence interval, it will also be within a 99% confidence interval.

(b) Decreasing the significance level (α) will increase the probability of making a Type 1 Error.

(c) Suppose the null hypothesis is $\mu = 5$ and we fail to reject H_0. Under this scenario, the true population mean is 5.

(d) If the alternative hypothesis is true, then the probability of making a Type 2 Error and the power of a test add up to 1.

(e) With large sample sizes, even small differences between the null value and the true value of the parameter, a difference often called the effect size , will be identified as statistically significant.

Chapter 5

Inference for numerical data

Chapter 4 introduced some primary tools of statistical inference—point estimates, interval estimates, and hypothesis tests. This chapter discusses settings where these tools are often used, including the analysis of paired observations and the comparison of two or more independent groups. The chapter also covers the important topic of estimating an appropriate sample size when a study is being designed. The chapter starts with introducing a new distribution, the t-distribution, which can be used for small sample sizes.

For labs, slides, and other resources, please visit
www.openintro.org/book/biostat

5.1   Single-sample inference with the t-distribution

The tools studied in Chapter 4 all made use of the t-statistic from a sample mean,

$$t = \frac{\bar{x} - \mu}{s/\sqrt{n}},$$

where the parameter μ is a population mean, $\bar{x}$ and s are the sample mean and standard deviation, and n is the sample size. Tests and confidence intervals were restricted to samples of at least 30 independent observations from a population where there was no evidence of strong skewness. This allowed for the Central Limit Theorem to be applied, justifying use of the normal distribution to calculate probabilities associated with the t-statistic.

In sample sizes smaller than 30, if the data are approximately symmetric and there are no large outliers, the t-statistic has what is called a t-distribution. When the normal distribution is used as the sampling distribution of the t-statistic, s is essentially being treated as a good replacement for the unknown population standard deviation σ. However, the sample standard deviation s, as an estimate of σ, has its own inherent variability like $\bar{x}$. The t density function adjusts for the variability in s by having more probability in the left and right tails than the normal distribution.

5.1.1   The t-distribution

Figure 5.1 shows a t-distribution and normal distribution. Like the standard normal distribution, the t-distribution is unimodal and symmetric about zero. However, the tails of a t-distribution are thicker than for the normal, so observations are more likely to fall beyond two standard deviations from the mean than under the normal distribution.[1] While the estimate of the standard error will be less accurate with smaller sample sizes, the thick tails of the t-distribution correct for the variability in s.

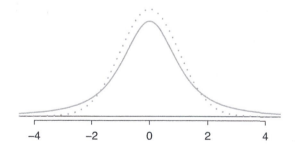

Figure 5.1: Comparison of a t-distribution (solid line) and a normal distribution (dotted line).

The t-distribution can be described as a family of symmetric distributions with a single parameter: degrees of freedom, which equals $n - 1$. Several t-distributions are shown in Figure 5.2. When there are more degrees of freedom, the t-distribution looks very much like the standard normal distribution. With degrees of freedom of 30 or more, the t-distribution is nearly indistinguishable from the normal distribution. Since the t-statistics in Chapter 4 were associated with sample sizes of at least 30, the degrees of freedom for the corresponding t-distributions were large enough to justify use of the normal distribution to calculate probabilities.

---

[1]The standard deviation of the t-distribution is actually a little more than 1. However, it is useful to think of the t-distribution as having a standard deviation of 1 in the context of using it to conduct inference.

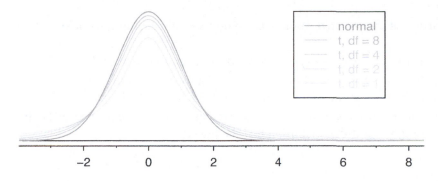

Figure 5.2: The larger the degrees of freedom, the more closely the *t*-distribution resembles the standard normal model.

DEGREES OF FREEDOM (DF)

The degrees of freedom characterize the shape of the *t*-distribution. The larger the degrees of freedom, the more closely the distribution approximates the normal model.

Probabilities for the *t*-distribution can be calculated either by using distribution tables or using statistical software. The use of software has become the preferred method because it is more accurate, allows for complete flexibility in the choice of *t*-values on the horizontal axis, and is not limited to a small range of degrees of freedom. The remainder of this section illustrates the use of a **t-table**, partially shown in Figure 5.3, in place of the normal probability table. A larger *t*-table is in Appendix B.2 on page 466. The R labs illustrate the use of software to calculate probabilities for the *t*-distribution. Readers intending to use software can skip to the next section.

one tail	0.100	0.050	0.025	0.010	0.005
two tails	0.200	0.100	0.050	0.020	0.010
df 1	3.08	6.31	12.71	31.82	63.66
2	1.89	2.92	4.30	6.96	9.92
3	1.64	2.35	3.18	4.54	5.84
⋮	⋮	⋮	⋮	⋮	⋮
17	1.33	1.74	2.11	2.57	2.90
18	**1.33**	**1.73**	**2.10**	**2.55**	**2.88**
19	1.33	1.73	2.09	2.54	2.86
20	1.33	1.72	2.09	2.53	2.85
⋮	⋮	⋮	⋮	⋮	⋮
400	1.28	1.65	1.97	2.34	2.59
500	1.28	1.65	1.96	2.33	2.59
∞	1.28	1.64	1.96	2.33	2.58

Figure 5.3: An abbreviated look at the *t*-table. Each row represents a different *t*-distribution. The columns describe the cutoffs for specific tail areas. The row with $df = 18$ has been **highlighted**.

Each row in the *t*-table represents a *t*-distribution with different degrees of freedom. The columns correspond to tail probabilities. For instance, for a *t*-distribution with $df = 18$, row 18 is used (highlighted in Figure 5.3). The value in this row that identifies the cutoff for an upper tail of 5% is found in the column where *one tail* is 0.050. This cutoff is 1.73. The cutoff for the lower 5% is -1.73; just like the normal distribution, all *t*-distributions are symmetric. If the area in each tail is 5%, then the area in two tails is 10%; thus, this column can also be described as the column where *two tails* is 0.100.

EXAMPLE 5.1

What proportion of the *t*-distribution with 18 degrees of freedom falls below -2.10?

Just like for a normal probability problem, it is advisable to start by drawing the distribution and shading the area below -2.10, as shown in Figure 5.4. From the table, identify the column containing the absolute value of -2.10; it is the third column. Since this is just the probability in one tail, examine the top line of the table; a one tail area for a value in the third column corresponds to 0.025. About 2.5% of the distribution falls below -2.10.

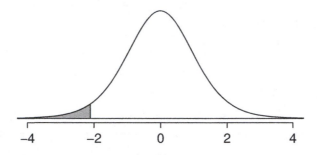

Figure 5.4: The *t*-distribution with 18 degrees of freedom. The area below -2.10 has been shaded.

EXAMPLE 5.2

A *t*-distribution with 20 degrees of freedom is shown in the left panel of Figure 5.5. Estimate the proportion of the distribution falling above 1.65 and below -1.65.

Identify the row in the *t*-table using the degrees of freedom: $df - 20$. Then, look for 1.65; the value is not listed, and falls between the first and second columns. Since these values bound 1.65, their tail areas will bound the tail area corresponding to 1.65. The two tail area of the first and second columns is between 0.100 and 0.200. Thus, between 10% and 20% of the distribution is more than 1.65 standard deviations from the mean. The precise area can be calculated using statistical software: 0.1146.

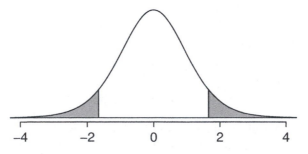

Figure 5.5: The *t*-distribution with 20 degrees of freedom, with the area further than 1.65 away from 0 shaded.

5.1.2 Using the t-distribution for tests and confidence intervals for a population mean

Chapter 4 provided formulas for tests and confidence intervals for population means in random samples large enough for the t-statistic to have a nearly normal distribution. In samples smaller than 30 from approximately symmetric distributions without large outliers, the t-statistic has a t-distribution with degrees of freedom equal to $n - 1$. Just like inference in larger samples, inference using the t-distribution also requires that the observations in the sample be independent. Random samples from very large populations always produce independent observations; in smaller populations, observations will be approximately independent as long as the size of the sample is no larger than 10% of the population.

Formulas for tests and intervals using the t–distribution are very similar to those using the normal distribution. For a sample of size n with sample mean $\overline{x}$ and standard deviation s, two-sided confidence intervals with confidence coefficient $100(1 - \alpha)\%$ have the form

$$\overline{x} \pm t_{\mathrm{df}}^{\star} \times \mathrm{SE},$$

where SE is the standard error of the sample mean ($s/\sqrt{n}$) and $t_{\mathrm{df}}^{\star}$ is the point on a t-distribution with $n - 1$ degrees of freedom and area $(1 - \alpha/2)$ to its left.

A one-sided interval with the same confidence coefficient will have the form

$$\overline{x} + t_{\mathrm{df}}^{\star} \times \mathrm{SE} \text{ (one-sided upper confidence interval), or}$$
$$\overline{x} - t_{\mathrm{df}}^{\star} \times \mathrm{SE} \text{ (one-sided lower confidence interval),}$$

except that in this case $t_{\mathrm{df}}^{\star}$ is the point on a t-distribution with $n - 1$ degrees of freedom and area $(1 - \alpha)$ to its left.

With the ability to conveniently calculate $t^{\star}$ for any sample size or associated α via computing software, the t-distribution can be used by default over the normal distribution. The rule of thumb that $n > 30$ qualifies as a large enough sample size to use the normal distribution dates back to when it was necessary to rely on distribution tables.

EXAMPLE 5.3

Dolphins are at the top of the oceanic food chain; as a consequence, dangerous substances such as mercury tend to be present in their organs and muscles at high concentrations. In areas where dolphins are regularly consumed, it is important to monitor dolphin mercury levels. This example uses data from a random sample of 19 Risso's dolphins from the Taiji area in Japan.[2] Calculate the 95% confidence interval for average mercury content in Risso's dolphins from the Taiji area using the data in Figure 5.6.

The observations are a simple random sample consisting of less than 10% of the population, so independence of the observations is reasonable. The summary statistics in Figure 5.6 do not suggest any skew or outliers; all observations are within 2.5 standard deviations of the mean. Based on this evidence, the approximate normality assumption seems reasonable.

Use the t-distribution to calculate the confidence interval:

$$\bar{x} \pm t_{df}^\star \times SE = \bar{x} \pm t_{18}^\star \times s/\sqrt{n}$$
$$= 4.4 \pm 2.10 \times 2.3/\sqrt{19}$$
$$= (3.29, 5.51) \ \mu g/\text{wet g}.$$

The $t^\star$ point can be read from the t-table on page 239, in the column with area totaling 0.05 in the two tails (third column) and the row with 18 degrees of freedom. Based on these data, one can be 95% confident the average mercury content of muscles in Risso's dolphins is between 3.29 and 5.51 μg/wet gram.

Alternatively, the $t^\star$ point can be calculated in R with the function qt, which returns a value of 2.1009.

n	$\bar{x}$	s	minimum	maximum
19	4.4	2.3	1.7	9.2

Figure 5.6: Summary of mercury content in the muscle of 19 Risso's dolphins from the Taiji area. Measurements are in μg/wet g (micrograms of mercury per wet gram of muscle).

---

[2]Taiji is a significant source of dolphin and whale meat in Japan. Thousands of dolphins pass through the Taiji area annually; assume that these 19 dolphins represent a simple random sample. Data reference: Endo T and Haraguchi K. 2009. High mercury levels in hair samples from residents of Taiji, a Japanese whaling town. Marine Pollution Bulletin 60(5):743-747.

GUIDED PRACTICE 5.4

The FDA's webpage provides some data on mercury content of various fish species.[3] From a sample of 15 white croaker (Pacific), a sample mean and standard deviation were computed as 0.287 and 0.069 ppm (parts per million), respectively. The 15 observations ranged from 0.18 to 0.41 ppm. Assume that these observations are independent. Based on summary statistics, does the normality assumption seem reasonable? If so, calculate a 90% confidence interval for the average mercury content of white croaker (Pacific).[4]

EXAMPLE 5.5

According to the EPA, regulatory action should be taken if fish species are found to have a mercury level of 0.5 ppm or higher. Conduct a formal significance test to evaluate whether the average mercury content of croaker white fish (Pacific) is different from 0.50 ppm. Use $\alpha = 0.05$.

The FDA regulatory guideline is a 'one-sided' statement; fish should not be eaten if the mercury level is larger than a certain value. However, without prior information on whether the mercury in this species tends to be high or low, it is best to do a two-sided test.

State the hypotheses: $H_0 : \mu = 0.5$ vs $H_A : \mu \neq 0.5$. Let $\alpha = 0.05$.

Calculate the t-statistic:

$$t = \frac{\bar{x} - \mu_0}{SE} = \frac{0.287 - 0.50}{0.069/\sqrt{15}} = -11.96$$

The probability that the absolute value of a t-statistic with 14 df is smaller than -11.96 is smaller than 0.01. Thus, $p < 0.01$. There is evidence to suggest at the $\alpha = 0.05$ significance level that the average mercury content of this fish species is lower than 0.50 ppm, since $\bar{x}$ is less than 0.50.

---

[3] www.fda.gov/food/foodborneillnesscontaminants/metals/ucm115644.htm

[4] There are no obvious outliers; all observations are within 2 standard deviations of the mean. If there is skew, it is not evident. There are no red flags for the normal model based on this (limited) information. $\bar{x} \pm t^\star_{14} \times SE \rightarrow 0.287 \pm 1.76 \times 0.0178 \rightarrow (0.256, 0.318)$. We are 90% confident that the average mercury content of croaker white fish (Pacific) is between 0.256 and 0.318 ppm.

5.2 Two-sample test for paired data

In the 2000 Olympics, was the use of a new wetsuit design responsible for an observed increase in swim velocities? In a study designed to investigate this question, twelve competitive swimmers swam 1500 meters at maximal speed, once wearing a wetsuit and once wearing a regular swimsuit.[5] The order of wetsuit versus swimsuit was randomized for each of the 12 swimmers. Figure 5.7 shows the average velocity recorded for each swimmer, measured in meters per second (m/s).[6]

	swimmer.number	wet.suit.velocity	swim.suit.velocity	velocity.diff
1	1	1.57	1.49	0.08
2	2	1.47	1.37	0.10
3	3	1.42	1.35	0.07
4	4	1.35	1.27	0.08
5	5	1.22	1.12	0.10
6	6	1.75	1.64	0.11
7	7	1.64	1.59	0.05
8	8	1.57	1.52	0.05
9	9	1.56	1.50	0.06
10	10	1.53	1.45	0.08
11	11	1.49	1.44	0.05
12	12	1.51	1.41	0.10

Figure 5.7: Paired Swim Suit Data

The swimsuit velocity data are an example of **paired data**, in which two sets of observations are uniquely paired so that an observation in one set matches an observation in the other; in this case, each swimmer has two measured velocities, one with a wetsuit and one with a swimsuit. A natural measure of the effect of the wetsuit on swim velocity is the difference between the measured maximum velocities (velocity.diff = wet.suit.velocity - swim.suit.velocity). Even though there are two measurements per swimmer, using the difference in velocities as the variable of interest allows for the problem to be approached like those in Section 5.1. Although it was not explicitly noted, the data used in Section 4.3.1 were paired; each respondent had both an actual and desired weight.

Suppose the parameter δ is the population average of the difference in maximum velocities during a 1500m swim if all competitive swimmers recorded swim velocities with each suit type. A hypothesis test can then be conducted with the null hypothesis that the mean population difference in swim velocities between suit types equals 0 (i.e., there is no difference in population average swim velocities), $H_0 : \delta = 0$, against the alternative that the difference is non-zero, $H_A : \delta \neq 0$.

STATING HYPOTHESES FOR PAIRED DATA

When testing a hypothesis about paired data, compare the groups by testing whether the population mean of the differences between the groups equals 0.

- For a two-sided test, $H_0 : \delta = 0$; $H_A : \delta \neq 0$.

- For a one-sided test, either $H_0 : \delta = 0$; $H_A : \delta > 0$ or $H_0 : \delta = 0$; $H_A : \delta < 0$.

---

[5]De Lucas et. al, The effects of wetsuits on physiological and biomechanical indices during swimming. *Journal of Science and Medicine in Sport*, 2000; 3(1): 1-8

[6]The data are available as swim in the oibiostat R package. The data are also used in Lock et. al *Statistics, Unlocking the Power of Data*, Wiley, 2013.

Some important assumptions are being made. First, it is assumed that the data are a random sample from the population. While the observations are likely independent, it is more difficult to justify that this sample of 12 swimmers is randomly drawn from the entire population of competitive swimmers. Nevertheless, it is often assumed in problems such as these that the participants are reasonably representative of competitive swimmers. Second, it is assumed that the population of differences is normally distributed. This is a small sample, one in which normality would be difficult to confirm. The dot plot for the difference in velocities in Figure 5.8 shows approximate symmetry.

Difference in Swim Velocities (m/s)

Figure 5.8: A dot plot of differences in swim velocities.

Let $\bar{x}_{\text{diff}}$ denote the sample average of the differences in maximum velocity, s_{diff} the sample standard deviation of the differences, and n the number of pairs in the dataset. The t-statistic used to test H_0 vs. H_A is:

$$\frac{\bar{x}_{\text{diff}} - \delta_0}{s_{\text{diff}}/\sqrt{n}},$$

where in this case $\delta_0 = 0$.[7]

EXAMPLE 5.6

Using the data in Figure 5.7, conduct a two-sided hypothesis test at $\alpha = 0.05$ to assess whether there is evidence to suggest that wetsuits have an effect on swim velocities during a 1500m swim.

The hypotheses are $H_0 : \delta = 0$ and $H_A : \delta \neq 0$. Let $\alpha = 0.05$.

Calculate the t-statistic:

$$t = \frac{\bar{x}_{\text{diff}} - \delta_0}{s_{\text{diff}}/\sqrt{n}} = \frac{0.078 - 0}{0.022/\sqrt{12}} = 12.32$$

The two-sided p-value is

$$p = P(T < -12.32) + P(T > 12.32),$$

where t has a t-distribution with $n - 1 = 11$ degrees of freedom. The t-table shows that $p < 0.01$. Software can be used to show that $p = 8.9 \times 10^{-8}$, a very small value indeed.

The data support the claim that the wetsuits changed swim velocity in a 1500m swim. The observed average increase of 0.078 m/s is significantly different than the null hypothesis of no change, and suggests that swim velocities are higher when swimmers wear wetsuits as opposed to swimsuits.

Calculating confidence intervals for paired data is also based on the differences between the values in each pair; the same approach as for single-sample data can be applied on the differences. For example, a two-sided 95% confidence interval for paired data has the form:

$$\left(\bar{x}_{\text{diff}} - t_{df}^{\star} \times \frac{s_{\text{diff}}}{\sqrt{n}}, \ \bar{x}_{\text{diff}} + t_{df}^{\star} \times \frac{s_{\text{diff}}}{\sqrt{n}} \right),$$

where $t^{\star}$ is the point on a t-distribution with $df = n - 1$ for n pairs, with area 0.025 to its right.

---

[7]This value is specified by the null hypothesis of no difference.

GUIDED PRACTICE 5.7

Using the data in Figure 5.7, calculate a 95% confidence interval for the average difference in swim velocities during a 1500m swim. Is the interval consistent with the results of the hypothesis test?[8]

The general approach when analyzing paired data is to first calculate the differences between the values in each pair, then use those differences in methods for confidence intervals and tests for a single sample. Any conclusion from an analysis should be stated in terms of the original paired measurements.

---

[8]Use the values of $\bar{x}_{\text{diff}}$ and s_{diff} as calculated previously: 0.078 and 0.022. The $t^\star$ value of 2.20 has $df = 11$ and 0.025 area to the right. The confidence interval is $(0.078 \pm \frac{0.022}{\sqrt{12}}) \rightarrow (0.064, 0.091)$ m/s. With 95% confidence, δ lies between 0.064 m/s and 0.09 m/s. The interval does not include 0 (no change), which is consistent with the result of the hypothesis test.

5.3 Two-sample test for independent data

Does treatment using embryonic stem cells (ESCs) help improve heart function following a heart attack? New and potentially risky treatments are sometimes tested in animals before studies in humans are conducted. In a 2005 paper in *Lancet*, Menard, et al. describe an experiment in which 18 sheep with induced heart attacks were randomly assigned to receive cell transplants containing either ESCs or inert material.[9] Various measures of cardiac function were measured 1 month after the transplant.

This design is typical of an intervention study. The analysis of such an experiment is an example of drawing inference about the difference in two population means, $\mu_1 - \mu_2$, when the data are independent, i.e., not paired. The point estimate of the difference, $\bar{x}_1 - \bar{x}_2$, is used to calculate a t-statistic that is the basis of confidence intervals and tests.

5.3.1 Confidence interval for a difference of means

Figure 5.9 contains summary statistics for the 18 sheep.[10] Percent change in heart pumping capacity was measured for each sheep. A positive value corresponds to increased pumping capacity, which generally suggests a stronger recovery from the heart attack. Is there evidence for a potential treatment effect of administering stem cells?

	n	$\bar{x}$	s
ESCs	9	3.50	5.17
control	9	-4.33	2.76

Figure 5.9: Summary statistics of the embryonic stem cell study.

[9]Menard C, et al., Transplantation of cardiac-committed mouse embryonic stem cells to infarcted sheep myocardium: a preclinical 2005; 366:1005-12, doi https://doi.org/10.1016/S0140-6736(05)67380-1
[10]The data are accessible as the dataset stem.cells in the openintro R package.

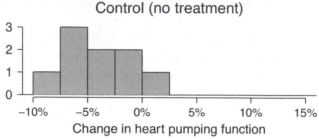

Figure 5.10: Histograms for both the embryonic stem cell group and the control group. Higher values are associated with greater improvement.

Figure 5.10 shows that the distributions of percent change do not have any prominent outliers, which would indicate a deviation from normality; this suggests that each sample mean can be modeled using a t-distribution. Additionally, the sheep in the study are independent of each other, and the sheep between groups are also independent. Thus, the t-distribution can be used to model the difference of the two sample means.

USING THE t-DISTRIBUTION FOR A DIFFERENCE IN MEANS

The t-distribution can be used for inference when working with the standardized difference of two means if (1) each sample meets the conditions for using the t-distribution and (2) the samples are independent.

A confidence interval for a difference of two means has the same basic structure as previously discussed confidence intervals:

$$(\bar{x}_1 - \bar{x}_2) \pm t_{df}^{\star} \times \text{SE} .$$

The following formula is used to calculate the standard error of $\bar{x}_1 - \bar{x}_2$. Since σ is typically unknown, the standard error is estimated by using s in place of σ.

$$SE_{\bar{x}_1 - \bar{x}_2} = \sqrt{\frac{\sigma_1^2}{n_1} + \frac{\sigma_2^2}{n_2}} \approx \sqrt{\frac{s_1^2}{n_1} + \frac{s_2^2}{n_2}} .$$

In this setting, the t-distribution has a somewhat complicated formula for the degrees of freedom that is usually calculated with software.[11] An alternative approach uses the smaller of $n_1 - 1$ and $n_2 - 1$ as the degrees of freedom.[12]

---

[11] See Section 5.6 for the formula.

[12] This technique for degrees of freedom is conservative with respect to a Type 1 Error; it is more difficult to reject the null hypothesis using this approach for degrees of freedom.

DISTRIBUTION OF A DIFFERENCE OF SAMPLE MEANS

The sample difference of two means, $\bar{x}_1 - \bar{x}_2$, can be modeled using the t-distribution and the standard error

$$SE_{\bar{x}_1 - \bar{x}_2} = \sqrt{\frac{s_1^2}{n_1} + \frac{s_2^2}{n_2}} \tag{5.8}$$

when each sample mean can itself be modeled using a t-distribution and the samples are independent. To calculate the degrees of freedom without using software, use the smaller of $n_1 - 1$ and $n_2 - 1$.

EXAMPLE 5.9

Calculate and interpret a 95% confidence interval for the effect of ESCs on the change in heart pumping capacity of sheep following a heart attack.

The point estimate for the difference is $\bar{x}_1 - \bar{x}_2 = \bar{x}_{\text{esc}} - \bar{x}_{\text{control}} = 7.83$.

The standard error is:

$$\sqrt{\frac{s_1^2}{n_1} + \frac{s_2^2}{n_2}} = \sqrt{\frac{5.17^2}{9} + \frac{2.76^2}{9}} = 1.95.$$

Since $n_1 = n_2 = 9$, use $df = 8$; $t_8^\star = 2.31$ for a 95% confidence interval. Alternatively, computer software can provide more accurate values: $df = 12.225, t^\star = 2.174$.

The confidence interval is given by:

$$(\bar{x}_1 - \bar{x}_2) \pm t_{df}^\star \times SE \quad \rightarrow \quad 7.83 \pm 2.31 \times 1.95 \quad \rightarrow \quad (3.38, 12.38).$$

With 95% confidence, the average amount that ESCs improve heart pumping capacity lies between 3.38% to 12.38%.[13] The data provide evidence for a treatment effect of administering stem cells.

5.3.2 Hypothesis tests for a difference in means

Is there evidence that newborns from mothers who smoke have a different average birth weight than newborns from mothers who do not smoke? The dataset births contains data from a random sample of 150 cases of mothers and their newborns in North Carolina over a year; there are 50 cases in the smoking group and 100 cases in the nonsmoking group.[14]

	fAge	mAge	weeks	weight	sexBaby	smoke
1	NA	13	37	5.00	female	nonsmoker
2	NA	14	36	5.88	female	nonsmoker
3	19	15	41	8.13	male	smoker
$\vdots$	$\vdots$	$\vdots$	$\vdots$	$\vdots$	$\vdots$	
150	45	50	36	9.25	female	nonsmoker

Figure 5.11: Four cases from the births dataset.

---

[13]From software, the confidence interval is (3.58, 12.08).

[14]This dataset is available in the openintro R package.

EXAMPLE 5.10

Evaluate whether it is appropriate to apply the *t*-distribution to the difference in sample means between the two groups.

Since the data come from a simple random sample and consist of less than 10% of all such cases, the observations are independent. While each distribution is strongly skewed, the large sample sizes of 50 and 100 allow for the use of the *t*-distribution to model each mean separately. Thus, the difference in sample means may be modeled using a *t*-distribution.

Figure 5.12: The top panel represents birth weights for infants whose mothers smoked. The bottom panel represents the birth weights for infants whose mothers who did not smoke. The distributions exhibit moderate-to-strong and strong skew, respectively.

A hypothesis test can be conducted to evaluate whether there is a relationship between mother's smoking status and average newborn birth weight. The null hypothesis represents the case of no difference between the groups, $H_0 : \mu_{ns} - \mu_s = 0$, where μ_{ns} represents the population mean of newborn birthweight for infants with mothers who did not smoke, and μ_s represents mean newborn birthweight for infants with mothers who smoked. Under the alternative hypothesis, there is some difference in average newborn birth weight between the groups, $H_A : \mu_{ns} - \mu_s \neq 0$. The hypotheses can also be written as $H_0 : \mu_{ns} = \mu_s$ and $H_A : \mu_{ns} \neq \mu_s$.

STATING HYPOTHESES FOR TWO-GROUP DATA

When testing a hypothesis about two independent groups, directly compare the two population means and state hypotheses in terms of μ_1 and μ_2.

- For a two-sided test, $H_0 : \mu_1 = \mu_2$; $H_A : \mu_1 \neq \mu_2$.

- For a one-sided test, either $H_0 : \mu_1 = \mu_2$; $H_A : \mu_1 > \mu_2$ or $H_0 : \mu_1 = \mu_2$; $H_A : \mu_1 < \mu_2$.

In this setting, the formula for a t-statistic is:

$$t = \frac{(\bar{x}_1 - \bar{x}_2) - (\mu_1 - \mu_2)}{SE_{\bar{x}_1 - \bar{x}_2}} = \frac{(\bar{x}_1 - \bar{x}_2) - (\mu_1 - \mu_2)}{\sqrt{\frac{s_1^2}{n_1} + \frac{s_2^2}{n_2}}}.$$

Under the null hypothesis of no difference between the groups, $H_0 : \mu_1 - \mu_2 = 0$, the formula simplifies to

$$t = \frac{(\bar{x}_1 - \bar{x}_2)}{\sqrt{\frac{s_1^2}{n_1} + \frac{s_2^2}{n_2}}}.$$

EXAMPLE 5.11

Using Figure 5.13, conduct a hypothesis test to evaluate whether there is evidence that newborns from mothers who smoke have a different average birth weight than newborns from mothers who do not smoke.

The hypotheses are $H_0 : \mu_1 = \mu_2$ and $H_A : \mu_1 \neq \mu_2$, where μ_1 represents the average newborn birth weight for nonsmoking mothers and μ_2 represents average newborn birth weight for mothers who smoke. Let $\alpha = 0.05$.

Calculate the t-statistic:

$$t = \frac{(\bar{x}_1 - \bar{x}_2)}{\sqrt{\frac{s_1^2}{n_1} + \frac{s_2^2}{n_2}}} = \frac{7.18 - 6.78}{\sqrt{\frac{1.60^2}{100} + \frac{1.43^2}{50}}} = 1.54.$$

Approximate the degrees of freedom as $50 - 1 = 49$. The t-score of 1.49 falls between the first and second columns in the $df = 49$ row of the t-table, so the two-sided p-value is between 0.10 and 0.20.[15]

This p-value is larger than the significance value, 0.05, so the null hypothesis is not rejected. There is insufficient evidence to state there is a difference in average birth weight of newborns from North Carolina mothers who did smoke during pregnancy and newborns from North Carolina mothers who did not smoke during pregnancy.

	smoker	nonsmoker
mean	6.78	7.18
st. dev.	1.43	1.60
samp. size	50	100

Figure 5.13: Summary statistics for the births dataset.

---

[15]From R, $df = 89.277$ and $p = 0.138$.

5.3.3   The paired test vs. independent group test

In the two-sample setting, students often find it difficult to determine whether a paired test or an independent group test should be used. The paired test applies only in situations where there is a natural pairing of observations between groups, such as in the swim data. Pairing can be obvious, such as the two measurements for each swimmer, or more subtle, such as measurements of respiratory function in twins, where one member of the twin pair is treated with an experimental treatment and the other with a control. In the case of two independent groups, there is no natural way to pair observations.

A common error is to overlook pairing in data and assume that two groups are independent. The swimsuit data can be used to illustrate the possible harm in conducting an independent group test rather than a paired test. In Section 5.2, the paired t-test showed a significant difference in the swim velocities between swimmers wearing wetsuits versus regular swimsuits. Suppose the analysis had been conducted without accounting for the fact that the measurements were paired.

The mean and standard deviation for the 12 wet suit velocities are 1.51 and 0.14 (m/sec), respectively, and 1.43 and 0.14 (m/sec) for the 12 swim suit velocities. A two-group test statistic is:

$$t = \frac{1.52 - 1.43}{\sqrt{0.14^2/12 + 0.14^2/12}} = 1.37.$$

If the degrees of freedom are approximated as $11 = 12 - 1$, the two-sided p-value as calculated from software is 0.20. According to this method, the null hypothesis of equal mean velocities for the two suit types would not be rejected.

It is not difficult to show that the numerator of the paired test (the average of the within swimmer differences) and the numerator of the two-group test (the difference of the average times for the two groups) are identical. The values of the test statistics differ because the denominators are different—specifically, the standard errors associated with each statistic are different. For the paired test statistic, the standard error uses the standard deviation of the within pair differences (0.22) and has value $0.022/\sqrt{12} = 0.006$. The two-group test statistic combines the standard deviations for the original measurements and has value $\sqrt{0.14^2/12 + 0.14^2/12} = 0.06$. The standard error for the two-group test is 10-fold larger than for the paired test.

This striking difference in the standard errors is caused by the much lower variability of the individual velocity differences compared to the variability of the original measurements. Due to the correlation between swim velocities for a single swimmer, the differences in the two velocity measurements for each swimmer are consistently small, resulting in low variability. Pairing has allowed for increased precision in estimating the difference between groups.

The swim suit data illustrates the importance of context, which distinguishes a statistical problem from a purely mathematical one. While both the paired and two-group tests are numerically feasible to calculate, without an apparent error, the context of the problem dictates that the correct approach is to use a paired test.

GUIDED PRACTICE 5.12

Ⓖ

Propose an experimental design for the embryonic stem cell study in sheep that would have required analysis with a paired t-test.[16]

---

[16]The experiment could have been done on pairs of siblings, with one assigned to the treatment group and one assigned to the control group. Alternatively, sheep could be matched up based on particular characteristics relevant to the experiment; for example, sheep could be paired based on similar weight or age. Note that in this study, a design involving two measurements taken on each sheep would be impractical.

5.3.4 Case study: discrimination in developmental disability support

Section 1.7.1 presented an analysis of the relationship between age, ethnicity, and amount of expenditures for supporting developmentally disabled residents in the state of California, using the dds.discr dataset. When the variable age is ignored, the expenditures per consumer is larger on average for White non-Hispanics than Hispanics, but Figure 1.53 showed that average differences by ethnicity were much smaller within age cohorts. This section demonstrates the use of t-tests to conduct a more formal analysis of possible differences in expenditure by ethnicity, both overall (i.e., ignoring age) and within age cohorts.

Comparing expenditures overall

When ignoring age, expenditures within the ethnicity groups Hispanic and White non-Hispanic show substantial right-skewing (Figure 1.45). A transformation is advisable before conducting a t-test. As shown in Figure 5.14, a natural log transformation effectively eliminates skewing.

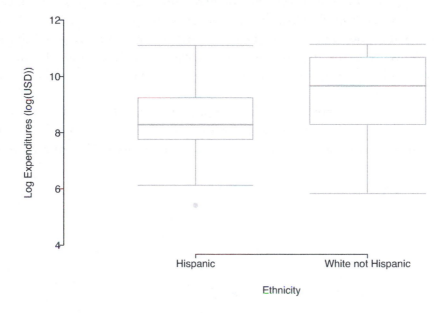

Figure 5.14: A plot of log(expenditures) by ethnicity.

Is there evidence of a difference in mean expenditures by ethnic group? Conduct a t-test of the null hypothesis $H_0 : \mu_1 = \mu_2$ versus the two-sided alternative $H_A : \mu_1 \neq \mu_2$, where μ_1 is the population mean log expenditure in Hispanics and μ_2 is the population mean log expenditure in White non-Hispanics.

	Ethnicity	n	$\overline{x}$	s
1	Hispanic	376	8.56	1.17
2	White non Hispanic	401	9.47	1.35

Figure 5.15: Summary statistics for the transformed variable log(expenditures) in the dds.discr data.

The summary statistics required to calculate the t-statistic are shown in Figure 5.15. The t-statistic for the test is

$$t = \frac{9.47 - 8.56}{\sqrt{1.35^2/401 + 1.17^2/376}} = 10.1.$$

The degrees of freedom of the test can be approximated as $376 - 1 = 375$; the p-value can be calculated using a normal approximation. Regardless of whether a t or normal distribution is used, the probability of a test statistic with absolute value larger than 10 is vanishingly small—the p-value is less than 0.001. When ignoring age, there is significant evidence of a difference in mean expenditures between Hispanics and White non-Hispanics. It appears that on average, White non-Hispanics receive a higher amount of developmental disability support from the state of California ($\bar{x}_1 < \bar{x}_2$).

However, as indicated in Section 1.7.1, this is a misleading result. The analysis as conducted does not account for the confounding effect of age, which is associated with both expenditures and ethnicity. As individuals age, they typically require more support from the government. In this dataset, White non-Hispanics tend to be older than Hispanics; this difference in age distribution contributes to the apparent difference in expenditures between two groups.

Comparing expenditures within age cohorts

One way to account for the effect of age is to compare mean expenditures within age cohorts. When comparing individuals of similar ages but different ethnic groups, are the differences in mean expenditures larger than would be expected by chance alone?

Figure 1.52 shows that the age cohort 13-17 is the largest among the Hispanic consumers, while the cohort 22-50 is the largest among White non-Hispanics. This section will examine the evidence against the null hypothesis of no difference in mean expenditures within these two cohorts.

Figure 5.16 shows that within both the age cohorts of 13-17 years and 22-50 years, the distribution of expenditures is reasonably symmetric; there is no need to apply a transformation before conducting a t-test. The skewing evident when age was ignored is due to the differing distributions of age within ethnicities.

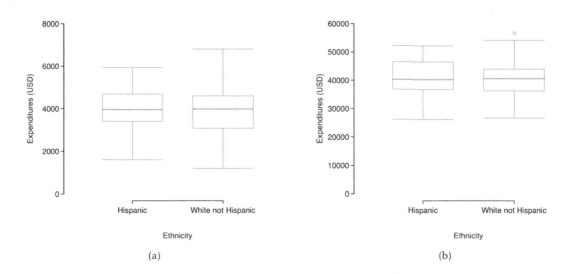

Figure 5.16: (a)A plot of expenditures by ethnicity in the age cohort 13 - 17. (b) A plot of expenditures by ethnicity in the age cohort 22 - 50.

Figure 5.17 contains the summary statistics for computing the test statistic to compare expenditures in the two groups within this age cohort. The test statistic has value $t = 0.318$, with degrees of freedom 66. The two-sided p-value is 0.75. There is not evidence of a difference between mean expenditures in Hispanics and White non-Hispanics ages 13-17.

	Ethnicity	n	$\overline{x}$	s
1	Hispanic	103	3955.28	938.82
2	White not Hispanic	67	3904.36	1071.02

Figure 5.17: Summary statistics for expenditures, Ages 13-17.

The analysis of the age cohort 22 - 50 years shows the same qualitative result. The t-statistic calculated from the summary statistics in Figure 5.18 has value $t = 0.659$ and p-value 0.51. Just as in the 13-17 age cohort, there is insufficient evidence to reject the null hypothesis of no difference between the means.

	Ethnicity	n	$\overline{x}$	s
1	Hispanic	43	40924.12	6467.09
2	White not Hispanic	133	40187.62	6081.33

Figure 5.18: Summary statistics for expenditures, Ages 22 - 50.

The inference-based analyses for these two age cohorts support the conclusions reached through the exploratory approach used in Section 1.7.1—comparing individuals of similar ages shows that there are not large differences between mean expenditures for White non-Hispanics versus Hispanics. An analysis that accounts for age as a confounding variable does not suggest there is evidence of ethnic discrimination in developmental disability support provided by the State of California.

5.3.5   Pooled standard deviation estimate

Occasionally, two populations will have standard deviations that are so similar that they can be treated as identical. For example, historical data or a well-understood biological mechanism may justify this strong assumption. In such cases, it can be more precise to use a pooled standard deviation to make inferences about the difference in population means.

The **pooled standard deviation** of two groups uses data from both samples to estimate the common standard deviation and standard error. If there are good reasons to believe that the population standard deviations are equal, an improved estimate of the group variances can be obtained by pooling the data from the two groups:

$$s^2_{\text{pooled}} = \frac{s_1^2(n_1 - 1) + s_2^2(n_2 - 1)}{n_1 + n_2 - 2},$$

where n_1 and n_2 are the sample sizes, and s_1 and s_2 represent the sample standard deviations. In this setting, the t-statistic uses s^2_{pooled} in place of s_1^2 and s_2^2 in the standard error formula, and the degrees of freedom for the t–statistic is the sum of the degrees of freedom for the two sample variances:

$$df = (n_1 - 1) + (n_2 - 1) = n_1 + n_2 - 2.$$

The t-statistic for testing the null hypothesis of no difference between population means becomes

$$t = \frac{\overline{x}_1 - \overline{x}_2}{s_{\text{pooled}}\sqrt{\frac{1}{n_1} + \frac{1}{n_2}}}.$$

The formula for the two-sided confidence interval for the difference in population means is

$$(\overline{x}_1 - \overline{x}_2) \pm t^{\star} \times s_{\text{pooled}}\sqrt{\frac{1}{n_1} + \frac{1}{n_2}},$$

where $t^{\star}$ is the point on a t-distribution with $n_1 + n_2 - 2$ degrees of freedom chosen according to the confidence coefficient.

The benefits of pooling the standard deviation are realized through obtaining a better estimate of the standard deviation for each group and using a larger degrees of freedom parameter for the t-distribution. Both of these changes may permit a more accurate model of the sampling distribution of $\overline{x}_1 - \overline{x}_2$, if the standard deviations of the two groups are indeed equal. In most applications, however, it is difficult to verify the assumption of equal population standard deviations, and thus safer to use the methods discussed in Sections 5.3.1 and 5.3.2.

5.4   Power calculations for a difference of means

Designing a study often involves many complex issues; perhaps the most important statistical issue in study design is the choice of an appropriate sample size. The **power** of a statistical test is the probability that the test will reject the null hypothesis when the alternative hypothesis is true; sample sizes are chosen to make that probability sufficiently large, typically between 80% and 90%.

Two competing considerations arise when choosing a sample size. The sample size should be sufficiently large to allow for important group differences to be detected in a hypothesis test. Practitioners often use the term 'detecting a difference' to mean correctly rejecting a null hypothesis, i.e., rejecting a null hypothesis when the alternative is true. If a study is so small that detecting a statistically significant difference is unlikely even when there are potentially important differences, enrolling participants might be unethical, since subjects could potentially be exposed to a dangerous experimental treatment. However, it is also unethical to conduct studies with an overly large sample size, since more participants than necessary would be exposed to an intervention with uncertain value. Additionally, collecting data is typically expensive and time consuming; it would be a waste of valuable resources to design a study with an overly large sample size.

This section begins by illustrating relevant concepts in the context of a hypothetical clinical trial, where the goal is to calculate a sufficient sample size for being 80% likely to detect practically important effects.[17] Afterwards, formulas are provided for directly calculating sample size, as well as references to software that can perform the calculations.

5.4.1   Reviewing the concepts of a test

EXAMPLE 5.13

A company would like to run a clinical trial with participants whose systolic blood pressures are between 140 and 180 mmHg. Suppose previously published studies suggest that the standard deviation of patient blood pressures will be about 12 mmHg, with an approximately symmetric distribution.[18] What would be the approximate standard error for $\bar{x}_{trmt} - \bar{x}_{ctrl}$ if 100 participants were enrolled in each treatment group?

The standard error is calculated as follows:

$$SE_{\bar{x}_{trmt} - \bar{x}_{ctrl}} = \sqrt{\frac{s_{trmt}^2}{n_{trmt}} + \frac{s_{ctrl}^2}{n_{ctrl}}} = \sqrt{\frac{12^2}{100} + \frac{12^2}{100}} = 1.70.$$

This may be an imperfect estimate of $SE_{\bar{x}_{trmt} - \bar{x}_{ctrl}}$, since the standard deviation estimate of 12 mmHg from prior data may not be correct. However, it is sufficient for getting started, and making an assumption like this is often the only available option.

---

[17]While sample size planning is also important for observational studies, those techniques are not discussed here.

[18]In many studies like this one, each participant's blood pressure would be measured at the beginning and end of the study, and the outcome measurement for the study would be the average difference in blood pressure in each of the treatment groups. For this hypothetical study, we assume for simplicity that blood pressure is measured at only the end of the study, and that the randomization ensures that blood pressures at the beginning of the study are equal (on average) between the two groups.

Since the degrees of freedom are greater than 30, the distribution of $\bar{x}_{trmt} - \bar{x}_{ctrl}$ will be approximately normal. Under the null hypothesis, the mean is 0 and the standard deviation is 1.70 (from the standard error).

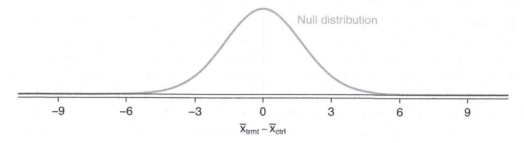

Figure 5.19: Null distribution for the t-statistic in Example 5.14.

EXAMPLE 5.14

For what values of $\bar{x}_{trmt} - \bar{x}_{ctrl}$ would the null hypothesis be rejected, using $\alpha = 0.05$?

If the observed difference is in the far left or far right tail of the null distribution, there is sufficient evidence to reject the null hypothesis. For $\alpha = 0.05$, H_0 is rejected if the difference is in the lower 2.5% or upper 2.5% tail:

Lower 2.5%: For the normal model, this is 1.96 standard errors below 0, so any difference smaller than $-1.96 \times 1.70 = -3.332$ mmHg.

Upper 2.5%: For the normal model, this is 1.96 standard errors above 0, so any difference larger than $1.96 \times 1.70 = 3.332$ mmHg.

The boundaries of these **rejection regions** are shown below. Note that if the new treatment is effective, mean blood pressure should be lower in the treatment group than in the control group; i.e., the difference should be in the lower tail.

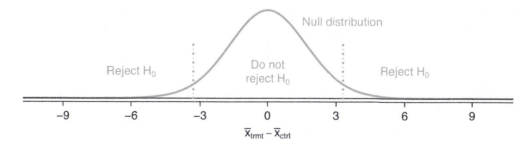

The next step is to perform some hypothetical calculations to determine the probability of rejecting the null hypothesis if the alternative hypothesis were true.

5.4.2 Computing the power for a 2-sample test

If there is a real effect from an intervention, and the effect is large enough to have practical value, the probability of detecting that effect is referred to as the **power**. Power can be computed for different sample sizes or different effect sizes.

There is no easy way to define when an effect size is large enough to be of value; this is not a statistical issue. For example, in a clinical trial, the scientifically significant effect is the incremental value of the intervention that would justify changing current clinical recommendations from an existing intervention to a new one. In such a setting, the effect size is usually determined from long discussions between the research team and study sponsors.

Suppose that for this hypothetical blood pressure medication study, the researchers are interested in detecting any effect on blood pressure that is 3 mmHg or larger than the standard medication. Here, 3 mmHg is the minimum **population effect size** of interest.

EXAMPLE 5.15

Suppose the study proceeded with 100 patients per treatment group and the new drug does reduce average blood pressure by an additional 3 mmHg relative to the standard medication. What is the probability of detecting this effect?

Determine the sampling distribution for $\bar{x}_{trmt} - \bar{x}_{ctrl}$ when the true difference is -3 mmHg; this has the same standard deviation of 1.70 as the null distribution, but the mean is shifted 3 units to the left. Then, calculate the fraction of the distribution for $\bar{x}_{trmt} - \bar{x}_{ctrl}$ that falls within the rejection region for the null distribution, as shown in Figure 5.20.

The probability of being in the left side of the rejection region ($x < -3.332$) can be calculated by converting to a Z-score and using either the normal probability table or statistical software.[19]

$$Z = \frac{-3.332 - (-3)}{1.7} = -0.20 \quad \rightarrow \quad P(Z \leq -0.20) = 0.4207.$$

The power for the test is about 42% when $\mu_{trmt} - \mu_{ctrl} = -3$ mm/Hg and each group has a sample size of 100.

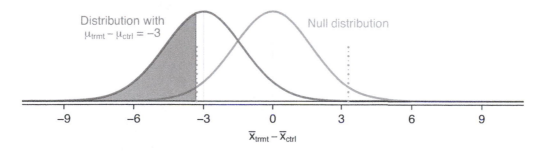

Figure 5.20: The rejection regions are outside of the dotted lines. Recall that the boundaries for $\alpha = 0.05$ were calculated to be ± 3.332 mmHg.

---

[19]The probability of being in the right side of the rejection region is negligible and can be ignored.

5.4.3 Determining a proper sample size

The last example demonstrated that with a sample size of 100 in each group, there is a probability of about 0.42 of detecting an effect size of 3 mmHg. If the study were conducted with this sample size, even if the new medication reduced blood pressure by 3 mmHg compared to the control group, there is a less than 50% chance of concluding that the medication is beneficial. Studies with low power are often inconclusive, and there are important reasons to avoid such a situation:

– Participants were subjected to a drug for a study that may have little scientific value.

– The company may have invested hundreds of millions of dollars in developing the new drug, and may now be left with uncertainty about its potential.

– Another clinical trial may need to be conducted to obtain a more conclusive answer as to whether the drug does hold any practical value, and that would require substantial time and expense.

To ensure a higher probability of detecting a clinically important effect, a larger sample size should be chosen. What about a study with 500 patients per group?

GUIDED PRACTICE 5.16

Calculate the power to detect a change of -3 mmHg using a sample size of 500 per group. Recall that the standard deviation of patient blood pressures was expected to be about 12 mmHg.[20]

(a) Determine the standard error.

(b) Identify the null distribution and rejection regions, as well as the alternative distribution when $\mu_{trmt} - \mu_{ctrl} = -3$.

(c) Compute the probability of rejecting the null hypothesis.

With a sample size of 500 per group, the power of the test is much larger than necessary. Not only does this lead to a study that would be overly expensive and time consuming, it also exposes more patients than necessary to the experimental drug.

Sample sizes are generally chosen such that power is around 80%, although in some cases 90% is the target. Other values may be reasonable for a specific context, but 80% and 90% are most commonly chosen as a good balance between high power and limiting the number of patients exposed to a new treatment (as well as reducing experimental costs).

---

[20](a) The standard error will now be $SE = \sqrt{\frac{12^2}{500} + \frac{12^2}{500}} = 0.76$.

(b) The null distribution, rejection boundaries, and alternative distribution are shown below. The rejection regions are the areas outside the two dotted lines at $\bar{x}_{trmt} - \bar{x}_{ctrl} \pm 0.76 \times 1.96 = \pm 1.49$.

(c) Compute the Z-score and find the tail area, $Z = \frac{-1.49 - (-3)}{0.76} = 1.99 \rightarrow P(Z \leq 1.99) = 0.9767$, which is the power of the test for a difference of 3 mmHg. With 500 patients per group, the study would be 97.7% likely to detect an effect size of 3 mmHg.

EXAMPLE 5.17

Identify the sample size that would lead to a power of 80%.

_____

The Z-score that defines a lower tail area of 0.80 is about $Z = 0.84$. In other words, 0.84 standard errors from -3, the mean of the alternative distribution.

For $\alpha = 0.05$, the rejection region always extends 1.96 standard errors from 0, the center of the null distribution.

The distance between the centers of the null and alternative distributions can be expressed in terms of the standard error:

$$(0.84 \times SE) + (1.96 \times SE) = 2.8 \times SE.$$

This quantity necessarily equals the minimum effect size of interest, 3 mmHg, which is the distance between -3 and 0. It is then possible to solve for n:

$$3 = 2.8 \times SE$$

$$3 = 2.8 \times \sqrt{\frac{12^2}{n} + \frac{12^2}{n}}$$

$$n = \frac{2.8^2}{3^2} \times \left(12^2 + 12^2\right) = 250.88$$

The study should enroll at least 251 patients per group for 80% power. Note that sample size should always be rounded up in order to achieve the desired power. Even if the calculation had yielded a number closer to 250 (e.g., 250.25), the study should still enroll 251 patients per grou, since having 250 patients per group would result in a power lower than 80%.

GUIDED PRACTICE 5.18

Suppose the targeted power is 90% and $\alpha = 0.01$. How many standard errors should separate the centers of the null and alternative distributions, where the alternative distribution is centered at the minimum effect size of interest? Assume the test is two-sided.[21]

_____

[21] Find the Z-score such that 90% of the distribution is below it: $Z = 1.28$. Next, find the cutoffs for the rejection regions: ±2.58. Thus, the centers of the null and alternative distributions should be about $1.28 + 2.58 = 3.86$ standard errors apart.

Figure 5.21 shows the power for sample sizes from 20 participants to 5,000 participants when $\alpha = 0.05$ and the true difference is -3 mmHg. While power increases with sample size, having more than 250-300 participants provides little additional value towards detecting an effect.

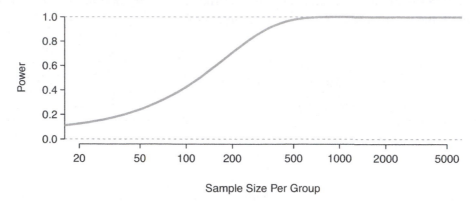

Figure 5.21: The curve shows the power for different sample sizes in the context of the blood pressure example when the true difference is -3.

5.4.4   Formulas for power and sample size

The previous sections have illustrated how power and sample size can be calculated from first principles, using the fundamental ideas behind distributions and testing. In practice, power and sample size calculations are so important that statistical software should be the method of choice; there are many commercially available and public domain programs for performing such calculations. However, hand calculations using formulas can provide quick estimates in the early stages of planning a study.

Use the following formula to calculate sample size for comparing two means, assuming each group will have n participants:

$$n = \frac{(\sigma_1^2 + \sigma_2^2)(z_{1-\alpha/2} + z_{1-\beta})^2}{\Delta^2}.$$

In this formula:

- μ_1, μ_2, σ_1, and σ_2 are the population means and standard deviations of the two groups.
- $\Delta = \mu_1 - \mu_2$ is the minimally important difference that investigators wish to detect.
- The null and alternative hypotheses are $H_0 : \Delta = 0$ (i.e., no difference between the means) and $H_A : \Delta \neq 0$, i.e., a two-sided alternative.
- The two-sided significance level is α, and $z_{1-\alpha/2}$ is the point on a standard normal distribution with area $1 - \alpha/2$ to its left and $\alpha/2$ area to its right.
- β is the probability of incorrectly failing to reject H_0 for a specified value of Δ; $1 - \beta$ is the power. The value $z_{1-\beta}$ is the point on a standard normal distribution with area $1 - \beta$ to its left.

For a study with sample size n per group, where Z is a normal random variable with mean 0 and standard deviation 1, power is given by:

$$\text{Power} = P\left(Z < -z_{1-\alpha/2} + \frac{\Delta}{\sqrt{\sigma_1^2/n + \sigma_2^2/n}}\right).$$

These formulas could have been used to do the earlier power and sample size calculations for the hypothetical study of blood pressure lowering medication. To calculate the sample size needed for 80% power in detecting a change of 3 mmHg, $\alpha = 0.05$, $1 - \beta = 0.80$, $\Delta = 3$ mmHg, and $\sigma_1 = \sigma_2 = 12$ mmHg. The formula yields a sample size n per group of

$$n = \frac{(12^2 + 12^2)(1.96 + 0.84)^2}{(-3.0)^2} = 250.88,$$

which can be rounded up to 251.

The formula for power can be used to verify the sample size of 251:

$$\begin{aligned}
\text{Power} &= P\left(Z < -1.96 + \frac{3}{\sqrt{12^2/251 + 12^2/251}}\right) \\
&= P(Z < 1.25) \\
&= 0.85.
\end{aligned}$$

The calculated power is slightly larger than 80% because of the rounding to 251.

The sample size calculations done before any data are collected are one of the most critical aspects of conducting a study. If an analysis is done incorrectly, it can be redone once the error is discovered. However, if data were collected for a sample size that is either too large or too small, it can be impossible to correct the error, especially in studies with human subjects. As a result, sample size calculations are nearly always done using software. For two-sample t-tests, the R function power.t.test is both freely available and easy to use.

5.5   Comparing means with ANOVA

In some settings, it is useful to compare means across several groups. It might be tempting to do pairwise comparisons between groups; for example, if there are three groups (A, B, C), why not conduct three separate t-tests (A vs. B, A vs. C, B vs. C)? Conducting multiple tests on the same data increases the rate of Type I error, making it more likely that a difference will be found by chance, even if there is no difference among the population means. Multiple testing is discussed further in Section 5.5.3.

Instead, the methodology behind a t-test can be generalized to a procedure called **analysis of variance (ANOVA)**, which uses a single hypothesis test to assess whether the means across several groups are equal. Strong evidence favoring the alternative hypothesis in ANOVA is described by unusually large differences among the group means.

H_0: The mean outcome is the same across all k groups. In statistical notation, $\mu_1 = \mu_2 = \cdots = \mu_k$ where μ_i represents the mean of the outcome for observations in category i.

H_A: At least one mean is different.

There are three conditions on the data that must be checked before performing ANOVA: 1) observations are independent within and across groups, 2) the data within each group are nearly normal, and 3) the variability across the groups is about equal.

EXAMPLE 5.19

Examine Figure 5.22. Compare groups I, II, and III. Is it possible to visually determine if the differences in the group centers is due to chance or not? Now compare groups IV, V, and VI. Do the differences in these group centers appear to be due to chance?

(E) It is difficult to discern a difference in the centers of groups I, II, and III, because the data within each group are quite variable relative to any differences in the average outcome. However, there appear to be differences in the centers of groups IV, V, and VI. For instance, group V appears to have a higher mean than that of the other two groups. The differences in centers for groups IV, V, and VI are noticeable because those differences are large relative to the variability in the individual observations within each group.

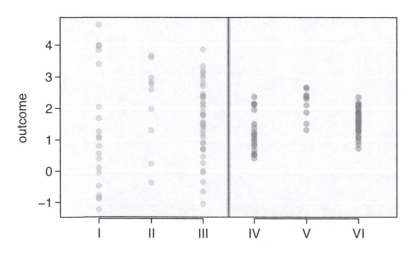

Figure 5.22: Side-by-side dot plot for the outcomes for six groups.

5.5.1   Analysis of variance (ANOVA) and the F-test

The famuss dataset was introduced in Chapter 1, Section 1.2.2. In the FAMuSS study, researchers examined the relationship between muscle strength and genotype at a location on the ACTN3 gene. The measure for muscle strength is percent change in strength in the non-dominant arm (ndrm.ch). Is there a difference in muscle strength across the three genotype categories (CC, CT, TT)?

GUIDED PRACTICE 5.20

The null hypothesis under consideration is the following: $\mu_{CC} = \mu_{CT} = \mu_{TT}$. Write the null and corresponding alternative hypotheses in plain language.[22]

Figure 5.23 provides summary statistics for each group. A side-by-side boxplot for the change in non-dominant arm strength is shown in Figure 5.24; Figure 5.25 shows the Q-Q plots by each genotype. Notice that the variability appears to be approximately constant across groups; nearly constant variance across groups is an important assumption that must be satisfied for using ANOVA. Based on the Q-Q plots, there is evidence of moderate right skew; the data do not follow a normal distribution very closely, but could be considered to 'loosely' follow a normal distribution.[23] It is reasonable to assume that the observations are independent within and across groups; it is unlikely that participants in the study were related, or that data collection was carried out in a way that one participant's change in arm strength could influence another's.

	CC	CT	TT
Sample size (n_i)	173	261	161
Sample mean ($\bar{x}_i$)	48.89	53.25	58.08
Sample SD (s_i)	29.96	33.23	35.69

Figure 5.23: Summary statistics of change in non-dominant arm strength, split by genotype.

Figure 5.24: Side-by-side box plot of the change in non-dominant arm strength for 595 participants across three groups.

---

[22]H_0: The average percent change in non-dominant arm strength is equal across the three genotypes. H_A: The average percent change in non-dominant arm strength varies across some (or all) groups.

[23]In a more advanced course, it can be shown that the ANOVA procedure still holds with deviations from normality when sample sizes are moderately large. Additionally, a more advanced course would discuss appropriate transformations to induce normality.

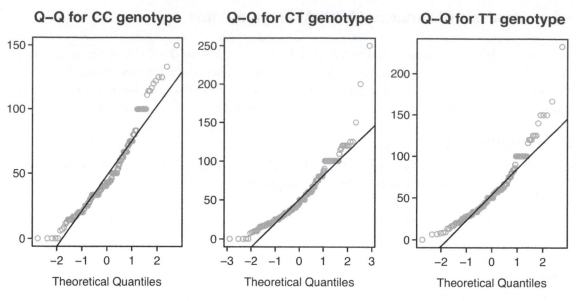

Figure 5.25: Q-Q plots of the change in non-dominant arm strength for 595 participants across three groups.

EXAMPLE 5.21

The largest difference between the sample means is between the CC and TT groups. Consider again the original hypotheses:

H_0: $\mu_{CC} = \mu_{CT} = \mu_{TT}$

H_A: The average percent change in non-dominant arm strength (μ_i) varies across some (or all) groups.

Why might it be inappropriate to run the test by simply estimating whether the difference of μ_{CC} and μ_{TT} is statistically significant at a 0.05 significance level?

It is inappropriate to informally examine the data and decide which groups to formally test. This is a form of **data fishing**; choosing the groups with the largest differences for the formal test will lead to an increased chance of incorrectly rejecting the null hypothesis (i.e., an inflation in the Type I error rate). Instead, all the groups should be tested using a single hypothesis test.

Analysis of variance focuses on answering one question: is the variability in the sample means large enough that it seems unlikely to be from chance alone? The variation between groups is referred to as the **mean square between groups** (MSG); the MSG is a measure of how much each group mean varies from the overall mean. Let $\overline{x}$ represent the mean of outcomes across all groups, where $\overline{x}_i$ is the mean of outcomes in a particular group i and n_i is the sample size of group i. The mean square between groups is:

$$MSG = \frac{1}{k-1} \sum_{i=1}^{k} n_i (\overline{x}_i - \overline{x})^2 = \frac{1}{df_G} SSG,$$

where SSG is the **sum of squares between groups**, $\sum_{i=1}^{k} n_i (\overline{x}_i - \overline{x})^2$, and $df_G = k - 1$ is the degrees of freedom associated with the MSG when there are k groups.

Under the null hypothesis, any observed variation in group means is due to chance and there is no real difference between the groups. In other words, the null hypothesis assumes that the groupings are non-informative, such that all observations can be thought of as belonging to a single group. If this scenario is true, then it is reasonable to expect that the variability between the group means should be equal to the variability observed within a single group. The **mean square error** (MSE) is a pooled variance estimate with associated degrees of freedom $df_E = n - k$ that provides a measure of variability within the groups. The mean square error is computed as:

$$MSE = \frac{1}{n-k} \sum_{i=1}^{k} (n_i - 1)s_i^2 = \frac{1}{df_E} SSE,$$

where the SSE is the **sum of squared errors**, n_i is the sample size of group i, and s_i is the standard deviation of group i.

Under the null hypothesis that all the group means are equal, any differences among the sample means are only due to chance; thus, the MSG and MSE should also be equal. ANOVA is based on comparing the MSG and MSE. The test statistic for ANOVA, the **F-statistic**, is the ratio of the between-group variability to the within-group variability:

$$F = \frac{MSG}{MSE}. \tag{5.22}$$

EXAMPLE 5.23

Calculate the F-statistic for the famuss data summarized in Figure 5.23. The overall mean $\bar{x}$ across all observations is 53.29.

First, calculate the MSG and MSE.

$$MSG = \frac{1}{k-1} \sum_{i=1}^{k} n_i (\bar{x}_i - \bar{x})^2$$

$$= \frac{1}{3-1}[(173)(48.89 - 53.29)^2 + (261)(53.25 - 53.29)^2 + (161)(58.08 - 53.29)^2]$$

$$= 3521.69$$

$$MSE = \frac{1}{n-k} \sum_{i=1}^{k} (n_i - 1)s_i^2$$

$$= \frac{1}{595-3}[(173-1)(29.96^2) + (261-1)(33.23^2) + (161-1)(35.69^2)]$$

$$= 1090.02$$

The F-statistic is the ratio:

$$\frac{MSG}{MSE} = \frac{3521.69}{1090.02} = 3.23.$$

A p-value can be computed from the F-statistic using an F-distribution, which has two associated parameters: df_1 and df_2. For the F-statistic in ANOVA, $df_1 = df_G$ and $df_2 = df_E$. An F distribution with 2 and 592 degrees of freedom, corresponding to the F-statistic for the genotype and muscle strength hypothesis test, is shown in Figure 5.26.

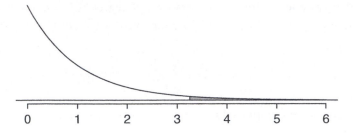

Figure 5.26: An F-distribution with $df_1 = 2$ and $df_2 = 592$. The tail area greater than $F = 3.23$ is shaded.

The larger the observed variability in the sample means (MSG) relative to the within-group variability (MSE), the larger F will be. Larger values of F represent stronger evidence against the null hypothesis. The upper tail of the distribution is used to compute a p-value, which is typically done using statistical software.

EXAMPLE 5.24

The p-value corresponding to the test statistic is equal to about 0.04. Does this provide strong evidence against the null hypothesis at significance level $\alpha = 0.05$?

The p-value is smaller than 0.05, indicating the evidence is strong enough to reject the null hypothesis at a significance level of 0.05. The data suggest that average change in strength in the non-dominant arm varies by participant genotype.

THE F-STATISTIC AND THE F-TEST

Analysis of variance (ANOVA) is used to test whether the mean outcome differs across two or more groups. ANOVA uses a test statistic F, which represents a standardized ratio of variability in the sample means relative to the variability within the groups. If H_0 is true and the model assumptions are satisfied, the statistic F follows an F distribution with parameters $df_1 = k-1$ and $df_2 = n-k$. The upper tail of the F-distribution is used to calculate the p-value.

5.5.2   Reading an ANOVA table from software

The calculations required to perform an ANOVA by hand are tedious and prone to human error. Instead, it is common to use statistical software to calculate the F-statistic and associated p-value. The results of an ANOVA can be summarized in a table similar to that of a regression summary, which will be discussed in Chapters 6 and 7.

Figure 5.27 shows an ANOVA summary to test whether the mean change in non-dominant arm strength varies by genotype. Many of these values should look familiar; in particular, the F-statistic and p-value can be retrieved from the last two columns.

	Df	Sum Sq	Mean Sq	F value	Pr(>F)
famuss$actn3.r577x	2	7043	3522	3.231	0.0402
Residuals	592	645293	1090		

Figure 5.27: ANOVA summary for testing whether the mean change in non-dominant arm strength varies by genotype at the actn3.r577x location on the ACTN3 gene.

5.5.3   Multiple comparisons and controlling Type I Error rate

Rejecting the null hypothesis in an ANOVA analysis only allows for a conclusion that there is evidence for a difference in group means. In order to identify the groups with different means, it is necessary to perform further testing. For example, in the famuss analysis, there are three comparisons to make: CC to CT, CC to TT, and CT to TT. While these comparisons can be made using two sample t-tests, it is important to control the Type I error rate. One of the simplest ways to reduce the overall probability of identifying a significant difference by chance in a multiple comparisons setting is to use the Bonferroni correction procedure.

In the Bonferroni correction procedure, the p-value from a two-sample t-test is compared to a modified significance level, $\alpha^\star$; $\alpha^\star = \alpha/K$, where K is the total number of comparisons being considered. For k groups, $K = \frac{k(k-1)}{2}$. When calculating the t-statistic, use the pooled estimate of standard deviation between groups (which equals $\sqrt{MSE}$); to calculate the p-value, use a t-distribution with df_2. It is typically more convenient to do these calculations using software.

BONFERRONI CORRECTION

The **Bonferroni correction** suggests that a more stringent significance level is appropriate when conducting multiple tests:

$$\alpha^\star = \alpha/K$$

where K is the number of comparisons being considered. For k groups, $K = \frac{k(k-1)}{2}$.

EXAMPLE 5.25

The ANOVA conducted on the famuss dataset showed strong evidence of differences in the mean strength change in the non-dominant arm between the three genotypes. Complete the three possible pairwise comparisons using the Bonferroni correction and report any differences.

Use a modified significance level of $\alpha^\star = 0.05/3 = 0.0167$. The pooled estimate of the standard deviation is $\sqrt{MSE} = \sqrt{1090.02} = 33.02$.

Genotype CC versus Genotype CT:

$$t = \frac{\overline{x}_1 - \overline{x}_2}{s_{\text{pooled}}\sqrt{\frac{1}{n_1} + \frac{1}{n_2}}} = \frac{48.89 - 53.25}{33.02\sqrt{\frac{1}{173} + \frac{1}{261}}} = -1.35.$$

This results in a p-value of 0.18 on $df = 592$. This p-value is larger than $\alpha^\star = 0.0167$, so there is not evidence of a difference in the means of genotypes CC and CT.

Genotype CC versus Genotype TT:

$$t = \frac{\overline{x}_1 - \overline{x}_2}{s_{\text{pooled}}\sqrt{\frac{1}{n_1} + \frac{1}{n_2}}} = \frac{48.89 - 58.08}{33.02\sqrt{\frac{1}{173} + \frac{1}{161}}} = -2.54.$$

This results in a p-value of 0.01 on $df = 592$. This p-value is smaller than $\alpha^\star = 0.0167$, so there is evidence of a difference in the means of genotypes CC and TT.

Genotype CT versus Genotype TT:

$$t = \frac{\overline{x}_1 - \overline{x}_2}{s_{\text{pooled}}\sqrt{\frac{1}{n_1} + \frac{1}{n_2}}} = \frac{53.25 - 58.08}{33.02\sqrt{\frac{1}{261} + \frac{1}{161}}} = -1.46.$$

This results in a p-value of 0.14 on $df = 592$. This p-value is larger than $\alpha^\star = 0.0167$, so there is not evidence of a difference in the means of genotypes CT and TT.

In summary, the mean percent strength change in the non-dominant arm for genotype CT individuals is not statistically distinguishable from those of genotype CC and TT individuals. However, there is evidence that mean percent strength change in the non-dominant arm differs between individuals of genotype CC and TT are different.

5.5.4 Reading the results of pairwise t-tests from software

Statistical software can be used to calculate the p-values associated with each possible pairwise comparison of the groups in ANOVA. The results of the pairwise tests are summarized in a table that shows the p-value for each two-group test.

Figure 5.28 shows the p-values from the three possible two-group t-tests comparing change in non-dominant arm strengths between individuals with genotypes CC, CT, and TT. For example, the table indicates that when comparing mean change in non-dominant arm strength between TT and CC individuals, the p-value is 0.01. This coheres with the calculations above, and these unadjusted p-values should be compared to $\alpha^\star = 0.0167$.

	CC	CT
CT	0.18	-
TT	0.01	0.14

Figure 5.28: Unadjusted p-values for pairwise comparisons testing whether the mean change in non-dominant arm strength varies by genotype at the actn3.r577x location on ACTN3 gene.

The use of statistical software makes it easier to apply corrections for multiple testing, such that it is not necessary to explicitly calculate the value of $\alpha^\star$. Figure 5.29 shows the Bonferroni-adjusted p-values from the three possible tests. When statistical software applies the Bonferroni correction, the unadjusted p-value is multiplied by K, the number of comparisons, allowing for the values to be directly compared to α, not $\alpha^\star$. Comparing an unadjusted p-value to α/K is equivalent to comparing the quantity ($K \times p$-value) to α.

	CC	CT
CT	0.54	-
TT	0.03	0.43

Figure 5.29: Bonferroni-adjusted p-values for pairwise comparisons testing whether the mean change in non-dominant arm strength varies by genotype at the actn3.r577x location on ACTN3 gene.

5.6 Notes

The material in this chapter is particularly important. For many applications, t-tests and Analysis of Variance (ANOVA) are an essential part of the core of statistics in medicine and the life sciences. The comparison of two or more groups is often the primary aim of experiments both in the laboratory and in studies with human subjects. More generally, the approaches to interpreting and drawing conclusions from testing demonstrated in this chapter are used throughout the rest of the text and, indeed, in much of statistics.

While it is important to master the details of the techniques of testing for differences in two or more groups, it is even more critical to not lose sight of the fundamental principles behind the tests. A statistically significant difference in group means does not necessarily imply that group membership is the reason for the observed association. A significant association does not necessarily imply causation, even if it is highly significant; confounding variables may be involved. In most cases, causation can only be inferred in controlled experiments when interventions have been assigned randomly. It is also essential to carefully consider the context of a problem. For instance, students often find the distinction between paired and independent group comparisons confusing; understanding the problem context is the only reliable way to choose the correct approach.

It is generally prudent to use the form of the t-test that does not assume equal standard deviations, but the power calculations described in Section 5.4 assume models with equal standard deviations. The formulas are simpler when standard deviations are equal, and software is more widely available for that case. The differences in sample sizes are usually minor and less important than assumptions about target differences or the values of the standard deviations. If the standard deviations are expected to be very different, then more specialized software for computing sample size and power should be used. The analysis done after the study has been completed should then use the t-test for unequal standard deviations.

Tests for significant differences are sometimes overused in science, with not enough attention paid to estimates and confidence intervals. Confidence intervals for the difference of two population means show a range of underlying differences in means that are consistent with the data, and often lead to insights not possible from only the test statistic and p-value. Wide confidence intervals may show that a non-significant test is the result of high variability in the test statistic, perhaps caused by a sample size that was too small. Conversely, a highly significant p-value may be the result of such a large sample size that the observed differences are not scientifically meaningful; that may be evident from confidence intervals with very narrow width.

Finally, the formula used to approximate degrees of freedom v for the independent two-group t-test that does not assume equal variance is

$$v = \frac{\left[(s_1^2/n_1) + (s_2^2/n_2)\right]^2}{\left[(s_1^2/n_1)^2/(n_1-1) + (s_2^2/n_2)^2/(n_2-1)\right]},$$

where n_1, s_1 are the sample size and standard deviation for the first sample, and n_2, s_2 are the corresponding values for the second sample. Since v is routinely provided in the output from statistical software, there is rarely any need to calculate it by hand. The approximate formula $df = \min(n_1 - 1, n_2 - 1)$ always produces a smaller value for degrees of freedom and hence a larger p-value.

The labs for this chapter are structured around particularly important problems in practice: comparing two groups, such as a treatment and control group (Lab 1); assessing before starting a study whether a sample size is large enough to make it likely that important differences will be detected (Lab 2); comparing more than two groups using analysis of variance (Lab 3); controlling error rates when looking at many comparisons in a dataset (Lab 4); and thinking about hypothesis testing in the larger context of reproducibility (Lab 5). The first four labs provide guidance on how to conduct and interpret specific types of analyses. Students may find the last lab particularly useful in understanding the distinction between a p-value and other probabilities relevant in an inferential setting, such as power.

5.7 Exercises

5.7.1 Single-sample inference with the t-distribution

5.1 Identify the critical t. An independent random sample is selected from an approximately normal population with unknown standard deviation. Find the degrees of freedom and the critical t-value ($t^\star$) for the given sample size and confidence level.

(a) $n = 6$, CL = 90%

(b) $n = 21$, CL = 98%

(c) $n = 29$, CL = 95%

(d) $n = 12$, CL = 99%

5.2 Find the p-value, Part I. An independent random sample is selected from an approximately normal population with an unknown standard deviation. Find the p-value for the given sets of alternative hypothesis and test statistic, and determine if the null hypothesis would be rejected at $\alpha = 0.05$.

(a) $H_A : \mu > \mu_0$, $n = 11$, $T = 1.91$

(b) $H_A : \mu < \mu_0$, $n = 17$, $T = -3.45$

(c) $H_A : \mu \neq \mu_0$, $n = 7$, $T = 0.83$

(d) $H_A : \mu > \mu_0$, $n = 28$, $T = 2.13$

5.3 Cutoff values. The following are cutoff values for the upper 5% of a t-distribution with either degrees of freedom 10, 50, or 100: 2.23, 1.98, and 2.01. Identify which value belongs to which distribution and explain your reasoning.

5.4 Find the p-value, Part II. An independent random sample is selected from an approximately normal population with an unknown standard deviation. Find the p-value for the given sets of alternative hypothesis and test statistic, and determine if the null hypothesis would be rejected at $\alpha = 0.01$.

(a) $H_A : \mu > 0.5$, $n = 26$, $T = 2.485$

(b) $H_A : \mu < 3$, $n = 18$, $T = 0.5$

5.5 Working backwards, Part I. A 95% confidence interval for a population mean, μ, is given as (18.985, 21.015). This confidence interval is based on a simple random sample of 36 observations. Calculate the sample mean and standard deviation. Assume that all conditions necessary for inference are satisfied. Use the t-distribution in any calculations.

5.6 Working backwards, Part II. A 90% confidence interval for a population mean is (65, 77). The population distribution is approximately normal and the population standard deviation is unknown. This confidence interval is based on a simple random sample of 25 observations. Calculate the sample mean, the margin of error, and the sample standard deviation.

5.7 Sleep habits of New Yorkers. New York is known as "the city that never sleeps". A random sample of 25 New Yorkers were asked how much sleep they get per night. Statistical summaries of these data are shown below. Do these data provide strong evidence that New Yorkers sleep less than 8 hours a night on average?

n	$\bar{x}$	s	min	max
25	7.73	0.77	6.17	9.78

(a) Write the hypotheses in symbols and in words.

(b) Check conditions, then calculate the test statistic, T, and the associated degrees of freedom.

(c) Find and interpret the p-value in this context. Drawing a picture may be helpful.

(d) What is the conclusion of the hypothesis test?

(e) If you were to construct a 90% confidence interval that corresponded to this hypothesis test, would you expect 8 hours to be in the interval?

5.8 Heights of adults. Researchers studying anthropometry collected body girth measurements and skeletal diameter measurements, as well as age, weight, height and gender, for 507 physically active individuals. The histogram below shows the sample distribution of heights in centimeters.[24]

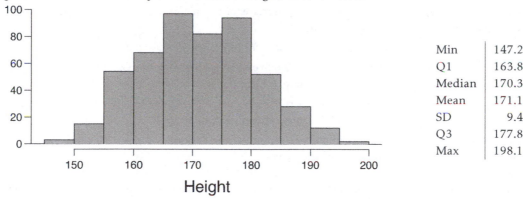

Min	147.2
Q1	163.8
Median	170.3
Mean	171.1
SD	9.4
Q3	177.8
Max	198.1

(a) What is the point estimate for the average height of active individuals? What about the median?

(b) What is the point estimate for the standard deviation of the heights of active individuals? What about the IQR?

(c) Is a person who is 1m 80cm (180 cm) tall considered unusually tall? And is a person who is 1m 55cm (155cm) considered unusually short? Explain your reasoning.

(d) The researchers take another random sample of physically active individuals. Would you expect the mean and the standard deviation of this new sample to be the ones given above? Explain your reasoning.

(e) The sample means obtained are point estimates for the mean height of all active individuals, if the sample of individuals is equivalent to a simple random sample. What measure do we use to quantify the variability of such an estimate? Compute this quantity using the data from the original sample under the condition that the data are a simple random sample.

5.9 Find the mean. You are given the following hypotheses:

$$H_0 : \mu = 60$$
$$H_A : \mu < 60$$

We know that the sample standard deviation is 8 and the sample size is 20. For what sample mean would the p-value be equal to 0.05? Assume that all conditions necessary for inference are satisfied.

---

[24]G. Heinz et al. "Exploring relationships in body dimensions". In: *Journal of Statistics Education* 11.2 (2003).

5.10 $t^\star$ vs. $z^\star$. For a given confidence level, $t^\star_{df}$ is larger than $z^\star$. Explain how $t^\star_{df}$ being slightly larger than $z^\star$ affects the width of the confidence interval.

5.11 Play the piano. Georgianna claims that in a small city renowned for its music school, the average child takes less than 5 years of piano lessons. We have a random sample of 20 children from the city, with a mean of 4.6 years of piano lessons and a standard deviation of 2.2 years.

(a) Evaluate Georgianna's claim using a hypothesis test.

(b) Construct a 95% confidence interval for the number of years students in this city take piano lessons, and interpret it in context of the data.

(c) Do your results from the hypothesis test and the confidence interval agree? Explain your reasoning.

5.12 Auto exhaust and lead exposure. Researchers interested in lead exposure due to car exhaust sampled the blood of 52 police officers subjected to constant inhalation of automobile exhaust fumes while working traffic enforcement in a primarily urban environment. The blood samples of these officers had an average lead concentration of 124.32 μg/l and a SD of 37.74 μg/l; a previous study of individuals from a nearby suburb, with no history of exposure, found an average blood level concentration of 35 μg/l.[25]

(a) Write down the hypotheses that would be appropriate for testing if the police officers appear to have been exposed to a higher concentration of lead.

(b) Explicitly state and check all conditions necessary for inference on these data.

(c) Test the hypothesis that the downtown police officers have a higher lead exposure than the group in the previous study. Interpret your results in context.

(d) Based on your preceding result, without performing a calculation, would a 99% confidence interval for the average blood concentration level of police officers contain 35 μg/l?

(e) Based on your preceding result, without performing a calculation, would a 99% confidence interval for this difference contain 0? Explain why or why not.

5.13 Car insurance savings. A market researcher wants to evaluate car insurance savings at a competing company. Based on past studies he is assuming that the standard deviation of savings is $100. He wants to collect data such that he can get a margin of error of no more than $10 at a 95% confidence level. How large of a sample should he collect?

---

[25]WI Mortada et al. "Study of lead exposure from automobile exhaust as a risk for nephrotoxicity among traffic policemen." In: *American journal of nephrology* 21.4 (2000), pp. 274–279.

5.7.2 Two-sample test for paired data

5.14 Air quality. Air quality measurements were collected in a random sample of 25 country capitals in 2013, and then again in the same cities in 2014. We would like to use these data to compare average air quality between the two years. Should we use a paired or non-paired test? Explain your reasoning.

5.15 Paired or not, Part I. In each of the following scenarios, determine if the data are paired.

(a) Compare pre- (beginning of semester) and post-test (end of semester) scores of students.

(b) Assess gender-related salary gap by comparing salaries of randomly sampled men and women.

(c) Compare artery thicknesses at the beginning of a study and after 2 years of taking Vitamin E for the same group of patients.

(d) Assess effectiveness of a diet regimen by comparing the before and after weights of subjects.

5.16 Paired or not, Part II. In each of the following scenarios, determine if the data are paired.

(a) We would like to know if Intel's stock and Southwest Airlines' stock have similar rates of return. To find out, we take a random sample of 50 days, and record Intel's and Southwest's stock on those same days.

(b) We randomly sample 50 items from Target stores and note the price for each. Then we visit Walmart and collect the price for each of those same 50 items.

(c) A school board would like to determine whether there is a difference in average SAT scores for students at one high school versus another high school in the district. To check, they take a simple random sample of 100 students from each high school.

5.17 Global warming, Part I. Let's consider a limited set of climate data, examining temperature differences in 1948 vs 2018. We sampled 197 locations from the National Oceanic and Atmospheric Administration's (NOAA) historical data, where the data was available for both years of interest. We want to know: were there more days with temperatures exceeding 90°F in 2018 or in 1948?[26] The difference in number of days exceeding 90°F (number of days in 2018 - number of days in 1948) was calculated for each of the 197 locations. The average of these differences was 2.9 days with a standard deviation of 17.2 days. We are interested in determining whether these data provide strong evidence that there were more days in 2018 that exceeded 90°F from NOAA's weather stations.

(a) Is there a relationship between the observations collected in 1948 and 2018? Or are the observations in the two groups independent? Explain.

(b) Write hypotheses for this research in symbols and in words.

(c) Check the conditions required to complete this test. A histogram of the differences is given to the right.

(d) Calculate the test statistic and find the p-value.

(e) Use $\alpha = 0.05$ to evaluate the test, and interpret your conclusion in context.

(f) What type of error might we have made? Explain in context what the error means.

(g) Based on the results of this hypothesis test, would you expect a confidence interval for the average difference between the number of days exceeding 90°F from 1948 and 2018 to include 0? Explain your reasoning.

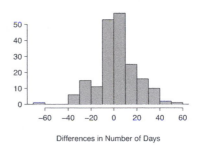

Differences in Number of Days

---

[26]NOAA, www.ncdc.noaa.gov/cdo-web/datasets, April 24, 2019.

5.18 High School and Beyond, Part I. The National Center of Education Statistics conducted a survey of high school seniors, collecting test data on reading, writing, and several other subjects. Here we examine a simple random sample of 200 students from this survey. Side-by-side box plots of reading and writing scores as well as a histogram of the differences in scores are shown below.

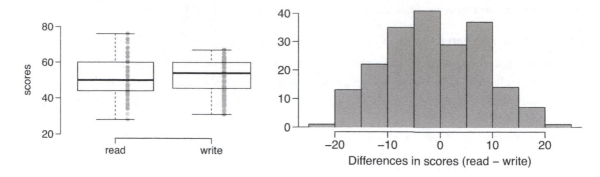

(a) Is there a clear difference in the average reading and writing scores?

(b) Are the reading and writing scores of each student independent of each other?

(c) The average observed difference in scores is $\bar{x}_{read-write} = -0.545$, and the standard deviation of the differences is 8.887 points. Do these data provide convincing evidence of a difference between the average scores on the two exams? Conduct a hypothesis test; interpret your conclusions in context.

(d) Based on the results of this hypothesis test, would you expect a confidence interval for the average difference between the reading and writing scores to include 0? Explain your reasoning.

5.19 Global warming, Part II. We considered the change in the number of days exceeding 90°F from 1948 and 2018 at 197 randomly sampled locations from the NOAA database in Exercise 5.17. The mean and standard deviation of the reported differences are 2.9 days and 17.2 days.

(a) Calculate a 90% confidence interval for the average difference between number of days exceeding 90°F between 1948 and 2018. We've already checked the conditions for you.

(b) Interpret the interval in context.

(c) Does the confidence interval provide convincing evidence that there were more days exceeding 90°F in 2018 than in 1948 at NOAA stations? Explain.

5.20 High school and beyond, Part II. We considered the differences between the reading and writing scores of a random sample of 200 students who took the High School and Beyond Survey in Exercise 5.18. The mean and standard deviation of the differences are $\bar{x}_{read-write} = -0.545$ and 8.887 points.

(a) Calculate a 95% confidence interval for the average difference between the reading and writing scores of all students.

(b) Interpret this interval in context.

(c) Does the confidence interval provide convincing evidence that there is a real difference in the average scores? Explain.

5.21 Gifted children. Researchers collected a simple random sample of 36 children who had been identified as gifted in a large city. The following histograms show the distributions of the IQ scores of mothers and fathers of these children. Also provided are some sample statistics.[27]

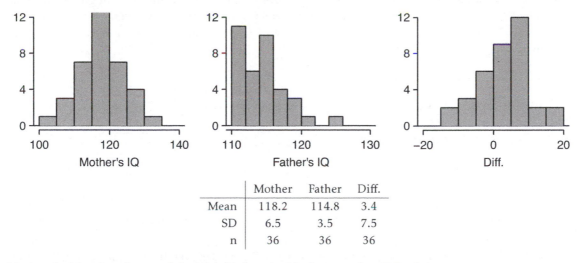

	Mother	Father	Diff.
Mean	118.2	114.8	3.4
SD	6.5	3.5	7.5
n	36	36	36

(a) Are the IQs of mothers and the IQs of fathers in this data set related? Explain.

(b) Conduct a hypothesis test to evaluate if the scores are equal on average. Make sure to clearly state your hypotheses, check the relevant conditions, and state your conclusion in the context of the data.

5.22 DDT exposure. Suppose that you are interested in determining whether exposure to the organochloride DDT, which has been used extensively as an insecticide for many years, is associated with breast cancer in women. As part of a study that investigated this issue, blood was drawn from a sample of women diagnosed with breast cancer over a six-year period and a sample of healthy control subjects matched to the cancer patients on age, menopausal status, and date of blood donation. Each woman's blood level of DDE (an important byproduct of DDT in the human body) was measured, and the difference in levels for each patient and her matched control calculated. A sample of 171 such differences has mean $\bar{d} = 2.7$ ng/mL and standard deviation $s_d = 15.9$ ng/mL. Differences were calculated as $DDE_{cancer} - DDE_{control}$.

(a) Test the null hypothesis that the mean blood levels of DDE are identical for women with breast cancer and for healthy control subjects. What do you conclude?

(b) Would you expect a 95% confidence interval for the true difference in population mean DDE levels to contain the value 0?

5.23 Blue-green eggshells. It is hypothesized that the blue-green color of the eggshells of many avian species represents an informational signal as to the health of the female that laid the eggs. To investigate this hypothesis, researchers conducted a study in which birds assigned to the treatment group were provided with supplementary food before and during laying; they predict that if eggshell coloration is related to female health at laying, females given supplementary food will lay more intensely blue-green eggs than control females. Nests were paired according to when nest construction began, and the study examined 16 nest pairs.

(a) The blue-green chroma (BGC) of eggs was measured on the day of laying; BGC refers to the proportion of total reflectance that is in the blue-green region of the spectrum, with a higher value representing a deeper blue-green color. In the food supplemented group, BGC chroma had $\bar{x} = 0.594$ and $s = 0.010$; in the control group, BGC chroma had $\bar{x} = 0.586$ and $s = 0.009$. A paired t-test resulted in $t = 2.28$ and $p = 0.038$. Interpret the results in the context of the data.

(b) In general, healthier birds are also known to lay heavier eggs. Egg mass was also measured for both groups. In the food supplemented group, egg mass had $\bar{x} = 1.70$ grams and $s = 0.11$ grams; in the control group, egg mass had $\bar{x} = 0.586$ grams and $s = 0.009$ grams. The test statistic from a paired t-test was 2.64 with p-value 0.019. Compute and interpret a 95% confidence interval for δ, the population mean difference in egg mass between the groups.

---

[27]F.A. Graybill and H.K. Iyer. *Regression Analysis: Concepts and Applications.* Duxbury Press, 1994, pp. 511–516.

5.7.3 Two-sample test for independent data

5.24 Diamond prices, Part I. A diamond's price is determined by various measures of quality, including carat weight. The price of diamonds increases as carat weight increases. While the difference between the size of a 0.99 carat diamond and a 1 carat diamond is undetectable to the human eye, the price difference can be substantial.[28]

	0.99 carats	1 carat
Mean	$ 44.51	$ 56.81
SD	$ 13.32	$ 16.13
n	23	23

(a) Use the data to assess whether there is a difference between the average standardized prices of 0.99 and 1 carat diamonds.

(b) Construct a 95% confidence interval for the average difference between the standardized prices of 0.99 and 1 carat diamonds.

5.25 Friday the 13th, Part I. In the early 1990's, researchers in the UK collected data on traffic flow, number of shoppers, and traffic accident related emergency room admissions on Friday the 13th and the previous Friday, Friday the 6th. The histograms below show the distribution of number of cars passing by a specific intersection on Friday the 6th and Friday the 13th for many such date pairs. Also given are some sample statistics, where the difference is the number of cars on the 6th minus the number of cars on the 13th.[29]

Friday the 6th

Friday the 13th

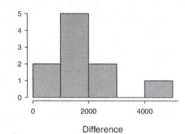

Difference

	6th	13th	Diff.
$\bar{x}$	128,385	126,550	1,835
s	7,259	7,664	1,176
n	10	10	10

(a) Are there any underlying structures in these data that should be considered in an analysis? Explain.

(b) What are the hypotheses for evaluating whether the number of people out on Friday the 6th is different than the number out on Friday the 13th?

(c) Check conditions to carry out the hypothesis test from part (b).

(d) Calculate the test statistic and the p-value.

(e) What is the conclusion of the hypothesis test?

(f) Interpret the p-value in this context.

(g) What type of error might have been made in the conclusion of your test? Explain.

---

[28] H. Wickham. *ggplot2: elegant graphics for data analysis.* Springer New York, 2009.
[29] T.J. Scanlon et al. "Is Friday the 13th Bad For Your Health?" In: *BMJ* 307 (1993), pp. 1584–1586.

5.26 Egg volume. In a study examining 131 collared flycatcher eggs, researchers measured various characteristics in order to study their relationship to egg size (assayed as egg volume, in mm^3). These characteristics included nestling sex and survival. A single pair of collared flycatchers generally lays around 6 eggs per breeding season; laying order of the eggs was also recorded.

(a) Is there evidence at the $\alpha = 0.10$ significance level to suggest that egg size differs between male and female chicks? If so, do heavier eggs tend to contain males or females? For male chicks, $\bar{x} = 1619.95$, $s = 127.54$, and $n = 80$. For female chicks, $\bar{x} = 1584.20$, $s = 102.51$, and $n = 48$. Sex was only recorded for eggs that hatched.

(b) Construct a 95% confidence interval for the difference in egg size between chicks that successfully fledged (developed capacity to fly) and chicks that died in the nest. From the interval, is there evidence of a size difference in eggs between these two groups? For chicks that fledged, $\bar{x} = 1605.87$, $s = 126.32$, and $n = 89$. For chicks that died in the nest, $\bar{x} = 1606.91$, $s = 103.46$, $n = 42$.

(c) Are eggs that are laid first a significantly different size compared to eggs that are laid sixth? For eggs laid first, $\bar{x} = 1581.98$, $s = 155.95$, and $n = 22$. For eggs laid sixth, $\bar{x} = 1659.62$, $s = 124.59$, and $n = 20$.

5.27 Friday the 13th, Part II. The Friday the 13th study reported in Exercise 5.25 also provides data on traffic accident related emergency room admissions. The distributions of these counts from Friday the 6th and Friday the 13th are shown below for six such paired dates along with summary statistics. You may assume that conditions for inference are met.

Friday the 6th

Friday the 13th

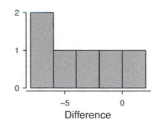

Difference

	6th	13th	diff
Mean	7.5	10.83	-3.33
SD	3.33	3.6	3.01
n	6	6	6

(a) Conduct a hypothesis test to evaluate if there is a difference between the average numbers of traffic accident related emergency room admissions between Friday the 6th and Friday the 13th.

(b) Calculate a 95% confidence interval for the difference between the average numbers of traffic accident related emergency room admissions between Friday the 6th and Friday the 13th.

(c) The conclusion of the original study states, "Friday 13th is unlucky for some. The risk of hospital admission as a result of a transport accident may be increased by as much as 52%. Staying at home is recommended." Do you agree with this statement? Explain your reasoning.

5.28 Avian influenza, Part I. In recent years, widespread outbreaks of avian influenza have posed a global threat to both poultry production and human health. One strategy being explored by researchers involves developing chickens that are genetically resistant to infection. In 2011, a team of investigators reported in *Science* that they had successfully generated transgenic chickens that are resistant to the virus. As a part of assessing whether the genetic modification might be hazardous to the health of the chicks, hatch weights between transgenic chicks and non-transgenic chicks were collected. Does the following data suggest that there is a difference in hatch weights between transgenic and non-transgenic chickens?

	transgenic chicks (g)	non-transgenic chicks (g)
$\bar{x}$	45.14	44.99
s	3.32	4.57
n	54	54

5.29 Chicken diet and weight, Part I. Chicken farming is a multi-billion dollar industry, and any methods that increase the growth rate of young chicks can reduce consumer costs while increasing company profits, possibly by millions of dollars. An experiment was conducted to measure and compare the effectiveness of various feed supplements on the growth rate of chickens. Newly hatched chicks were randomly allocated into six groups, and each group was given a different feed supplement. Below are some summary statistics from this data set along with box plots showing the distribution of weights by feed type.[30]

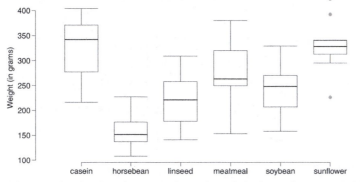

	Mean	SD	n
casein	323.58	64.43	12
horsebean	160.20	38.63	10
linseed	218.75	52.24	12
meatmeal	276.91	64.90	11
soybean	246.43	54.13	14
sunflower	328.92	48.84	12

(a) Describe the distributions of weights of chickens that were fed linseed and horsebean.

(b) Do these data provide strong evidence that the average weights of chickens that were fed linseed and horsebean are different? Use a 5% significance level.

(c) What type of error might we have committed? Explain.

(d) Would your conclusion change if we used $\alpha = 0.01$?

5.30 Fuel efficiency of manual and automatic cars, Part I. Each year the US Environmental Protection Agency (EPA) releases fuel economy data on cars manufactured in that year. Below are summary statistics on fuel efficiency (in miles/gallon) from random samples of cars with manual and automatic transmissions manufactured in 2012. Do these data provide strong evidence of a difference between the average fuel efficiency of cars with manual and automatic transmissions in terms of their average city mileage? Assume that conditions for inference are satisfied.[31]

City MPG		
	Automatic	Manual
Mean	16.12	19.85
SD	3.58	4.51
n	26	26

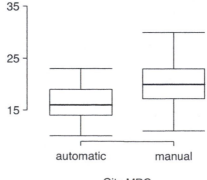

City MPG

---

[30]Chicken Weights by Feed Type, from the datasets package in R..
[31]U.S. Department of Energy, Fuel Economy Data, 2012 Datafile.

5.31 Chicken diet and weight, Part II. Casein is a common weight gain supplement for humans. Does it have an effect on chickens? Using data provided in Exercise 5.29, test the hypothesis that the average weight of chickens that were fed casein is different than the average weight of chickens that were fed soybean. If your hypothesis test yields a statistically significant result, discuss whether or not the higher average weight of chickens can be attributed to the casein diet. Assume that conditions for inference are satisfied.

5.32 Fuel efficiency of manual and automatic cars, Part II. The table provides summary statistics on highway fuel economy of cars manufactured in 2012 (from Exercise 5.30). Use these statistics to calculate a 98% confidence interval for the difference between average highway mileage of manual and automatic cars, and interpret this interval in the context of the data.[32]

		Hwy MPG
	Automatic	Manual
Mean	22.92	27.88
SD	5.29	5.01
n	26	26

5.33 Gaming and distracted eating. A group of researchers are interested in the possible effects of distracting stimuli during eating, such as an increase or decrease in the amount of food consumption. To test this hypothesis, they monitored food intake for a group of 44 patients who were randomized into two equal groups. The treatment group ate lunch while playing solitaire, and the control group ate lunch without any added distractions. Patients in the treatment group ate 52.1 grams of biscuits, with a standard deviation of 45.1 grams, and patients in the control group ate 27.1 grams of biscuits, with a standard deviation of 26.4 grams. Do these data provide convincing evidence that the average food intake (measured in amount of biscuits consumed) is different for the patients in the treatment group? Assume that conditions for inference are satisfied.[33]

5.34 Placebos without deception. While placebo treatment can influence subjective symptoms, it is typically believed that patient response to placebo requires concealment or deception; in other words, a patient must believe that they are receiving an effective treatment in order to experience the benefits of being treated with an inert substance. Researchers recruited patients suffering from irritable bowel syndrome (IBS) to test whether placebo responses are neutralized by awareness that the treatment is a placebo.

Patients were randomly assigned to either the treatment arm or control arm. Those in the treatment arm were given placebo pills, which were described as "something like sugar pills, which have been shown in rigorous clinical testing to produce significant mind-body self-healing processes". Those in the control arm did not receive treatment. At the end of the study, all participants answered a questionnaire called the IBS Global Improvement Scale (IBS-GIS) which measures whether IBS symptoms have improved; higher scores are indicative of more improvement.

At the end of the study, the 37 participants in the open placebo group had IBS-GIS scores with $\bar{x} = 5.0$ and $s = 1.5$, while the 43 participants in the no treatment group had IBS-GIS scores with $\bar{x} = 3.9$ and $s = 1.3$.

Based on an analysis of the data, summarize whether the study demonstrates evidence that placebos administered without deception may be an effective treatment for IBS.

---

[32]U.S. Department of Energy, Fuel Economy Data, 2012 Datafile.

[33]R.E. Oldham-Cooper et al. "Playing a computer game during lunch affects fullness, memory for lunch, and later snack intake". In: *The American Journal of Clinical Nutrition* 93.2 (2011), p. 308.

5.35 Prison isolation experiment, Part I. Subjects from Central Prison in Raleigh, NC, volunteered for an experiment involving an "isolation" experience. The goal of the experiment was to find a treatment that reduces subjects' psychopathic deviant T scores. This score measures a person's need for control or their rebellion against control, and it is part of a commonly used mental health test called the Minnesota Multiphasic Personality Inventory (MMPI) test. The experiment had three treatment groups:

(1) Four hours of sensory restriction plus a 15 minute "therapeutic" tape advising that professional help is available.

(2) Four hours of sensory restriction plus a 15 minute "emotionally neutral" tape on training hunting dogs.

(3) Four hours of sensory restriction but no taped message.

Forty-two subjects were randomly assigned to these treatment groups, and an MMPI test was administered before and after the treatment. Distributions of the differences between pre and post treatment scores (pre - post) are shown below, along with some sample statistics. Use this information to independently test the effectiveness of each treatment. Make sure to clearly state your hypotheses, check conditions, and interpret results in the context of the data.[34]

Treatment 1

Treatment 2

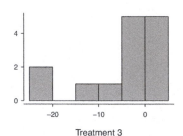

Treatment 3

	Tr 1	Tr 2	Tr 3
Mean	6.21	2.86	-3.21
SD	12.3	7.94	8.57
n	14	14	14

5.7.4 Power calculations for a difference of means

5.36 Email outreach efforts. A medical research group is recruiting people to complete short surveys about their medical history. For example, one survey asks for information on a person's family history in regards to cancer. Another survey asks about what topics were discussed during the person's last visit to a hospital. So far, as people sign up, they complete an average of just 4 surveys, and the standard deviation of the number of surveys is about 2.2. The research group wants to try a new interface that they think will encourage new enrollees to complete more surveys, where they will randomize each enrollee to either get the new interface or the current interface. How many new enrollees do they need for each interface to detect an effect size of 0.5 surveys per enrollee, if the desired power level is 80%?

5.37 Increasing corn yield. A large farm wants to try out a new type of fertilizer to evaluate whether it will improve the farm's corn production. The land is broken into plots that produce an average of 1,215 pounds of corn with a standard deviation of 94 pounds per plot. The owner is interested in detecting any average difference of at least 40 pounds per plot. How many plots of land would be needed for the experiment if the desired power level is 90%? Assume each plot of land gets treated with either the current fertilizer or the new fertilizer.

---

[34]Prison isolation experiment, stat.duke.edu/resources/datasets/prison-isolation.

5.7.5 Comparing means with ANOVA

5.38 Fill in the blank. When doing an ANOVA, you observe large differences in means between groups. Within the ANOVA framework, this would most likely be interpreted as evidence strongly favoring the _____ hypothesis.

5.39 Chicken diet and weight, Part III. In Exercises 5.29 and 5.31 we compared the effects of two types of feed at a time. A better analysis would first consider all feed types at once: casein, horsebean, linseed, meat meal, soybean, and sunflower. The ANOVA output below can be used to test for differences between the average weights of chicks on different diets.

	Df	Sum Sq	Mean Sq	F value	Pr(>F)
feed	5	231,129.16	46,225.83	15.36	0.0000
Residuals	65	195,556.02	3,008.55		

Conduct a hypothesis test to determine if these data provide convincing evidence that the average weight of chicks varies across some (or all) groups. Make sure to check relevant conditions. Figures and summary statistics are shown below.

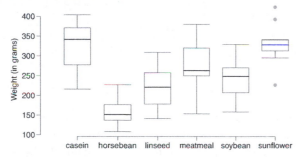

	Mean	SD	n
casein	323.58	64.43	12
horsebean	160.20	38.63	10
linseed	218.75	52.24	12
meatmeal	276.91	64.90	11
soybean	246.43	54.13	14
sunflower	328.92	48.84	12

5.40 Teaching descriptive statistics. A study compared five different methods for teaching descriptive statistics. The five methods were traditional lecture and discussion, programmed textbook instruction, programmed text with lectures, computer instruction, and computer instruction with lectures. 45 students were randomly assigned, 9 to each method. After completing the course, students took a 1-hour exam.

(a) What are the hypotheses for evaluating if the average test scores are different for the different teaching methods?

(b) What are the degrees of freedom associated with the F-test for evaluating these hypotheses?

(c) Suppose the p-value for this test is 0.0168. What is the conclusion?

5.41 Coffee, depression, and physical activity. Caffeine is the world's most widely used stimulant, with approximately 80% consumed in the form of coffee. Participants in a study investigating the relationship between coffee consumption and exercise were asked to report the number of hours they spent per week on moderate (e.g., brisk walking) and vigorous (e.g., strenuous sports and jogging) exercise. Based on these data the researchers estimated the total hours of metabolic equivalent tasks (MET) per week, a value always greater than 0. The table below gives summary statistics of MET for women in this study based on the amount of coffee consumed.[35]

	≤ 1 cup/week	2-6 cups/week	1 cup/day	2-3 cups/day	≥ 4 cups/day	Total
Mean	18.7	19.6	19.3	18.9	17.5	
SD	21.1	25.5	22.5	22.0	22.0	
n	12,215	6,617	17,234	12,290	2,383	50,739

Caffeinated coffee consumption

(a) Write the hypotheses for evaluating if the average physical activity level varies among the different levels of coffee consumption.

(b) Check conditions and describe any assumptions you must make to proceed with the test.

(c) Below is part of the output associated with this test. Fill in the empty cells.

	Df	Sum Sq	Mean Sq	F value	Pr(>F)
coffee					0.0003
Residuals		25,564,819			
Total		25,575,327			

(d) What is the conclusion of the test?

5.42 Student performance across discussion sections. A professor who teaches a large introductory statistics class (197 students) with eight discussion sections would like to test if student performance differs by discussion section, where each discussion section has a different teaching assistant. The summary table below shows the average final exam score for each discussion section as well as the standard deviation of scores and the number of students in each section.

	Sec 1	Sec 2	Sec 3	Sec 4	Sec 5	Sec 6	Sec 7	Sec 8
n_i	33	19	10	29	33	10	32	31
$\bar{x}_i$	92.94	91.11	91.80	92.45	89.30	88.30	90.12	93.35
s_i	4.21	5.58	3.43	5.92	9.32	7.27	6.93	4.57

The ANOVA output below can be used to test for differences between the average scores from the different discussion sections.

	Df	Sum Sq	Mean Sq	F value	Pr(>F)
section	7	525.01	75.00	1.87	0.0767
Residuals	189	7584.11	40.13		

Conduct a hypothesis test to determine if these data provide convincing evidence that the average score varies across some (or all) groups. Check conditions and describe any assumptions you must make to proceed with the test.

---

[35]M. Lucas et al. "Coffee, caffeine, and risk of depression among women". In: *Archives of internal medicine* 171.17 (2011), p. 1571.

5.43 GPA and major. Undergraduate students taking an introductory statistics course at Duke University conducted a survey about GPA and major. The side-by-side box plots show the distribution of GPA among three groups of majors. Also provided is the ANOVA output.

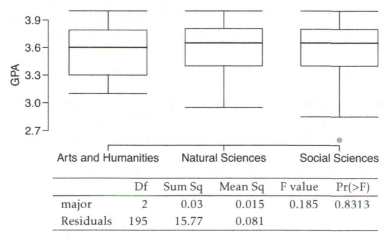

	Df	Sum Sq	Mean Sq	F value	Pr(>F)
major	2	0.03	0.015	0.185	0.8313
Residuals	195	15.77	0.081		

(a) Write the hypotheses for testing for a difference between average GPA across majors.

(b) What is the conclusion of the hypothesis test?

(c) How many students answered these questions on the survey, i.e. what is the sample size?

5.44 Work hours and education. The General Social Survey collects data on demographics, education, and work, among many other characteristics of US residents.[36] Using ANOVA, we can consider educational attainment levels for all 1,172 respondents at once. Below are the distributions of hours worked by educational attainment and relevant summary statistics that will be helpful in carrying out this analysis.

	Educational attainment					
	Less than HS	HS	Jr Coll	Bachelor's	Graduate	Total
Mean	38.67	39.6	41.39	42.55	40.85	40.45
SD	15.81	14.97	18.1	13.62	15.51	15.17
n	121	546	97	253	155	1,172

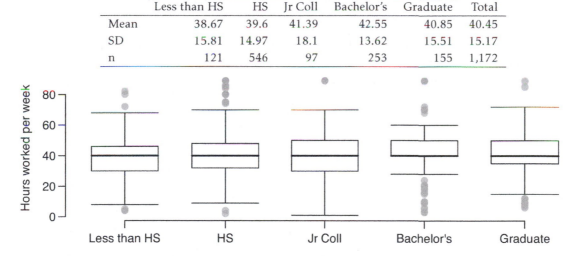

(a) Write hypotheses for evaluating whether the average number of hours worked varies across the five groups.

(b) Check conditions and describe any assumptions you must make to proceed with the test.

(c) Below is part of the output associated with this test. Fill in the empty cells.

	Df	Sum Sq	Mean Sq	F value	Pr(>F)
degree			501.54		0.0682
Residuals		267,382			
Total					

(d) What is the conclusion of the test?

---

[36]National Opinion Research Center, General Social Survey, 2010.

5.45 True / False: ANOVA, Part I. Determine if the following statements are true or false in ANOVA, and explain your reasoning for statements you identify as false.

(a) As the number of groups increases, the modified significance level for pairwise tests increases as well.

(b) As the total sample size increases, the degrees of freedom for the residuals increases as well.

(c) The constant variance condition can be somewhat relaxed when the sample sizes are relatively consistent across groups.

(d) The independence assumption can be relaxed when the total sample size is large.

5.46 Child care hours. The China Health and Nutrition Survey aims to examine the effects of the health, nutrition, and family planning policies and programs implemented by national and local governments.[37] It, for example, collects information on number of hours Chinese parents spend taking care of their children under age 6. The side-by-side box plots below show the distribution of this variable by educational attainment of the parent. Also provided below is the ANOVA output for comparing average hours across educational attainment categories.

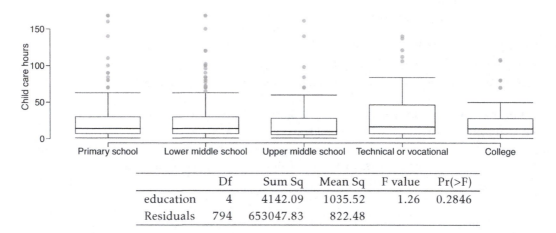

	Df	Sum Sq	Mean Sq	F value	Pr(>F)
education	4	4142.09	1035.52	1.26	0.2846
Residuals	794	653047.83	822.48		

(a) Write the hypotheses for testing for a difference between the average number of hours spent on child care across educational attainment levels.

(b) What is the conclusion of the hypothesis test?

---

[37] UNC Carolina Population Center, China Health and Nutrition Survey, 2006.

5.47 Prison isolation experiment, Part II. Exercise 5.35 introduced an experiment that was conducted with the goal of identifying a treatment that reduces subjects' psychopathic deviant T scores, where this score measures a person's need for control or his rebellion against control. In Exercise 5.35 you evaluated the success of each treatment individually. An alternative analysis involves comparing the success of treatments. The relevant ANOVA output is given below.

	Df	Sum Sq	Mean Sq	F value	Pr(>F)
treatment	2	639.48	319.74	3.33	0.0461
Residuals	39	3740.43	95.91		

$$s_{pooled} = 9.793 \text{ on } df = 39$$

(a) What are the hypotheses?

(b) What is the conclusion of the test? Use a 5% significance level.

(c) If in part (b) you determined that the test is significant, conduct pairwise tests to determine which groups are different from each other. If you did not reject the null hypothesis in part (b), recheck your answer.

5.48 True / False: ANOVA, Part II. Determine if the following statements are true or false, and explain your reasoning for statements you identify as false.

If the null hypothesis that the means of four groups are all the same is rejected using ANOVA at a 5% significance level, then ...

(a) we can then conclude that all the means are different from one another.

(b) the standardized variability between groups is higher than the standardized variability within groups.

(c) the pairwise analysis will identify at least one pair of means that are significantly different.

(d) the appropriate α to be used in pairwise comparisons is $0.05 / 4 = 0.0125$ since there are four groups.

Chapter 6

Simple linear regression

The relationship between two numerical variables can be visualized using a scatterplot in the xy-plane. The **predictor** or **explanatory variable** is plotted on the horizontal axis, while the **response variable** is plotted on the vertical axis.[1]

This chapter explores simple linear regression, a technique for estimating a straight line that best fits data on a scatterplot.[2] A line of best fit functions as a linear model that can not only be used for prediction, but also for inference. Linear regression should only be used with data that exhibit linear or approximately linear relationships.

For example, scatterplots in Chapter 1 illustrated the linear relationship between height and weight in the NHANES data, with height as a predictor of weight. Adding a best-fitting line to these data using regression techniques would allow for prediction of an individual's weight based on their height. The linear model could also be used to investigate questions about the population-level relationship between height and weight, since the data are a random sample from the population of adults in the United States.

Not all relationships in data are linear. For example, the scatterplot in Figure 1.28 of Chapter 1 shows a highly non-linear relationship between between annual per capita income and life expectancy for 165 countries in 2011. Relationships are called **strong relationships** if the pattern of the dependence between the predictor and response variables is clear, even if it is nonlinear as in Figure 1.28. A **weak relationship** is one in which the points in the scatterplot are so diffuse as to make it difficult to discern any relationship. Figure 1.29 in Chapter 1 showed relationships progressing from weak to strong moving from left to right in the top and bottom panels. Each of the relationships shown in the second panels from the left are **moderate relationships**. Finally, changing the scale of measurement of one or both variables, such as changing age from age in years to age in months, simply stretches or compresses one or both axes and does not change the nature of the relationship. If a relationship is linear it will remain so, and with a simple change of scale, a nonlinear relationship will remain nonlinear.

---

[1] Sometimes, the predictor variable is referred to as the independent variable, and the response variable referred to as the dependent variable.

[2] Although the response variable in linear regression is necessarily numerical, the predictor variable can be numerical or categorical.

The next chapter covers multiple regression, a statistical model used to estimate the relationship between a single numerical response variable and several predictor variables.

For labs, slides, and other resources, please visit
www.openintro.org/book/biostat

6.1 Examining scatterplots

Various demographic and cardiovascular risk factors were collected as a part of the Prevention of REnal and Vascular END-stage Disease (PREVEND) study, which took place in the Netherlands. The initial study population began as 8,592 participants aged 28-75 years who took a first survey in 1997-1998.[3] Participants were followed over time; 6,894 participants took a second survey in 2001-2003, and 5,862 completed the third survey in 2003-2006. In the third survey, measurement of cognitive function was added to the study protocol. Data from 4,095 individuals who completed cognitive testing are in the prevend dataset, available in the R package oibiostat.

As adults age, cognitive function changes over time, largely due to various cerebrovascular and neurodegenerative changes. It is thought that cognitive decline is a long-term process that may start as early as 45 years of age.[4] The Ruff Figural Fluency Test (RFFT) is one measure of cognitive function that provides information about cognitive abilities such as planning and the ability to switch between different tasks. The test consists of drawing as many unique designs as possible from a pattern of dots, under timed conditions; scores range from 0 to 175 points (worst and best score, respectively).

RFFT scores for a random sample of 500 individuals are shown in Figure 6.1, plotted against age at enrollment, which is measured in years. The variables Age and RFFT are negatively associated; older participants tend to have lower cognitive function. There is an approximately linear trend observable in the data, which suggests that adding a line could be useful for summarizing the relationship between the two variables.

It is important to avoid adding straight lines to non-linear data, such as in Figure 1.28.

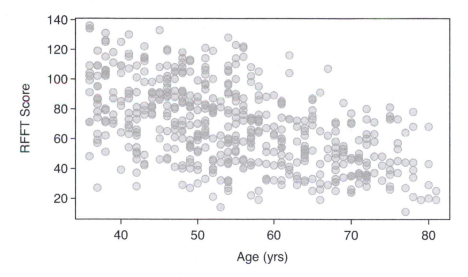

Figure 6.1: A scatterplot showing age vs. RFFT. Age is the predictor variable, while RFFT score is the response variable.

---

[3]Participants were selected from the city of Groningen on the basis of their urinary albumin excretion; urinary albumin excretion is known to be associated with abnormalities in renal function.

[4]Joosten H, et al. Cardiovascular risk profile and cognitive function in young, middle-aged, and elderly subjects. Stroke. 2013;44:1543-1549, https://doi.org/10.1161/STROKEAHA.111.000496

The following conditions should be true in a scatterplot for a line to be considered a reasonable approximation to the relationship in the plot and for the application of the methods of inference discussed later in the chapter:

1 **Linearity.** The data shows a linear trend. If there is a nonlinear trend, an advanced regression method should be applied; such methods are not covered in this text. Occasionally, a transformation of the data will uncover a linear relationship in the transformed scale.

2 **Constant variability.** The variability of the response variable about the line remains roughly constant as the predictor variable changes.

3 **Independent observations.** The (x, y) pairs are independent; i.e., the value of one pair provides no information about other pairs. Be cautious about applying regression to sequential observations in time (**time series** data), such as height measurements taken over the course of several years. Time series data may have a complex underlying structure, and the relationship between the observations should be accounted for in a model.

4 **Residuals that are approximately normally distributed.** This condition can be checked only after a line has been fit to the data and will be explained in Section 6.3.1, where the term residual is defined. In large datasets, it is sufficient for the residuals to be approximately symmetric with only a few outliers. This condition becomes particularly important when inferences are made about the line, as discussed in Section 6.4.

GUIDED PRACTICE 6.1

Figure 6.2 shows the relationship between clutch.volume and body.size in the frog data. The plot also appears as Figure 1.26 in Chapter 1. Are the first three conditions met for linear regression?[5]

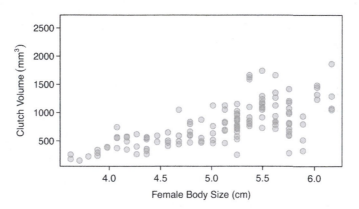

Figure 6.2: A plot of clutch.volume versus body.size in the frog data.

---

[5]No. While the relationship appears linear and it is reasonable to assume the observations are independent (based on information about the frogs given in Chapter 1), the variability in clutch.volume is noticeably less for smaller values of body.size than for larger values.

6.2 Estimating a regression line using least squares

Figure 6.3 shows the scatterplot of age versus RFFT score, with the **least squares regression line** added to the plot; this line can also be referred to as a **linear model** for the data. An RFFT score can be predicted for a given age from the equation of the regression line:

$$\widehat{RFFT} = 137.55 - 1.26(\text{age}).$$

The vertical distance between a point in the scatterplot and the predicted value on the regression line is the **residual** for the observation represented by the point; observations below the line have negative residuals, while observations above the line have positive residuals. The size of a residual is usually discussed in terms of its absolute value; for example, a residual of -13 is considered larger than a residual of 5.

For example, consider the predicted RFFT score for an individual of age 56. According to the linear model, this individual has a predicted score of $137.550 - 1.261(56) = 66.934$ points. In the data, however, there is a participant of age 56 with an RFFT score of 72; their score is about 5 points higher than predicted by the model (this observation is shown on the plot with a "×").

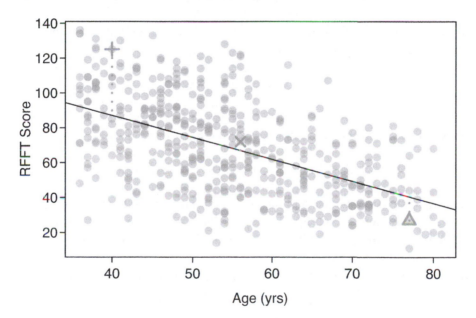

Figure 6.3: A scatterplot showing age (horizontal axis) vs. RFFT (vertical axis) with the regression line added to the plot. Three observations are marked in the figure; the one marked by a "+" has a large residual of about +38, the one marked by a "×" has a small residual of about +5, and the one marked by a "△" has a moderate residual of about -13. The vertical dotted lines extending from the observations to the regression line represent the residuals.

RESIDUAL: DIFFERENCE BETWEEN OBSERVED AND EXPECTED

The residual of the ith observation (x_i, y_i) is the difference of the observed response (y_i) and the response predicted based on the model fit $(\widehat{y_i})$:

$$e_i = y_i - \widehat{y_i}$$

The value $\widehat{y_i}$ is calculated by plugging x_i into the model equation.

The **least squares regression line** is the line which minimizes the sum of the squared residuals for all the points in the plot. Let $\hat{y}_i$ be the predicted value for an observation with value x_i for the explanatory variable. The value $e_i = y_i - \hat{y}_i$ is the residual for a data point (x_i, y_i) in a scatterplot with n pairs of points. The least squares line is the line for which

$$e_1^2 + e_2^2 + \cdots + e_n^2 \tag{6.2}$$

is smallest.

For a general population of ordered pairs (x, y), the **population regression model** is

$$y = \beta_0 + \beta_1 x + \varepsilon.$$

The term ε is a normally distributed 'error term' that has mean 0 and standard deviation σ. Since $E(\varepsilon) = 0$, the model can also be written

$$E(Y|x) = \beta_0 + \beta_1 x,$$

where the notation $E(Y|x)$ denotes the expected value of Y when the predictor variable has value x.[6] For the PREVEND data, the population regression line can be written as

$$\text{RFFT} = \beta_0 + \beta_1(\text{age}) + \varepsilon, \text{ or as } E(\text{RFFT}|\text{age}) = \beta_0 + \beta_1(\text{age}).$$

The term β_0 is the vertical intercept for the line (often referred to simply as the intercept) and β_1 is the slope. The notation b_0 and b_1 are used to represent the point estimates of the parameters β_0 and β_1. The point estimates b_0 and b_1 are estimated from data; β_0 and β_1 are parameters from the population model for the regression line.

b_0, b_1

Sample estimates of β_0, β_1

The regression line can be written as $\hat{y} = b_0 + b_1(x)$, where $\hat{y}$ represents the predicted value of the response variable. The slope of the least squares line, b_1, is estimated by

$$b_1 = \frac{s_y}{s_x} r, \tag{6.3}$$

where r is the correlation between the two variables, and s_x and s_y are the sample standard deviations of the explanatory and response variables, respectively. The intercept for the regression line is estimated by

$$b_0 = \overline{y} - b_1 \overline{x}. \tag{6.4}$$

Typically, regression lines are estimated using statistical software.

---

[6]The error term ε can be thought of as a population parameter for the residuals (e). While ε is a theoretical quantity that refers to the deviation between an observed value and $E(Y|x)$, a residual is calculated as the deviation between an observed value and the prediction from the linear model.

EXAMPLE 6.5

From the summary statistics displayed in Figure 6.4 for prevend.samp, calculate the equation of the least-squares regression line for the PREVEND data.

$$b_1 = \frac{s_y}{s_x}r = \frac{27.40}{11.60}(-0.534) = -1.26$$

$$b_0 = \overline{y} - b_1\overline{x} = 68.40 - (-1.26)(54.82) = 137.55.$$

The results agree with the equation shown at the beginning of this section:

$$\widehat{RFFT} = 137.55 - 1.26(age).$$

	Age (yrs)	RFFT score
mean	$\overline{x} = 54.82$	$\overline{y} = 68.40$
standard deviation	$s_x = 11.60$	$s_y = 27.40$
		$r = -0.534$

Figure 6.4: Summary statistics for age and RFFT from prevend.samp.

GUIDED PRACTICE 6.6

Figure 6.5 shows the relationship between height and weight in a sample from the NHANES dataset introduced in Chapter 1. Calculate the equation of the regression line given the summary statistics: $\overline{x} = 168.78, \overline{y} = 83.83, s_x = 10.29, s_y = 21.04, r = 0.410.$[7]

Figure 6.5: A plot of Height versus Weight in nhanes.samp.adult.500, with a least-squares regression line

GUIDED PRACTICE 6.7

Predict the weight in pounds for an adult who is 5 feet, 11 inches tall. 1 cm = .3937 in; 1 lb = 0.454 kg.[8]

---

[7]The equation of the line is $\widehat{weight} = -57.738 + 0.839(height)$, where height is in centimeters and weight is in kilograms.

[8]5 feet, 11 inches equals 71/.3937 = 180.34 centimeters. From the regression equation, the predicted weight is −57.738 + 0.839(180.34) = 93.567 kilograms. In pounds, this weight is 93.567/0.454 = 206.280.

6.3 Interpreting a linear model

A least squares regression line functions as a statistical model that can be used to estimate the relationship between an explanatory and response variable. While the calculations for constructing a regression line are relatively simple, interpreting the linear model is not always straightforward. In addition to discussing the mathematical interpretation of model parameters, this section also addresses methods for assessing whether a linear model is an appropriate choice, interpreting categorical predictors, and identifying outliers.

The slope parameter of the regression line specifies how much the line rises (positive slope) or declines (negative slope) for one unit of change in the explanatory variable. In the PREVEND data, the line decreases by 1.26 points for every increase of 1 year. However, it is important to clarify that RFFT score *tends* to decrease as age increases, with *average* RFFT score decreasing by 1.26 points for each additional year of age. As visible from the scatter of the data around the line, the line does not perfectly predict RFFT score from age; if this were the case, all the data would fall exactly on the line.

When interpreting the slope parameter, it is also necessary to avoid phrasing indicative of a causal relationship, since the line describes an association from data collected in an observational study. From these data, it is not possible to conclude that increased age causes a decline in cognitive function.[9]

Mathematically, the intercept on the vertical axis is a predicted value on the line when the explanatory variable has value 0. In biological or medical examples, 0 is rarely a meaningful value of the explanatory variable. For example, in the PREVEND data, the linear model predicts a score of 137.55 when age is 0—however, it is nonsensical to predict an RFFT score for a newborn infant.

In fact, least squares lines should never be used to extrapolate values outside the range of observed values. Since the PREVEND data only includes participants between ages 36 and 81, it should not be used to predict RFFT scores for people outside that age range. The nature of a relationship may change for very small or very large values of the explanatory variable; for example, if participants between ages 15 and 25 were studied, a different relationship between age and RFFT scores might be observed. Even making predictions for values of the explanatory variable slightly larger than the minimum or slightly smaller than the maximum can be dangerous, since in many datasets, observations near the minimum or maximum values (of the explanatory variable) are sparse.

Linear models are useful tools for summarizing a relationship between two variables, but it is important to be cautious about making potentially misleading claims based on a regression line. The following subsection discusses two commonly used approaches for examining whether a linear model can reasonably be applied to a dataset.

6.3.1 Checking residuals from a linear model

Recall that there are four assumptions that must be met for a linear model to be considered reasonable: linearity, constant variability, independent observations, normally distributed residuals. In the PREVEND data, the relationship between RFFT score and age appears approximately linear, and it is reasonable to assume that the data points are independent. To check the assumptions of constant variability around the line and normality of the residuals, it is helpful to consult residual plots and normal probability plots (Section 3.3.7).[10]

---

[9]Similarly, avoid language such as increased age *leads to* or *produces* lower RFFT scores.

[10]While simple arithmetic can be used to calculate the residuals, the size of most datasets makes hand calculations impractical. The plots here are based on calculations done in R.

Examining patterns in residuals

There are a variety of residual plots used to check the fit of a least squares line. The plots shown in this text are scatterplots in which the residuals are plotted on the vertical axis against predicted values from the model on the horizontal axis. Other residual plots may instead show values of the explanatory variable or the observed response variable on the horizontal axis. When a least squares line fits data very well, the residuals should scatter about the horizontal line $y = 0$ with no apparent pattern.

Figure 6.6 shows three residual plots from simulated data; the plots on the left show data plotted with the least squares regression line, and the plots on the right show residuals on the y-axis and predicted values on the x-axis. A linear model is a particularly good fit for the data in the first row, where the residual plot shows random scatter above and below the horizontal line. In the second row, the original data cycles below and above the regression line; this nonlinear pattern is more evident in the residual plot. In the last row, the variability of the residuals is not constant; the residuals are slightly more variable for larger predicted values.

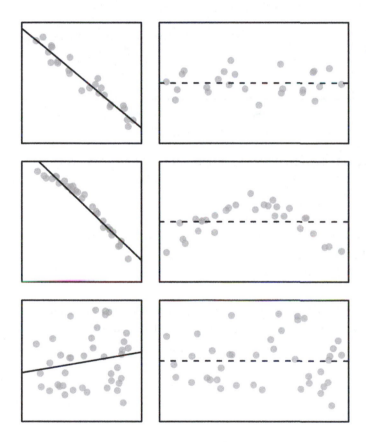

Figure 6.6: Sample data with their best fitting lines (left) and their corresponding residual plots (right).

Figure 6.7 shows a residual plot from the estimated linear model $\widehat{RFFT} = 137.55 - 1.26(age)$. While the residuals show scatter around the line, there is less variability for lower predicted RFFT scores. A data analyst might still decide to use the linear model, with the knowledge that predictions of high RFFT scores may not be as accurate as for lower scores. Reading a residual plot critically can reveal weaknesses about a linear model that should be taken into account when interpreting model results. More advanced regression methods beyond the scope of this text may be more suitable for these data.

Figure 6.7: Residual plot for the model in Figure 6.3 using `prevend.samp`.

EXAMPLE 6.8

Figure 6.8 shows a residual plot for the model predicting weight from height using the sample of 500 adults from the NHANES data, `nhanes.samp.adult.500`. Assess whether the constant variability assumption holds for the linear model.

The residuals above the line are more variable, taking on more extreme values than those below the line. Larger than expected residuals imply that there are many large weights that are underpredicted; in other words, the model is less accurate at predicting relatively large weights.

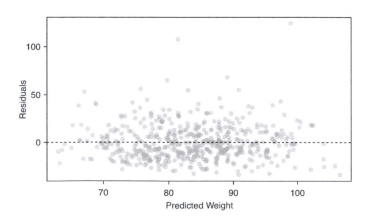

Figure 6.8: A residual plot from the linear model for height versus weight in `nhanes.samp.adult.500`.

Checking normality of the residuals

The normal probability plot, introduced in Section 3.3.7, is best suited for checking normality of the residuals, since normality can be difficult to assess using histograms alone. Figure 6.9 shows both the histogram and normal probability plot of the residuals after fitting a least squares regression to the age versus RFFT data.

Figure 6.9: A histogram and normal probability plot of the residuals from the linear model for RFFT versus Age in prevend.samp.

The normal probability plot shows that the residuals are nearly normally distributed, with only slight deviations from normality in the left and right tails.

GUIDED PRACTICE 6.9

G

Figure 6.10 shows a histogram and normal probability plot for the linear model to predict weight from height in nhanes.samp.adult.500. Evaluate the normality of the residuals.[11]

Figure 6.10: A histogram and normal probability plot of the residuals from the linear model for height versus weight in nhanes.samp.adult.500.

---

[11] The data are roughly normal, but there are deviations from normality in the tails, particularly the upper tail. There are some relatively large observations, which is evident from the residual plot shown in Figure 6.8.

6.3.2   Using R^2 to describe the strength of a fit

The correlation coefficient r measures the strength of the linear relationship between two variables. However, it is more common to measure the strength of a linear fit using r^2, which is commonly written as R^2 in the context of regression.[12]

The quantity R^2 describes the amount of variation in the response that is explained by the least squares line. While R^2 can be easily calculated by simply squaring the correlation coefficient, it is easier to understand the interpretation of R^2 by using an alternative formula:

$$R^2 = \frac{\text{variance of predicted } y\text{-values}}{\text{variance of observed } y\text{-values}}.$$

It is possible to show that R^2 can also be written

$$R^2 = \frac{s_y^2 - s_{\text{residuals}}^2}{s_y^2}.$$

In the linear model predicting RFFT scores from age, the predicted values on the least squares line are the values of RFFT that are 'explained' by the linear model. The variability of the residuals about the line represents the remaining variability after the prediction; i.e., the variability unexplained by the model. For example, if a linear model perfectly captured all the data, then the variance of the predicted y-values would be equal to the variance of the observed y-values, resulting in $R^2 = 1$. In the linear model for $\widehat{RFFT}$, the proportion of variability explained is

$$R^2 = \frac{s_{RFFT}^2 - s_{\text{residuals}}^2}{s_{RFFT}^2} = \frac{750.52 - 536.62}{750.52} = \frac{213.90}{750.52} = 0.285,$$

about 29%. This is equal to the square of the correlation coefficient, $r^2 = -0.534^2 = 0.285$.

Since R^2 in simple linear regression is simply the square of the correlation coefficient between the predictor and the response, it does not add a new tool to regression. It becomes much more useful in models with several predictors, where it has the same interpretation as the proportion of variability explained by a model but is no longer the square of any one of the correlation coefficients between the individual responses and the predictor. Those models are discussed in Chapter 7.

GUIDED PRACTICE 6.10

In the NHANES data, the variance of Weight is 442.53 kg^2 and the variance of the residuals is 368.1. What proportion of the variability in the data is explained by the model?[13]

GUIDED PRACTICE 6.11

If a linear model has a very strong negative relationship with a correlation of -0.97, how much of the variation in the response is explained by the explanatory variable?[14]

---

[12]In software output, R^2 is usually labeled **R-squared**.

[13]About 16.8%: $\frac{s_{\text{weight}}^2 - s_{\text{residuals}}^2}{s_{\text{weight}}^2} = \frac{442.53 - 368.1}{442.53} = \frac{74.43}{442.53} = 0.168$

[14]About $R^2 = (-0.97)^2 = 0.94$ or 94% of the variation is explained by the linear model.

6.3.3 Categorical predictors with two levels

Although the response variable in linear regression is necessarily numerical, the predictor variable may be either numerical or categorical. This section explores the association between a country's infant mortality rate and whether or not 50% of the population has access to adequate sanitation facilities.

The World Development Indicators (WDI) is a database of country-level variables (i.e., indicators) recording outcomes for a variety of topics, including economics, health, mortality, fertility, and education.[15] The dataset wdi.2011 contains a subset of variables on 165 countries from the year 2011.[16] The infant mortality rate in a country is recorded as the number of deaths in the first year of life per 1,000 live births. Access to sanitation is recorded as the percentage of the population with adequate disposal facilities for human waste. Due to the availability of death certificates, infant mortality is measured reasonably accurately throughout the world. However, it is more difficult to obtain precise measurements of the percentage of a population with access to adequate sanitation facilities; instead, considering whether half the population has such access may be a more reliable measure. The analysis presented here is based on 163 of the 165 countries; the values for access to sanitation are missing for New Zealand and Turkmenistan.

Figure 6.11(a) shows that infant mortality rates are highly right-skewed, with a relatively small number of countries having high infant mortality rates. In 13 countries, infant mortality rates are higher than 70 deaths per thousand live births. Figure 6.11(b) shows infant mortality after a log transformation; the following analysis will use the more nearly symmetric transformed version of inf.mortality.

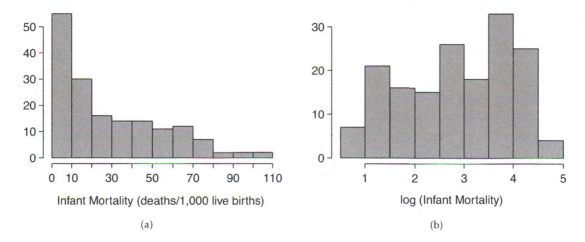

Figure 6.11: (a) Histogram of infant mortality, measured in deaths per 1,000 live births in the first year of life. (b) Histogram of the log-transformed infant mortality.

---

[15]http://data.worldbank.org/data-catalog/world-development-indicators

[16]The data were collected by a Harvard undergraduate in the Statistics department, and are accessible via the oibiostat package.

Figure 6.12 shows a scatterplot of log(inf.mortality) against the categorical variable for sanitation access, coded 1 if at least 50% of the population has access to adequate sanitation, and 0 otherwise. Since there are only two values of the predictor, the values of infant mortality are stacked above the two predictor values 0 and 1.[17]

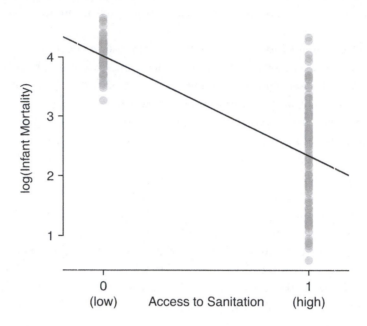

Figure 6.12: Country-level infant mortality rates, divided into low access ($x = 0$) and high access ($x = 1$) to sanitation. The least squares regression line is also shown.

The least squares regression line has the form

$$\log(\overline{\texttt{inf.mortality}}) = b_0 + b_1(\texttt{sanit.access}). \tag{6.12}$$

The estimated least squares regression line has intercept and slope parameters of 4.018 and -1.681, respectively. While the scatterplot appears unlike those for two numerical variables, the interpretation of the parameters remains unchanged. The slope, -1.681, is the estimated change in the logarithm of infant mortality when the categorical predictor changes from low access to sanitation facilities to high access. The intercept term 4.018 is the estimated log infant mortality for the set of countries where less than 50% of the population has access to adequate sanitation facilities (sanit.access = 0).

---

[17]Typically, side-by-side boxplots are used to display the relationship between a numerical variable and a categorical variable. In a regression context, it can be useful to use a scatterplot instead, in order to see the variability around the regression line.

Using the model in Equation 6.12, the prediction equation can be written

$$\log(\widehat{\texttt{inf.mortality}}) = 4.018 - 1.681(\texttt{sanit.access}).$$

Exponentiating both sides of the equation yields

$$\widehat{\texttt{inf.mortality}} = e^{4.018 - 1.681(\texttt{sanit.access})}.$$

When sanit.access = 0, the equation simplifies to $e^{4.018} = 55.590$ deaths among 1,000 live births; this is the estimated infant mortality rate in the countries with low access to sanitation facilities. When sanit.access = 1, the estimated infant mortality rate is $e^{4.018-1.681(1)} = e^{2.337} = 10.350$ deaths per 1,000 live births. The infant mortality rate drops by a factor of 0.186; i.e., the mortality rate in the high access countries is approximately 20% of that in the low access countries.[18]

EXAMPLE 6.13

Check the assumptions of constant variability around the regression line and normality of the residuals in the model for the relationship between the transformed infant mortality variable and access to sanitation variable. Residual plots are shown in Figure 6.13.

While the normal probability plot does show that the residuals are approximately normally distributed, the residual plot reveals that variability is far from constant around the two predictors. Another method for assessing the relationship between the two groups is advisable; this is discussed further in Section 6.4.

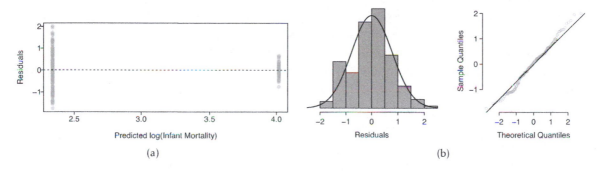

(a)   (b)

Figure 6.13: (a) Residual plot of log(inf.mortality) and sanit.access. (b) Histogram and normal probability plot of the residuals.

6.3.4 Outliers in regression

Depending on their position, data points in a scatterplot have varying degrees of contribution to the estimated parameters of a regression line. Points that are at particularly low or high values of the predictor (x) variable are said to have **high leverage**, and have a large influence on the estimated intercept and slope of the regression line; observations with x values closer to the center of the distribution of x do not have a large effect on the slope.

---

[18]When examining event rates in public health, associations are typically measured using rate ratios rather than rate differences.

A data point in a scatterplot is considered an **outlier in regression** if its value for the response (y) variable does not follow the general linear trend in the data. Outliers that sit at extreme values of the predictor variable (i.e., have high leverage) have the potential to contribute disproportionately to the estimated parameters of a regression line. If an observation does have a strong effect on the estimates of the line, such that estimates change substantially when the point is omitted, the observation is **influential**. These terms are formally defined in advanced regression courses.

This section examines the relationship between infant mortality and number of doctors, using data for each state and the District of Columbia.[19] Infant mortality is measured as the number of infant deaths in the first year of life per 1,000 live births, and number of doctors is recorded as number of doctors per 100,000 members of the population. Figure 6.14 shows scatterplots with infant mortality on the y-axis and number of doctors on the x-axis.

One point in Figure 6.14(a), marked in red, is clearly distant from the main cluster of points. This point corresponds to the District of Columbia, where there were approximately 807.2 doctors per 100,000 members of the population, and the infant mortality rate was 11.3 per 1,000 live births. Since 807.2 is a high value for the predictor variable, this observation has high leverage. It is also an outlier; the other points exhibit a downward sloping trend as the number of doctors increases, but this point, with an unusually high y-value paired with a high x-value, does not follow the trend.

Figure 6.14(b) illustrates that the DC observation is influential. Not only does the observation simply change the numerical value of the slope parameter, it reverses the direction of the linear trend; the regression line fitted with the complete dataset has a positive slope, but the line re-fitted without the DC observation has a negative slope. The large number of doctors per population is due to the presence of several large medical centers in an area with a population that is much smaller than a typical state.

It seems natural to ask whether or not an influential point should be removed from a dataset, but that may not be the right question. Instead, it is usually more important to assess whether the influential point might be an error in the data, or whether it belongs in the dataset. In this case, the District of Columbia has certain characteristics that may make comparisons with other states inappropriate; this is one argument in favor of excluding the DC observation from the data.

Generally speaking, if an influential point arises from random sampling from a large population and is not a data error, it should be left in the dataset, since it probably represents a small subset of the population from which the data were sampled.

GUIDED PRACTICE 6.14

Once the influential DC point is removed, assess whether it is appropriate to use linear regression on these data by checking the four assumptions behind least squares regression: linearity, constant variability, independent observations, and approximate normality of the residuals. Refer to the residual plots shown in Figure 6.15.[20]

---

[19]Data are from the Statistical Abstract of the United States, published by the US Census Bureau. Data are for 2010, and available as census.2010 in the oibiostat package.

[20]The scatterplot in Figure 6.14(b) does not show any nonlinear trends. Similarly, Figure 6.15(a) does not indicate any nonlinear trends or noticeable difference in the variability of the residuals, although it does show that there are relatively few observations for low values of predicted infant mortality. From Figure 6.15(b), the residuals are approximately normally distributed. Infant mortality across the states reflects a complex mix of different levels of income, access to health care, and individual state initiatives in health care; these and other state-specific features probably act independently across the states, although there is some dependence from federal influence such as funding for pre-natal care. Overall, independence seems like a reasonable assumption.

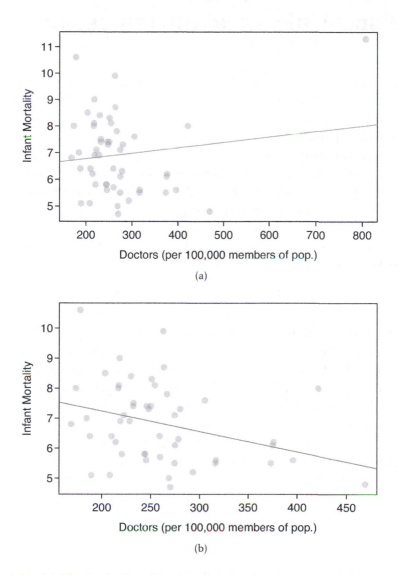

Figure 6.14: (a) Plot including District of Columbia data point. (b) Plot without influential District of Columbia data point.

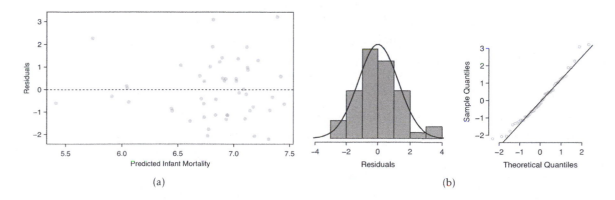

Figure 6.15: (a) Residual plot of inf.mortality and doctors. (b) Histogram and normal probability plot of the residuals.

6.4 Statistical inference with regression

The previous sections in this chapter have focused on linear regression as a tool for summarizing trends in data and making predictions. These numerical summaries are analogous to the methods discussed in Chapter 1 for displaying and summarizing data. Regression is also used to make inferences about a population.

The same ideas covered in Chapters 4 and 5 about using data from a sample to draw inferences about population parameters apply with regression. Previously, the goal was to draw inference about the population parameter μ; in regression, the population parameter of interest is typically the slope parameter β_1. Inference about the intercept term is rare, and limited to the few problems where the vertical intercept has scientific meaning.[21]

Inference in regression relies on the population linear model for the relationship between an explanatory variable X and a response variable Y given by

$$Y = \beta_0 + \beta_1 X + \varepsilon, \tag{6.15}$$

where ε is assumed to have a normal distribution with mean 0 and standard deviation σ ($\varepsilon \sim N(0, \sigma)$). This population model specifies that a response Y has value $\beta_0 + \beta_1 X$ plus a random term that pushes Y symmetrically above or below the value specified by the line.[22]

The set of ordered pairs (x_i, y_i) used when fitting a least squares regression line are assumed to have been sampled from a population in which the relationship between the explanatory and response variables follows Equation 6.15. Under this assumption, the slope and intercept values of the least squares regression line, b_0 and b_1, are estimates of the population parameters β_0 and β_1; b_0 and b_1 have sampling distributions, just as $\overline{X}$ does when thought of as an estimate of a population mean μ. A more advanced treatment of regression would demonstrate that the sampling distribution of b_1 is normal with mean $E(b_1) = \beta_1$ and standard deviation

$$\sigma_{b_1} = \frac{\sigma}{\sqrt{\sum (x_i - \overline{x})^2}}.$$

The sampling distribution of b_0 has mean $E(b_0) = \beta_0$ and standard deviation

$$\sigma_{b_0} = \sigma \sqrt{\frac{1}{n} + \frac{\overline{x}^2}{\sum (x_i - \overline{x})^2}}.$$

In both of these expressions, σ is the standard deviation of ε.

Hypothesis tests and confidence intervals for regression parameters have the same basic form as tests and intervals about population means. The test statistic for a null hypothesis $H_0 : \beta_1 = \beta_1^0$ about a slope parameter is

$$t = \frac{b_1 - \beta_1^0}{\text{s.e.}(b_1)},$$

where the formula for s.e.(b_1) is given below. In this setting, t has a t-distribution with $n - 2$ degrees of freedom, where n is the number of ordered pairs used to estimate the least squares line.

---

[21] In some applications of regression, the predictor x is replaced by $x^* = x - \overline{x}$. In that case, the vertical intercept is the value of the line when $x^* = 0$, or $x = \overline{x}$.

[22] Since $E(\varepsilon) = 0$, this model can also be written as $Y \sim N(\mu_x)$, with $\mu_x = E(Y) = \beta_0 + \beta_1 X$. The term ε is the population model for the observed residuals e_i in regression.

Typically, hypothesis testing in regression involves tests of whether the x and y variables are associated; in other words, whether the slope is significantly different from 0. In these settings, the null hypothesis is that there is no association between the explanatory and response variables, or $H_0 : \beta_1 = 0 = \beta_1^0$, in which case

$$t = \frac{b_1}{\text{s.e.}(b_1)}.$$

The hypothesis is rejected in favor of the two-sided alternative $H_A : \beta_1 \neq 0$ with significance level α when $|t| \geq t_{df}^\star$, where $t_{df}^\star$ is the point on a t-distribution with $n - 2$ degrees of freedom that has $\alpha/2$ area to its right (i.e., when $p \leq \alpha$).

A two-sided confidence interval for β_1 is given by

$$b_1 \pm \text{s.e.}(b_1) \times t_{df}^\star.$$

Tests for one-sided alternatives and one-sided confidence intervals make the usual adjustments to the rejection rule and confidence interval, and p-values are interpreted just as in Chapters 4 and 5.

Formulas for calculating standard errors

Statistical software is typically used to obtain t-statistics and p-values for inference with regression, since using the formulas for calculating standard error can be cumbersome.

The standard errors of b_0 and b_1 used in confidence intervals and hypothesis tests replace σ with s, the standard deviation of the residuals from a fitted line. Formally,

$$s = \sqrt{\frac{\sum e_i^2}{n-2}} = \sqrt{\frac{\sum (y_i - \hat{y}_i)^2}{n-2}}. \qquad (6.16)$$

The term s^2 is often called the mean squared error from the regression, and s the root mean squared error.

The two standard errors are

$$\text{s.e.}(b_1) = \frac{s}{\sqrt{\sum (x_i - \overline{x})^2}} \qquad \text{and} \qquad \text{s.e.}(b_0) = s\sqrt{\frac{1}{n} + \frac{\overline{x}^2}{\sum (x_i - \overline{x})^2}}.$$

EXAMPLE 6.17

Is there evidence of a significant association between number of doctors per 100,000 members of the population in a state and infant mortality rate?

The numerical output that R returns is shown in Figure 6.16.[23]

The question implies that the District of Columbia should not be included in the analysis. The assumptions for applying a least squares regression have been verified in Exercise 6.14. Whenever possible, formal inference should be preceded by a check of the assumptions for regression.

The null and alternative hypotheses are $H_0 : \beta_1 = 0$ and $H_A : \beta_1 \neq 0$.

The estimated slope of the least squares line is -0.0068, with standard error 0.0028. The t-statistic equals -2.40, and the probability that the absolute value of a t-statistic with $50 - 2 = 48$ degrees of freedom is smaller than -2.40 or larger than 2.40 is 0.021.

Since $p = 0.021 < 0.05$, the data support the alternative hypothesis that the number of physicians is associated with infant mortality at the 0.05 significance level. The sign of the slope implies that the association is negative; states with more doctors tend to have lower rates of infant mortality.

	Estimate	Std. Error	t value	Pr(>|t|)
(Intercept)	8.5991	0.7603	11.31	0.0000
Doctors Per 100,000	-0.0068	0.0028	-2.40	0.0206

Figure 6.16: Summary of regression output from R for the model predicting infant mortality from number of doctors, using the census.2010 dataset.

Care should be taken in interpreting the above results. The R^2 for the model is 0.107; the model explains only about 10% of the state-to-state variability in infant mortality, which suggests there are several other factors affecting infant mortality that are not accounted for in the model.[24] Additionally, an important implicit assumption being made in this example is that data from the year 2010 are representative; in other words, that the relationship between number of physicians and infant mortality is constant over time, and that the data from 2010 can be used to make inference about other years.

Note that it would be incorrect to make claims of causality from these data, such as stating that an additional 100 physicians (per 100,000 residents) would lead to a decrease of 0.68 in the infant mortality rate.

GUIDED PRACTICE 6.18

Calculate a 95% two-sided confidence interval for the slope parameter β_1 in the state-level infant mortality data.[25]

---

[23]Other software packages, such as Stata or Minitab, provide similar information but with slightly different labeling.

[24]Calculations of the R^2 value are not shown here.

[25]The $t^\star$ value for a t-distribution with 48 degrees of freedom is 2.01, and the standard error of b_1 is 0.0028. The 95% confidence interval is $-0.0068 \pm 2.01(0.0028) = (-0.0124, -0.0012)$.

Connection to two-group hypothesis testing

Conducting a regression analysis with a numerical response variable and a categorical predictor with two levels is analogous to conducting a two-group hypothesis test.

For example, Section 6.3.3 shows a regression model that compares the average infant mortality rate in countries with low access to sanitation facilities versus high access.[26] In other words, the purpose of the analysis is to compare mean infant mortality rate between the two groups: countries with low access versus countries with high access. Recall that the slope parameter b_1 is the difference between the means of log(mortality rate). A test of the null hypothesis $H_0 : \beta_1 = 0$ in the context of a categorical predictor with two levels is a test of whether the two means are different, just as for the two-group null hypothesis, $H_0 : \mu_1 = \mu_2$.

When the pooled standard deviation assumption (Section 5.3.5) is used, the t-statistic and p-value from a two-group hypothesis test are equivalent to that returned from a regression model.

Figure 6.17 shows the R output from a regression model in the wdi.2011 data, in which sanit.access = 1 for countries where at least 50% of the population has access to adequate sanitation and 0 otherwise. The abbreviated R output from two-group t-tests are shown in Figure 6.18. The version of the t-test that does not assume equal standard deviations and uses non-integer degrees of freedom is often referred to as the Welch test.

	Estimate	Std. Error	t value	Pr(>|t|)
(Intercept)	4.0184	0.1100	36.52	< 0.001
High Access	-1.6806	0.1322	-12.72	0.001

Figure 6.17: Regression of log(infant mortality) versus sanitation access.

Test	df	t value	Pr(>|t|)
Two-group t-test	161	12.72	< 0.001
Welch two-group t-test	155.82	17.36	< 0.001

Figure 6.18: Results from the independent two-group t-test, under differing assumptions about standard deviations between groups, for mean log(infant mortality) between sanitation access groups.

The sign of the t-statistic differs because for the two-group test, the difference in mean log(infant mortality) was calculated by subtracting the mean in the high access group from the mean in the low access group; in the regression model, the negative sign reflects the reduction in mean log(infant mortality) when changing from low access to high access. Since the t-distribution is symmetric, the two-sided p-value is equal. In this case, p is a small number less than 0.001, as calculated from a t-distribution with $163 - 2 = 161$ degrees of freedom (recall that 163 countries are represented in the dataset). The degrees of freedom for the pooled two-group test and linear regression are equivalent.

Example 6.13 showed that the constant variability assumption does not hold for these data. As a result, it might be advisable for a researcher interested in comparing the infant mortality rates between these two groups to conduct a two-group hypothesis test without using the pooled standard deviation assumption. Since this test uses a different formula for calculating the standard error of the difference in means, the t-statistic is different; additionally, the degrees of freedom are not equivalent. In this particular example, there is not a noticeable effect on the p-value.

---

[26]Recall that a log transformation was used on the infant mortality rate.

6.5   Interval estimates with regression

Section 6.4 introduced interval estimates for regression parameters, such as the population slope β_1. An estimated regression line can also be used to construct interval estimates for the regression line itself and to calculate prediction intervals for a new observation.

6.5.1   Confidence intervals

As initially discussed in Section 6.2, the estimated regression line for the association between RFFT score and age from the 500 individuals in prevend.samp is

$$\widehat{\text{RFFT}} = 137.55 - 1.26(\text{age}).$$

Figure 6.19 shows the summary output from R when the regression model is fit. R also provides the value of R^2 as 0.285 and the value of s, the estimated standard deviation of the residuals, as 23.2.

	Estimate	Std. Error	t value	Pr(>|t|)
(Intercept)	137.55	5.02	27.42	0.000
Age	-1.26	0.09	-14.09	0.000
				$df = 498$

Figure 6.19: Summary of regression output from R for the model predicting RFFT score from age, using the prevend.samp dataset.

A confidence interval for the slope parameter β_1 is centered at the point estimate b_1, with a width based on the standard error for the slope. For this model, the 95% confidence interval for age is $-1.26 \pm (1.96)(0.09) = (-1.44, -1.09)$ years.[27] With 95% confidence, each additional year of age is associated with between a 1.1 and 1.4 point lower RFFT score.

A confidence interval can also be calculated for a specific point on a least squares line. Consider a specific value of the predictor variable, x^*, such as 60 years of age. At age 60 years, the predicted value of RFFT score is $137.55 - 1.26(60) = 61.95$ points. The fitted line suggests that individuals from this population who are 60 years of age score, on average, about 62 points on the RFFT. Each point on the estimated regression line represents the predicted average RFFT score for a certain age.

More generally, the population model for a regression line is $E(Y|x) = \beta_0 + \beta_1 x$, and at a value x^* of the predictor x, the fitted regression line

$$\widehat{E(Y|x^*)} = b_0 + b_1 x^*$$

estimates the mean of Y for members of the population with predictor value x^*.

Thus, each point on a fitted regression line represents a point estimate for $E(Y|x^*)$. The corresponding interval estimate for $E(Y|x^*)$ measures the uncertainty in the estimated mean of Y at predictor value x^*, just as how an interval estimate for the population slope β_1 represents the uncertainty around b_1.

---

[27]The critical value 1.96 is used here because at degrees of freedom 498, the t-distribution is very close to a normal distribution. From software, $t^\star_{0.975, df=498} = 1.9647$.

The confidence interval for $E(Y|x^*)$ is computed using the standard error of the estimated mean of the regression model at a value of the predictor:

$$\text{s.e.}(\widehat{E(Y|x^*)}) = \sqrt{s^2\left(\frac{1}{n} + \frac{(x^* - \overline{x})^2}{\sum(x_i - \overline{x})^2}\right)} = s\sqrt{\frac{1}{n} + \frac{(x^* - \overline{x})^2}{\sum(x_i - \overline{x})^2}}.$$

In this expression, s is given by Equation 6.16, the usual estimate of σ, the standard deviation of the error term ϵ in the population linear model $Y = \beta_0 + \beta_1 X + \epsilon$.

The standard error of an estimated mean in regression is rarely calculated by hand; with all but the smallest datasets, the calculations are long and best left to software. When necessary, it can be calculated from basic features of the data and summary statistics.

Consider computing a 95% confidence interval for $E(RFFT|age = 60)$.

- The sample size is $n = 500$.

- $s = 23.2$ appears in the regression output.

- The sample mean $\overline{x}$ of the predictor is $\overline{age} = 54.8$ years.

- $(x^* - \overline{x})^2$ is the squared distance between the predictor value of interest and the sample mean of the predictors: $(60 - 54.8)^2 = 27.04$.

- The sum $\sum(x_i - \overline{x})^2$ is the numerator in the calculation of the variance of the predictor, and equals $(n-1)\text{Var}(x) = (499)(134.4445) = 67,088$.

Using these values, the standard error of the estimated mean RFFT score at age 60 is

$$\text{s.e.}(\widehat{E(RFFT|age} = 60)) = 23.2\sqrt{\frac{1}{500} + \frac{27.04}{67,088}} = 1.14.$$

Thus, a 95% confidence interval for the estimated mean is $61.95 \pm (1.96)(1.14) = (59.72, 64.18)$ points. With 95% confidence, the interval $(59.72, 64.18)$ points contains the average RFFT score of a 60-year-old individual.

It is also possible to calculate approximate confidence intervals for the estimated mean at a specific value of a predictor. When $x^* = \overline{x}$, the second term in the square root will be 0, and the standard error of the estimated mean at the average value $\overline{x}$ will have the simple form $s/\sqrt{n}$. For values close to $\overline{x}$, approximating the standard error as $s/\sqrt{n}$ is often sufficient. In the PREVEND data, 60 years is reasonably close to the average age 54.8 years, and the approximate value of the standard error is $23.2/\sqrt{500} = 1.03$. For values x^* that are more distant from the mean, the second term in the square root cannot be reasonably ignored.

The approximate form of the standard error for the mean at a predictor value, $s/\sqrt{n}$, makes it easier to see that for large n, the standard error approaches 0; thus, the confidence interval narrows as sample size increases, allowing the estimates to become more precise. This behavior is identical to the confidence interval for a simple mean, as one would expect. It is possible to show algebraically that the confidence intervals at any value of the predictor become increasingly narrow as the sample size increases.

6.5.2 Prediction intervals

After fitting a regression line, a **prediction interval** is used to estimate a range of values for a new observation of the response variable Y with predictor value x^*; that is, an observation not in the data used to estimate the line. The point estimate $\widehat{Y|x^*} = b_0 + b_1 x^*$ is the same as $\widehat{E(Y|x^*)}$, but the corresponding interval estimate is wider than a confidence interval for the mean.

The width of the interval reflects both the uncertainty in the estimate of the mean, $\widehat{E(Y|x^*)}$, and the inherent variability of the response variable. The standard error for a predicted value $\widehat{Y|x^*}$ at predictor x^* is

$$\text{s.e.}(\widehat{Y|x^*}) = \sqrt{s^2 + s^2\left(\frac{1}{n} + \frac{(x^*-\overline{x})^2}{\sum(x_i-\overline{x})^2}\right)} = s\sqrt{1 + \frac{1}{n} + \frac{(x^*-\overline{x})^2}{\sum(x_i-\overline{x})^2}}.$$

The increased variability when estimating $Y|x^*$ versus $E(Y|x^*)$ is accounted for by the additional s^2 term inside the square root.

The standard error for a prediction can also be calculated from summary statistics; the calculation is similar to that for the standard error for a mean. From the values of the summary statistics,

$$\text{s.e.}(\widehat{RFFT|age} = 60) = 23.2\sqrt{1 + \frac{1}{500} + \frac{27.04}{67,088}} = 23.23.$$

The 95% prediction interval is $61.95 \pm (1.96)(23.23) = (16.42, 107.68)$ points. These data and the model suggest that with 95% confidence, a newly selected 60-year-old will score between 16 and 108 points on the RFFT. This interval is wider than the confidence interval for the mean RFFT score (at age 60 years).

Just as with confidence intervals, an approximate prediction interval for a predictor near the average of the predictors can be constructed by considering the case when $x^* = \overline{x}$ and the standard error reduces to $s\sqrt{1 + 1/n}$. This approximate standard error shows why prediction intervals are wider than confidence intervals and do not become narrower as sample size increases. For large sample sizes, the term $1/n$ is close to 0, and the standard error is close to s, the standard deviation of the residuals about the line. Even when the mean is estimated perfectly, a prediction interval will reflect the variability in the data (specifically, the variability in the response variable).

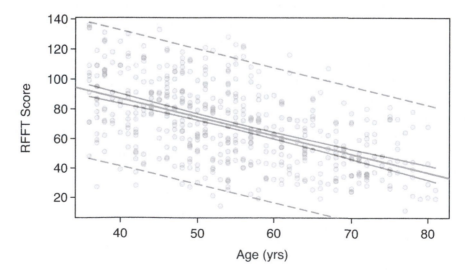

Figure 6.20: A scatterplot showing RFFT versus age, with the regression line in blue. The confidence intervals are marked by solid red lines, while the prediction intervals are shown in dashed red lines.

Figure 6.20 visually demonstrates confidence intervals and prediction intervals in the PRE-VEND example; the regression line is shown in blue, while the 95% confidence intervals and prediction intervals for each value of age are shown in red. At any value of age, the width of the interval estimate at that value is represented by the vertical distance between the two red lines. For example, the width of the confidence interval at age 60 years is represented by the distance between the points (60, 59.72) and (60, 64.18); the solid red lines pass through the upper and lower confidence bounds calculated in the earlier example. Similarly, the dashed red lines that represent prediction intervals pass through (60, 16.42) and (60, 107.68), the 95% upper and lower bounds for the predicted RFFT score of a 60-year-old.

The plot shows how the confidence intervals are most narrow at the mean age, 54.8 years, and become wider at values of age further from the mean. The prediction intervals are always wider than the confidence intervals. While the mean can be estimated with relative precision along the regression line, prediction intervals reflect the scatter of RFFT scores about the line (which is directly related to the inherent variability of RFFT scores within the study participants). While larger sample sizes can lead to narrower confidence intervals, the width of the prediction intervals will remain essentially unchanged unless the sampling scheme is changed in a way that reduces the variability of the response. A sample of individuals restricted to ages 60 - 70 years, for example, would be expected to have less variable RFFT scores, which would allow for narrower prediction intervals.

The distinction between confidence and prediction intervals is important and often overlooked. A clinical practitioner interested in the expected outcomes of a test generally should rely on confidence intervals for the mean along a regression line. The PREVEND data suggest that 60 year olds will score on average about 62 points on the test, and the average score is between 59.7 points and 62.2 points. When the RFFT is administered to a new 60 year old, however, the likely range of responses will be between 16.4 and 107.7. The prediction interval is wide because the scores on the test are quite variable.

6.6 Notes

This chapter provides only an introduction to simple linear regression; the next chapter, Chapter 7, expands on the principles of simple regression to models with more than one predictor variable.

When fitting a simple regression, be sure to visually assess whether the model is appropriate. Nonlinear trends or outliers are often obvious in a scatterplot with the least squares line plotted. If outliers are evident, the data source should be consulted when possible, since outliers may be indicative of errors in data collection. It is also important to consider whether observed outliers belong to the target population of inference, and assess whether the outliers should be included in the analysis.

There are several variants of residual plots used for model diagnostics. The ones shown in Section 6.3.1, which plot the predicted values on the horizontal axis, easily generalize to settings with multiple predictors, since there is always a single predicted value even when there is more than one predictor. If the only model used is a simple regression, plotting residuals against predictor values may make it easier to identify a case with a notable residual. Additionally, data analysts will sometimes plot residuals against case number of the predictor, since runs of large or small residuals may indicate that adjacent cases are correlated.

The R^2 statistic is widely used in the social sciences, where the unexplained variability in the data is typically much larger than the variability captured or explained by a model. It is important to be aware of what information R^2 does and does not provide. Even though a model may have a low proportion of explained variability, regression coefficients in the model can still be highly statistically significant. The R^2 should not be interpreted as a measure of the quality of the fit of the model. It is possible for R^2 to be large even when the data do not show a linear relationship.

Linear regression models are often estimated after an investigator has noticed a linear relationship in data, and experienced investigators can often guess correctly that regression coefficients will be significant before calculating a p-value. Unlike with two-sample hypothesis tests, regression models are rarely specified in advance at the design stage. In practice, it is best to be skeptical about a small p-value in a regression setting, and wait to see whether the observed statistically significant relationship can be confirmed in an independent dataset. The issue of model validation and assessing whether results of a regression analysis will generalize to other datasets is often discussed at length in advanced courses.

In more advanced texts, substantial attention is devoted to the subtleties of fitting straight line models. For instance, there are strategies for adjusting an analysis when one or more of the assumptions for regression do not hold. There are also specific methods to numerically assess the leverage or influence that each observation has on a fitted model.

Lab 1 explores the relationship between cognitive function and age in adults by fitting and interpreting a straight line to these variables in the PREVEND dataset, in addition to discussing the statistical model for least squares regression and residual plots used to assess the assumptions for linear regression. The lab is a useful reminder that least squares regression is much more than the mechanics of finding a line that best fits a dataset. Lab 2 uses simulated data to explore the quantity R^2. Lab 3 explores the use of binary categorical predictor variables in regression and shows how two-sample t-tests can be calculated using linear regression, in addition to introducing inference in a regression context. Categorical predictor variables are common in medicine and the life sciences.

6.7 Exercises

6.7.1 Examining scatterplots

6.1 Identify relationships, Part I. For each of the six plots, identify the strength of the relationship (e.g. weak, moderate, or strong) in the data and whether fitting a linear model would be reasonable.

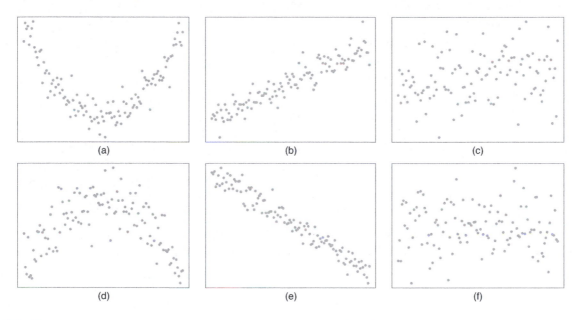

6.2 Identify relationships, Part II. For each of the six plots, identify the strength of the relationship (e.g. weak, moderate, or strong) in the data and whether fitting a linear model would be reasonable.

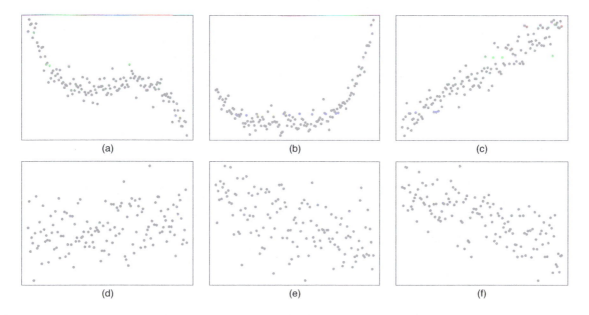

6.7.2 Estimating a regression line using least squares

6.3 Body measurements, Part I. Researchers studying anthropometry collected body girth measurements and skeletal diameter measurements, as well as age, weight, height and gender for 507 physically active individuals.[28] The scatterplot below shows the relationship between height and shoulder girth (over deltoid muscles), both measured in centimeters.

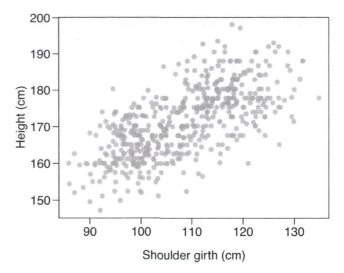

(a) Describe the relationship between shoulder girth and height.

(b) How would the relationship change if shoulder girth was measured in inches while the units of height remained in centimeters?

6.4 Body measurements, Part II. The scatterplot below shows the relationship between weight measured in kilograms and hip girth measured in centimeters from the data described in Exercise 6.3.

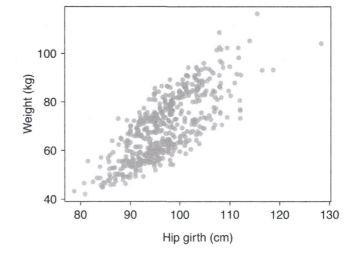

(a) Describe the relationship between hip girth and weight.

(b) How would the relationship change if weight was measured in pounds while the units for hip girth remained in centimeters?

[28]G. Heinz et al. "Exploring relationships in body dimensions". In: *Journal of Statistics Education* 11.2 (2003).

6.5 Over-under, Part I. Suppose we fit a regression line to predict the shelf life of an apple based on its weight. For a particular apple, we predict the shelf life to be 4.6 days. The apple's residual is -0.6 days. Did we over or under estimate the shelf-life of the apple? Explain your reasoning.

6.6 Over-under, Part II. Suppose we fit a regression line to predict the number of incidents of skin cancer per 1,000 people from the number of sunny days in a year. For a particular year, we predict the incidence of skin cancer to be 1.5 per 1,000 people, and the residual for this year is 0.5. Did we over or under estimate the incidence of skin cancer? Explain your reasoning.

6.7 Murders and poverty, Part I. The following regression output is for predicting annual murders per million from percentage living in poverty in a random sample of 20 metropolitan areas.

	Estimate	Std. Error	t value	Pr(>|t|)
(Intercept)	-29.901	7.789	-3.839	0.001
poverty%	2.559	0.390	6.562	0.000
$s = 5.512$	$R^2 = 70.52\%$		$R^2_{adj} = 68.89\%$	

(a) Write out the linear model.

(b) Interpret the intercept.

(c) Interpret the slope.

(d) Calculate the correlation coefficient.

6.8 Cats, Part I. The following regression output is for predicting the heart weight (in g) of cats from their body weight (in kg). The coefficients are estimated using a dataset of 144 domestic cats.

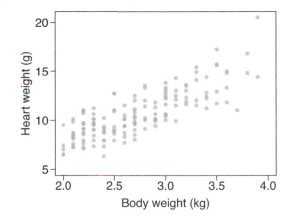

	Estimate	Std. Error	t value	Pr(>|t|)
(Intercept)	-0.357	0.692	-0.515	0.607
body wt	4.034	0.250	16.119	0.000
$s = 1.452$	$R^2 = 64.66\%$		$R^2_{adj} = 64.41\%$	

(a) Write out the linear model.

(b) Interpret the intercept.

(c) Interpret the slope.

(d) Calculate the correlation coefficient.

6.9 Age and RFFT score, Part I. A linear model fit to RFFT scores and age for 500 randomly sampled individuals from the PREVEND data has equation $\widehat{RFFT} = 137.55 - 1.26(Age)$.

(a) Interpret the slope and intercept values in the context of the data; i.e., explain the linear model in terms that a non-statistician would understand. Comment on whether the intercept value has any interpretive meaning in this setting.

(b) Based on the linear model, how much does RFFT score differ, on average, between an individual who is 60 years old versus an individual who is 50 years old?

(c) According to the linear model, what is the average RFFT score for an individual who is 70 years old?

(d) Examine Figure 6.1. Is it valid to use the linear model to estimate RFFT score for an individual who is 20 years old? Explain your answer.

6.10 Guppies, Part I. Guppies are small, brightly colored tropical fish often seen in freshwater fish aquariums. A study was conducted in 147 male guppies to examine the relationship between coloration and heterozygosity; heterozygosity refers to the condition of having different alleles at a given genetic locus. The guppies were randomly sampled from a river in the wild.

In an initial stage of the study, researchers examined whether length and height are linearly associated. The mean length is 1261.21 cm, with standard deviation 95.62 cm. The mean height is 201.75 cm, with standard deviation 20.68. The correlation between length and height is 0.85.

(a) From a visual inspection, does it seem like the line is a reasonable fit for the data?

(b) Write the equation of the regression line for predicting length from height.

(c) Estimate the predicted mean length of a guppy with height 180 cm.

6.7.3 Interpreting a linear model

6.11 Visualize the residuals. The scatterplots shown below each have a superimposed regression line. If we were to construct a residual plot (residuals versus x) for each, describe what those plots would look like.

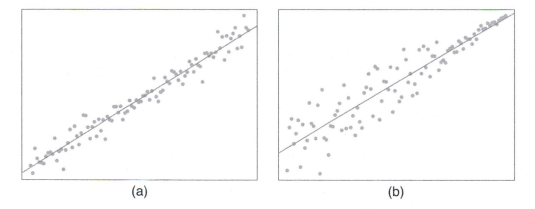

(a)   (b)

6.12 Trends in the residuals. Shown below are two plots of residuals remaining after fitting a linear model to two different sets of data. Describe important features and determine if a linear model would be appropriate for these data. Explain your reasoning.

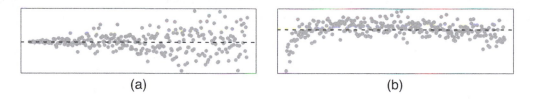

(a)   (b)

6.13 Guppies, Part II. Exercise 6.10 showed a plot of length versus height for 147 male guppies with a least squares regression line.

(a) Identify two points that have relatively high leverage and discuss whether these points seem to be particularly influential.

(b) Based on the plot, comment on whether it is appropriate to use R^2 as a metric for describing the strength of the model fit.

(c) The R^2 for this model is 0.718. Interpret this value in the context of the data.

6.14   Nutrition at Starbucks, Part I. The scatterplot below shows the relationship between the number of calories and amount of carbohydrates (in grams) Starbucks food menu items contain.[29] Since Starbucks only lists the number of calories on the display items, we are interested in predicting the amount of carbs a menu item has based on its calorie content.

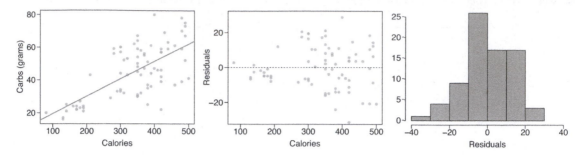

(a) Describe the relationship between number of calories and amount of carbohydrates (in grams) that Starbucks food menu items contain.

(b) In this scenario, what are the explanatory and response variables?

(c) Why might we want to fit a regression line to these data?

(d) Do these data meet the conditions required for fitting a least squares line?

6.15   Nutrition at Starbucks, Part II. Exercise 6.14 introduced a data set on nutrition information on Starbucks food menu items. Based on the scatterplot and the residual plot provided, describe the relationship between the protein content and calories of these menu items, and determine if a simple linear model is appropriate to predict amount of protein from the number of calories.

[29]Source: Starbucks.com, collected on March 10, 2011, www.starbucks.com/menu/nutrition.

6.16 Body measurements, Part III. Exercise 6.3 introduces data on shoulder girth and height of a group of individuals. The mean shoulder girth is 107.20 cm with a standard deviation of 10.37 cm. The mean height is 171.14 cm with a standard deviation of 9.41 cm. The correlation between height and shoulder girth is 0.67.

(a) Write the equation of the regression line for predicting height.

(b) Interpret the slope and the intercept in this context.

(c) Calculate R^2 of the regression line for predicting height from shoulder girth, and interpret it in the context of the application.

(d) A randomly selected student from your class has a shoulder girth of 100 cm. Predict the height of this student using the model.

(e) The student from part (d) is 160 cm tall. Calculate the residual, and explain what this residual means.

(f) A one year old has a shoulder girth of 56 cm. Would it be appropriate to use this linear model to predict the height of this child?

6.17 Outliers, Part I. Identify the outliers in the scatterplots shown below, and determine what type of outliers they are. Explain your reasoning.

(a)

(b)

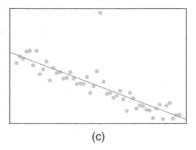

(c)

6.18 Outliers, Part II. Identify the outliers in the scatterplots shown below and determine what type of outliers they are. Explain your reasoning.

(a)

(b)

(c)

6.19 Guppies, Part III. The residual plots below are for the linear model fit in Exercise 6.10 predicting length from height for 147 male guppies.

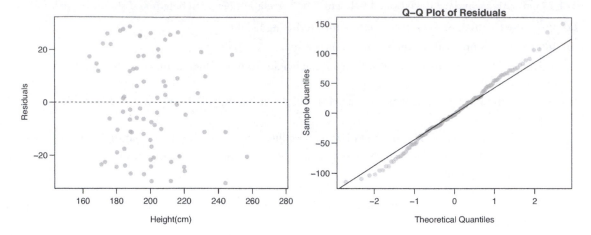

(a) From a plot of residual values versus predicted values, are the assumptions of linearity and constant variability satisfied? Explain your answer.

(b) Is it reasonable to assume that the observations were independent, based on the description of the study? Explain your answer.

(c) Are the residuals approximately normally distributed? Explain your answer.

6.20 Guppies, Part IV. Multilocus heterozygosity (MLH) is reflective of genetic quality; according to sexual selection research, it is thought that sexual ornamentation functions as a visual indicator of fitness. By selecting males with features such as bright coloration, females can improve the chances of reproductive success.

Male guppies are covered in a mixture of colored spots; orange coloration is consistently preferred by females. Heterozygosity was assessed by genotyping 9 loci and calculating the proportion of loci that are heterozygous. The research question of interest is whether MLH and orange color are linearly associated.

(a) Based on the plot of MLH versus relative orange area, describe the nature of the association in language accessible to a general audience.

(b) Comment on whether the assumptions of linearity and constant variability are reasonably met.

(c) Comment on whether the residuals are approximately normally distributed.

6.21 Diamond prices, Part II. Exercise 5.24 introduced data on the price of diamonds based on whether a diamond is 0.99 carats or 1 carat. Based on the summary statistics, write an estimated model equation predicting price from a binary indicator of carat weight. Be sure to clearly define the variables used in the model.

	0.99 carats	1 carat
Mean	$ 44.51	$ 56.81
SD	$ 13.32	$ 16.13
n	23	23

6.22 Avian influenza, Part II. Exercise 5.28 introduced data from an analysis investigating whether hatch weights between transgenic and non-transgenic chicks differ.

	transgenic chicks (g)	non-transgenic chicks (g)
$\bar{x}$	45.14	44.99
s	3.32	4.57
n	54	54

(a) Write an estimated least squares regression line for a model predicting hatch weight from chick type, where non-transgenic chicks are the reference group; i.e., the group for which the binary predictor takes on value 0.

(b) Write an estimated least squares regression line for a model predicting hatch weight from chick type, where transgenic chicks are the reference group.

6.7.4 Statistical inference with regression

6.23 Body measurements, Part IV. The scatterplot and least squares summary below show the relationship between weight measured in kilograms and height measured in centimeters of 507 physically active individuals.

	Estimate	Std. Error	t value	Pr(>|t|)
(Intercept)	-105.0113	7.5394	-13.93	0.0000
height	1.0176	0.0440	23.13	0.0000

(a) Describe the relationship between height and weight.

(b) Write the equation of the regression line. Interpret the slope and intercept in context.

(c) Do the data provide strong evidence that an increase in height is associated with an increase in weight? State the null and alternative hypotheses, report the p-value, and state your conclusion.

(d) The correlation coefficient for height and weight is 0.72. Calculate R^2 and interpret it in context.

6.24 Beer and blood alcohol content. Many people believe that gender, weight, drinking habits, and many other factors are much more important in predicting blood alcohol content (BAC) than simply considering the number of drinks a person consumed. Here we examine data from sixteen student volunteers at Ohio State University who each drank a randomly assigned number of cans of beer. These students were evenly divided between men and women, and they differed in weight and drinking habits. Thirty minutes later, a police officer measured their blood alcohol content (BAC) in grams of alcohol per deciliter of blood.[30] The scatterplot and regression table summarize the findings.

	Estimate	Std. Error	t value	Pr(>|t|)
(Intercept)	-0.0127	0.0126	-1.00	0.3320
beers	0.0180	0.0024	7.48	0.0000

(a) Describe the relationship between the number of cans of beer and BAC.

(b) Write the equation of the regression line. Interpret the slope and intercept in context.

(c) Do the data provide strong evidence that drinking more cans of beer is associated with an increase in blood alcohol? State the null and alternative hypotheses, report the p-value, and state your conclusion.

(d) The correlation coefficient for number of cans of beer and BAC is 0.89. Calculate R^2 and interpret it in context.

(e) Suppose we visit a bar, ask people how many drinks they have had, and also take their BAC. Do you think the relationship between number of drinks and BAC would be as strong as the relationship found in the Ohio State study?

---

[30]J. Malkevitch and L.M. Lesser. *For All Practical Purposes: Mathematical Literacy in Today's World*. WH Freeman & Co, 2008.

6.25 Husbands and wives, Part I. The scatterplot below summarizes husbands' and wives' heights in a random sample of 170 married couples in Britain, where both partners' ages are below 65 years. Summary output of the least squares fit for predicting wife's height from husband's height is also provided in the table.

	Estimate	Std. Error	t value	Pr(>|t|)
(Intercept)	43.5755	4.6842	9.30	0.0000
height_husband	0.2863	0.0686	4.17	0.0000

(a) Is there strong evidence that taller men marry taller women? State the hypotheses and include any information used to conduct the test.

(b) Write the equation of the regression line for predicting wife's height from husband's height.

(c) Interpret the slope and intercept in the context of the application.

(d) Given that $R^2 = 0.09$, what is the correlation of heights in this data set?

(e) You meet a married man from Britain who is 5'9" (69 inches). What would you predict his wife's height to be? How reliable is this prediction?

(f) You meet another married man from Britain who is 6'7" (79 inches). Would it be wise to use the same linear model to predict his wife's height? Why or why not?

(g) Is there statistically significant evidence of an association between husband height and wife height based on these data? Explain your answer.

(h) Would you expect a 95% confidence interval for husband height to contain 0? Explain your answer.

6.26 Helmets and lunches. The scatterplot shows the relationship between socioeconomic status measured as the percentage of children in a neighborhood receiving reduced-fee lunches at school (lunch) and the percentage of bike riders in the neighborhood wearing helmets (helmet). The average percentage of children receiving reduced-fee lunches is 30.8% with a standard deviation of 26.7% and the average percentage of bike riders wearing helmets is 38.8% with a standard deviation of 16.9%.

(a) If the R^2 for the least-squares regression line for these data is 72%, what is the correlation between lunch and helmet?

(b) Calculate the slope and intercept for the least-squares regression line for these data.

(c) Interpret the intercept of the least-squares regression line in the context of the application.

(d) Interpret the slope of the least-squares regression line in the context of the application.

(e) What would the value of the residual be for a neighborhood where 40% of the children receive reduced-fee lunches and 40% of the bike riders wear helmets? Interpret the meaning of this residual in the context of the application.

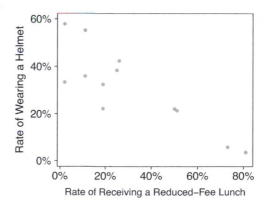

6.27 Husbands and wives, Part II. Exercise 6.25 presents a scatterplot displaying the relationship between husbands' and wives' ages in a random sample of 170 married couples in Britain, where both partners' ages are below 65 years. Given below is summary output of the least squares fit for predicting wife's age from husband's age.

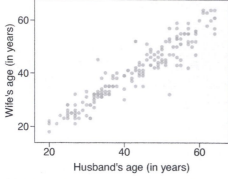

	Estimate	Std. Error	t value	Pr(>|t|)
(Intercept)	1.5740	1.1501	1.37	0.1730
age_husband	0.9112	0.0259	35.25	0.0000
				$df = 168$

(a) We might wonder, is the age difference between husbands and wives consistent across ages? If this were the case, then the slope parameter would be $\beta_1 = 1$. Use the information above to evaluate if there is strong evidence that the difference in husband and wife ages differs for different ages.

(b) Write the equation of the regression line for predicting wife's age from husband's age.

(c) Interpret the slope and intercept in context.

(d) Given that $R^2 = 0.88$, what is the correlation of ages in this data set?

(e) You meet a married man from Britain who is 55 years old. What would you predict his wife's age to be? How reliable is this prediction?

(f) You meet another married man from Britain who is 85 years old. Would it be wise to use the same linear model to predict his wife's age? Explain.

6.28 Guppies, Part V. Exercise 6.20 introduced a linear model for predicting relative orange area from proportion of loci that are heterozygous (MLH). Relative orange area refers to the percentage of the body that is orange (rather than a different color).

	Estimate	Std. Error	t value	Pr(>|t|)
(Intercept)	0.037	0.010	3.68	0.0033
MLH	0.051	0.018	2.77	0.0063
				$df = 145$

(a) Write the estimated model equation.

(b) What is the predicted mean relative orange area for a guppy that is heterozygous at 8 out of 9 loci?

(c) Based on the linear model, how much does mean relative orange area differ between a guppy that is heterozygous at 2 loci versus 4 loci (out of 9 total)?

(d) Conduct a hypothesis test to determine whether relative orange area is significantly associated with MLH. Do the results suggest that more elaborate sexual ornaments are associated with increased heterozygosity? Explain.

(e) Compute and interpret a 95% confidence interval for the slope parameter β_1.

6.29 Age and RFFT score, Part II. The following regression output is for predicting RFFT score of 500 randomly sampled individuals from the PREVEND data based on age (years).

	Estimate	Std. Error	t value	Pr(>|t|)
(Intercept)	137.55	5.02	27.42	0.000
Age	-1.26	0.09	-14.09	0.000

$$df = 498$$

(a) Do these data provide statistically significant evidence at the $\alpha = 0.01$ significance level that age is associated with RFFT score? State the null and alternative hypotheses, report the relevant p-value, and state your conclusion.

(b) Compute and interpret a 99% confidence interval for the population slope.

6.30 Avian influenza, Part III. Exercise 5.28 introduced data from an analysis investigating whether hatch weights between transgenic and non-transgenic chicks differ. Based on the results from conducting the two-group test, explain whether the 95% confidence interval for the β_1 parameter in a model predicting hatch weight from a group indicator would contain 0.

6.7.5 Interval estimates with regression

6.31 Husbands and wives, Part III. Exercise 6.27 introduces data from a random sample of 170 married couples in Britain, where both partners' ages are below 65 years, and fits a model predicting wife's age from husband's age. Wife's age has a mean of 40.68 years, with standard deviation 11.41 years. Husband's age has a mean of 42.92 years, with standard deviation 11.76 years. From software, the residual standard error is $s = 3.95$.

(a) Use the summary statistics to calculate a 95% confidence interval for the average age of wives whose husbands are 55 years old.

(b) You meet a married man from Britain who is 55 years old. Predict his wife's age and give a 95% prediction interval for her age.

(c) Repeat parts (a) and (b) using the approximate formulas for the appropriate standard errors.

6.32 Guppies, Part VI. The relationship between length and height for 147 male guppies was introduced in Exercise 6.10, which used the summary statistics to calculate the equation of the least squares line for length as a function of height and estimate the mean length of an adult male guppy with height 180 cm. The estimated residual standard error from this model is $s = 50.93$.

(a) Use the summary statistics given in Exercise 6.10 to construct a 95% confidence interval for the estimated mean length when height is 180 cm.

(b) Use a prediction interval based on the summary statistics to estimate the lengths for a new 180 cm guppy that would be more than two standard deviations above and below the estimated mean.

(c) Use the approximate formulas for the standard error for a mean and for a prediction to recalculate the intervals in parts (a) and (b).

Chapter 7

Multiple linear regression

In most practical settings, more than one explanatory variable is likely to be associated with a response. This chapter discusses how the ideas behind simple linear regression can be extended to a model with multiple predictor variables.

There are several applications of multiple regression. One of the most common applications in a clinical setting is estimating an association between a response variable and primary predictor of interest while adjusting for possible confounding variables. Sections 7.1 and 7.2 introduce the multiple regression model by examining the possible association between cognitive function and the use of statins after adjusting for potential confounders. Section 7.8 discusses another application of multiple regression—constructing a model that effectively explains the observed variation in the response variable.

The other sections in the chapter outline general principles of multiple regression, including the statistical model, methods for assessing quality of model fit, categorical predictors with more than two levels, interaction, and the connection between ANOVA and regression. The methods used to conduct hypothesis tests and construct confidence intervals for regression coefficients extend naturally from simple to multiple linear regression, so the section on the statistical model for multiple regression can be treated as optional.

For labs, slides, and other resources, please visit
www.openintro.org/book/biostat

7.1   Introduction to multiple linear regression

Statins are a class of drugs widely used to lower cholesterol. There are two main types of cholesterol: low density lipoprotein (LDL) and high density lipoprotein (HDL).[1] Research suggests that adults with elevated LDL may be at risk for adverse cardiovascular events such as a heart attack or stroke. In 2013, a panel of experts commissioned by the American College of Cardiology and the American Heart Association recommended that statin therapy be considered in individuals who either have any form of atherosclerotic cardiovascular disease[2] or have LDL cholesterol levels $\geq$ 190 mg/dL, individuals with Type II diabetes ages 40 to 75 with LDL between 70 to 189 mg/dL, and non-diabetic individuals ages of 40 to 75 with a predicted probability of future clogged arteries of at least 0.075.[3]

Health policy analysts have estimated that if the new guidelines were to be followed, almost half of Americans ages 40 to 75 and nearly all men over 60 would be prescribed a statin. However, some physicians have raised the question of whether treatment with a statin might be associated with an increased risk of cognitive decline.[4,5] Older adults are at increased risk for cardiovascular disease, but also for cognitive decline. A study by Joosten, et al. examined the association of statin use and other variables with cognitive ability in an observational cohort of 4,095 participants from the Netherlands who were part of the larger PREVEND study introduced in Section 6.1.[6] The analyses presented in this chapter are based on a random sample of 500 participants from the cohort.[7]

The investigators behind the Joosten study anticipated an issue in the analysis—statins are used more often in older adults than younger adults, and older adults suffer a natural cognitive decline. Age is a potential **confounder** in this setting. If age is not accounted for in the analysis, it may seem that cognitive decline is more common among individuals prescribed statins, simply because those prescribed statins are simply older and more likely to have reduced cognitive ability than those not prescribed statins.

---

[1] Total cholesterol level is the sum of LDL and HDL levels.

[2] i.e., arteries thickening and hardening with plaque

[3] Stone NJ, et al. 2013 ACC/AHA Guideline on the Treatment of Blood Cholesterol to Reduce Atherosclerotic Cardiovascular Risk in Adults Circulation. 2014;129:S1-S45. DOI: 10.1161/01.cir.0000437738.63853.7a

[4] Muldoon, Matthew F., et al. Randomized trial of the effects of simvastatin on cognitive functioning in hypercholesterolemic adults. The American journal of medicine 117.11 (2004): 823-829.

[5] King, Deborah S., et al. Cognitive impairment associated with atorvastatin and simvastatin. Pharmacotherapy: The Journal of Human Pharmacology and Drug Therapy 23.12 (2003): 1663-1667.

[6] Joosten H, Visser ST, van Eersel ME, Gansevoort RT, Bilo HJG, et al. (2014) Statin Use and Cognitive Function: Population-Based Observational Study with Long-Term Follow- Up. PLoS ONE 9(12): e115755. doi:10.1371/ journal.pone.0115755

[7] The random sample is accessible as prevend.samp in the oibiostat R package.

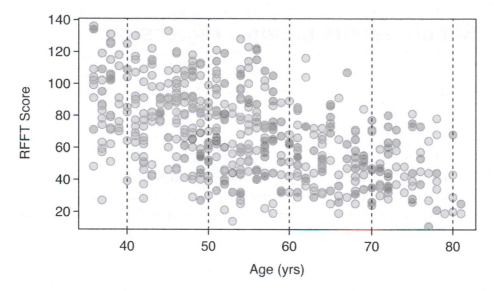

Figure 7.1: A scatterplot showing age vs. RFFT in prevend.samp. Statin users are represented with red points; participants not using statins are shown as blue points.

Figure 7.1 visually demonstrates why age is a potential confounder for the association between statin use and cognitive function, where cognitive function is measured via the Ruff Figural Fluency Test (RFFT). Scores range from 0 (worst) to 175 (best). The blue points indicate individuals not using statins, while red points indicate statin users. First, it is clear that age and statin use are associated, with statin use becoming more common as age increases; the red points are more prevalent on the right side of the plot. Second, it is also clear that age is associated with lower RFFT scores; ignoring the colors, the point cloud drifts down and to the right. However, a close inspection of the plot suggests that for ages in relatively small ranges (e.g., ages 50-60), statin use may not be strongly associated with RFFT score—there are approximately as many red dots with low RFFT scores as with high RFFT scores in a given age range. In other words, for subsets of participants with approximately similar ages, statin use may not be associated with RFFT. Multiple regression provides a way to estimate the association of statin use with RFFT while adjusting for age; i.e., accounting for the underlying relationship between age and statin use.

7.2   Simple versus multiple regression

A simple linear regression model can be fit for an initial examination of the association between statin use and RFFT score,

$$E(\text{RFFT}) = \beta_0 + \beta_{\text{Statin}}(\text{Statin}).$$

RFFT scores in prevend.samp are approximately normally distributed, ranging between approximately 10 and 140, with no obvious outliers (Figure 7.2(a)). The least squares regression line shown in Figure 7.2(b) has a negative slope, which suggests a possible negative association.

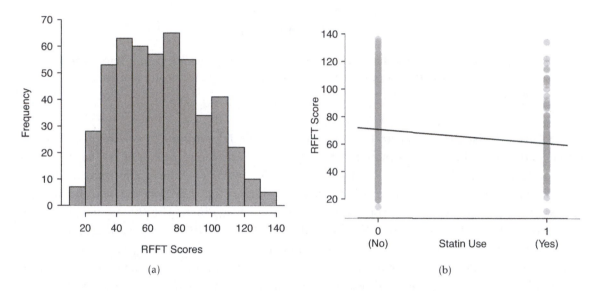

Figure 7.2: (a) Histogram of RFFT scores. (b) Scatterplot of RFFT score versus statin use in prevend.samp. The variable Statin is coded 1 for statin users, and 0 otherwise.

Figure 7.3 gives the parameter estimates of the least squares line, and indicates that the association between RFFT score and statin use is highly significant. On average, statin users score approximately 10 points lower on the RFFT. However, even though the association is statistically significant, it is potentially misleading since the model does not account for the underlying relationship between age and statin use. The association between age and statin use visible from Figure 7.1 is even more apparent in Figure 7.4, which shows that the median age of statin users is about 10 years higher than the median age of individuals not using statins.

	Estimate	Std. Error	t value	Pr(>|t|)
(Intercept)	70.7143	1.3808	51.21	0.0000
Statin	-10.0534	2.8792	-3.49	0.0005

Figure 7.3: R summary output for the simple regression model of RFFT versus statin use in prevend.samp.

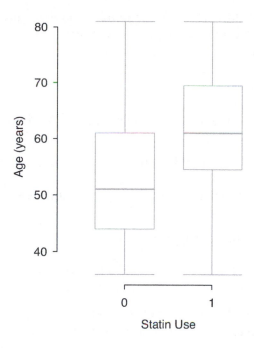

Figure 7.4: Boxplot of age by statin use in prevend.samp. The variable Statin is coded 1 for statin users, and 0 otherwise.

Multiple regression allows for a model that incorporates both statin use and age,

$$E(\text{RFFT}) = \beta_0 + \beta_{\text{Statin}}(\text{Statin}) + \beta_{\text{Age}}(\text{Age}).$$

In statistical terms, the association between RFFT and Statin is being estimated after adjusting for Age. This is an example of one of the more important applications of multiple regression: estimating an association between a response variable and primary predictor of interest while adjusting for possible confounders. In this setting, statin use is the primary predictor of interest.

The principles and assumptions behind the multiple regression model are introduced more formally in Section 7.4, along with the method used to estimate the coefficients. Figure 7.5 shows the parameter estimates for the model from R.

	Estimate	Std. Error	t value	Pr(>|t|)
(Intercept)	137.8822	5.1221	26.92	0.0000
Statin	0.8509	2.5957	0.33	0.7432
Age	-1.2710	0.0943	-13.48	0.0000

Figure 7.5: R summary output for the multiple regression model of RFFT versus statin use and age in prevend.samp.

EXAMPLE 7.1

Using the parameter estimates in Figure 7.5, write the prediction equation for the linear model. How does the predicted RFFT score for a 67-year-old not using statins compare to that of an individual of the same age who does use statins?

The equation of the linear model is

$$\widehat{RFFT} = 137.8822 + 0.8509(Statin) - 1.2710(Age).$$

The predicted RFFT score for a 67-year-old not using statins (Statin = 0) is

$$\widehat{RFFT} = 137.8822 + (0.8509)(0) - (1.2710)(67) = 52.7252.$$

The predicted RFFT score for a 67-year-old using statins (Statin = 1) is

$$\widehat{RFFT} = 137.8822 + (0.8509)(1) - (1.2710)(67) = 53.5761.$$

The two calculations differ only by the value of the coefficient β_{Statin}, 0.8509.[8] Thus, for two individuals who are the same age, the model predicts that RFFT score will be 0.8509 higher in the individual taking statins; statin use is associated with a small increase in RFFT score.

EXAMPLE 7.2

Suppose two individuals are both taking statins; one individual is 50 years of age, while the other is 60 years of age. Compare their predicted RFFT scores.

From the model equation, the coefficient of age β_{Age} is -1.2710; an increase in one unit of age (i.e., one year) is associated with a decrease in RFFT score of -1.2710, when statin use is the same. Thus, the individual who is 60 years of age is predicted to have an RFFT score that is about 13 points lower $((-1.2710)(10) = -12.710)$ than the individual who is 50 years of age.

This can be confirmed numerically:

The predicted RFFT score for a 50-year-old using statins is

$$\widehat{RFFT} = 137.8822 + (0.8509)(1) - (1.2710)(50) = 75.1831.$$

The predicted RFFT score for a 60-year-old using statins is

$$\widehat{RFFT} = 137.8822 + (0.8509)(1) - (1.2710)(60) = 62.4731.$$

The scores differ by $62.4731 - 75.1831 = -12.710$.

---

[8]In most cases, predictions do not need to be calculated to so many significant digits, since the coefficients are only estimates. This example uses the additional precision to illustrate the role of the coefficients.

GUIDED PRACTICE 7.3

What does the intercept represent in this model? Does the intercept have interpretive value?[9]

As in simple linear regression, t-statistics can be used to test hypotheses about the slope coefficients; for this model, the two null hypotheses are $H_0 : \beta_{Statin} = 0$ and $H_0 : \beta_{Age} = 0$. The p-values for the tests indicate that at significance level $\alpha = 0.05$, the association between RFFT score and statin use is not statistically significant, but the association between RFFT score and age is significant.

In a clinical setting, the interpretive focus lies on reporting the nature of the association between the primary predictor and the response and specifying which confounders have been adjusted for. The results of the analysis might be summarized as follows—

Although the use of statins appeared to be associated with lower RFFT scores when no adjustment was made for possible confounders, statin use is not significantly associated with RFFT score in a regression model that adjusts for age.

The results shown in Figure 7.5 do not provide information about either the quality of the model fit or its value as a prediction model. The next section describes the residual plots that can be used to check model assumptions and the use of R^2 to estimate how much of the variability in the response variable is explained by the model.

There is an important aspect of these data that should not be overlooked. The data do not come from a study in which participants were followed as they aged; i.e., a longitudinal study. Instead, this study was a cross-sectional study, in which patient age, statin use, and RFFT score were recorded for all participants during a short time interval. While the results of the study support the conclusion that older patients tend to have lower RFFT scores, they cannot be used to conclude that scores decline with age in individuals; there were no repeated measurements of RFFT taken as individual participants aged. Older patients come from an earlier birth cohort, and it is possible, for instance, that younger participants have more post-secondary school education or better health practices generally; such a cohort effect may have some explanatory effect on the observed association. The details of how a study is designed and how data are collected should always be taken into account when interpreting study results.

---

[9]The intercept represents an individual with value 0 for both Statin and Age; i.e., an individual not using statins with age of 0 years. It is not reasonable to predict RFFT score for a newborn, or to assess statin use; the intercept is meaningless and has no interpretive value.

7.3   Evaluating the fit of a multiple regression model

7.3.1   Using residuals to check model assumptions

The assumptions behind multiple regression are essentially the same as the four assumptions listed in Section 6.1 for simple linear regression. The assumption of linearity is extended to multiple regression by assuming that when only one predictor variable changes, it is linearly related to the change in the response variable. Assumption 2 becomes the slightly more general assumption that the residuals have approximately constant variance. Assumptions 3 and 4 do not change; it is assumed that the observations on each case are independent and the residuals are approximately normally distributed.

Since it is not possible to make a scatterplot of a response variable against several simultaneous predictors, residual plots become even more essential as tools for checking modeling assumptions.

To assess the linearity assumption, examine plots of residuals against each of the predictors. These plots might show an nonlinear trend that could be corrected with a transformation. The scatterplot of residual values versus age in Figure 7.6 shows no apparent nonlinear trends. It is not necessary to assess linearity against a categorical predictor, since a line drawn through two points (i.e., the means of the two groups) is necessarily linear.

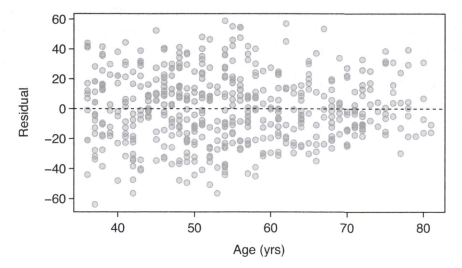

Figure 7.6: Residuals versus age in the model for RFFT vs statins and age in the PREVEND data.

Since each case has one predicted value and one residual, regardless of the number of predictors, residuals can still be plotted against predicted values to assess the constant variance assumption. The scatterplot in the left panel of Figure 7.7 shows that the variance of the residuals is slightly smaller for lower predicted values of RFFT, but is otherwise approximately constant.

Just as in simple regression, normal probability plots can be used to check the normality assumption of the residuals. The normal probability plot in the right panel of Figure 7.7 shows that the residuals from the model are reasonably normally distributed, with only slight departures from normality in the tails.

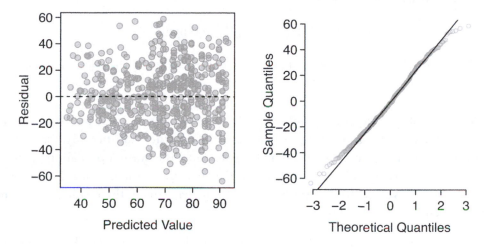

Figure 7.7: Residual plots from the linear model for RFFT versus statin use and age in prevend.samp.

EXAMPLE 7.4

Section 1.7 featured a case study examining the evidence for ethnic discrimination in the amount of financial support offered by the State of California to individuals with developmental disabilities. Although an initial look at the data suggested an association between expenditures and ethnicity, further analysis suggested that age is a confounding variable for the relationship.

A multiple regression model can be fit to these data to model the association between expenditures, age, and ethnicity in a subset that only includes data from Hispanics and White non-Hispanics. Two residual plots from the model fit for

$$E(\text{expenditures}) = \beta_0 + \beta_{\text{ethnicity}}(\text{ethnicity}) + \beta_{\text{age}}(\text{age})$$

are shown in Figure 7.8. From these plots, assess whether a linear regression model is appropriate for these data.

The model assumptions are clearly violated. The residual versus fitted plot shows obvious patterns; the residuals do not scatter randomly about the $y = 0$ line. Additionally, the variance of the residuals is not constant around the $y = 0$ line. As shown in the normal probability plot, the residuals show marked departures from normality, particularly in the upper tail; although this skewing may be partially resolved with a log transformation, the patterns in the residual versus fitted plot are more problematic.

Recall that a residual is the difference between an observed value and expected value; for an observation i, the residual equals $y_i - \hat{y}_i$. Positive residuals occur when a model's predictions are smaller that the observed values, and vice versa for negative residuals. In the residual versus fitted plot, it can be seen that in the middle range of predicted values, the model consistently under-predicts expenditures; on the upper and lower ends, the model over-predicts. This is a particularly serious issue with the model fit.

A single linear regression model is not appropriate for these data. For a more detailed examination of the model residuals, refer to Chapter 7, Lab 2. With some subsetting according to age cohort, it can be reasonable to use linear regression for modeling these data.

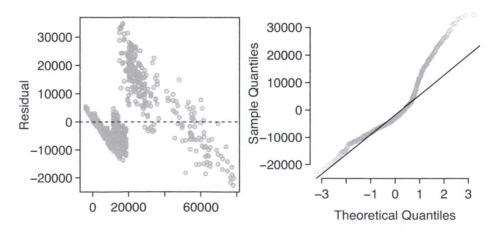

Figure 7.8: Residual versus fitted values plot and residual normal probability plot from the linear model for expenditures versus ethnicity and age for a subset of dds.discr.

7.3.2 Using R^2 and adjusted R^2 with multiple regression

Section 6.3.2 provided two definitions of the R^2 statistic—it is the square of the correlation coefficient r between a response and the single predictor in simple linear regression, and equivalently, it is the proportion of the variation in the response variable explained by the model. In statistical terms, the second definition can be written as

$$R^2 = \frac{\text{Var}(y_i) - \text{Var}(e_i)}{\text{Var}(y_i)} = 1 - \frac{\text{Var}(e_i)}{\text{Var}(y_i)},$$

where y_i and e_i denote the response and residual values for the ith case.

The first definition cannot be used in multiple regression, since there is a correlation coefficient between each predictor and the response variable. However, since there is a single set of residuals, the second definition remains applicable.

Although R^2 can be calculated directly from the equation, it is rarely calculated by hand since statistical software includes R^2 as a standard part of the summary output for a regression model.[10] In the model with response RFFT and predictors Statin and Age, $R^2 = 0.2852$. The model explains almost 29% of the variability in RFFT scores, a considerable improvement over the model with Statin alone ($R^2 = 0.0239$).

Adding a variable to a regression model always increases the value of R^2. Sometimes that increase is large and clearly important, such as when age is added to the model for RFFT scores. In other cases, the increase is small, and may not be worth the added complexity of including another variable. The **adjusted R-squared** is often used to balance predictive ability with complexity in a multiple regression model. Like R^2, the adjusted R^2 is routinely provided in software output.

ADJUSTED R^2 AS A TOOL FOR MODEL ASSESSMENT

The **adjusted R^2** is computed as

$$R^2_{adj} = 1 - \frac{\text{Var}(e_i)/(n-p-1)}{\text{Var}(y_i)/(n-1)} = 1 - \frac{\text{Var}(e_i)}{\text{Var}(y_i)} \times \frac{n-1}{n-p-1},$$

where n is the number of cases used to fit the model and p is the number of predictor variables in the model.

Essentially, the adjusted R^2 imposes a penalty for including additional predictors that do not contribute much towards explaining the observed variation in the response variable. The value of the adjusted R^2 in the model with both Statin and Age is 0.2823, which is essentially the same as the R^2 value of 0.2852. The additional predictor Age considerably increases the strength of the model, resulting in only a small penalty to the R^2 value.

While the adjusted R^2 is useful as a statistic for comparing models, it does not have an inherent interpretation like R^2. Students often confuse the interpretation of R^2 and adjusted R^2; while the two are similar, adjusted R^2 is *not* the proportion of variation in the response variable explained by the model. The use of adjusted R^2 for model selection will be discussed in Section 7.8.

---

[10]In R and other software, R^2 is typically labeled 'multiple R-squared'.

7.4   The general multiple linear regression model

This section provides a compact summary of the multiple regression model and contains more mathematical detail than most other sections; the next section, Section 7.5, discusses categorical predictors with more than two levels. The ideas outlined in this section and the next are illustrated with an extended analysis of the PREVEND data in Section 7.6.

7.4.1   Model parameters and least squares estimation

For multiple regression, the data consist of a response variable Y and p explanatory variables $X_1, X_2, \ldots, X_p$. Instead of the simple regression model

$$Y = \beta_0 + \beta_1 X + \varepsilon,$$

multiple regression has the form

$$Y = \beta_0 + \beta_1 X_1 + \beta_2 X_2 + \beta_3 X_3 + \cdots + \beta_p X_p + \varepsilon,$$

or equivalently

$$E(Y) = \beta_0 + \beta_1 X_1 + \beta_2 X_2 + \beta_3 X_3 + \cdots + \beta_p X_p,$$

since the normally distributed error term ε is assumed to have mean 0. Each predictor x_i has an associated coefficient β_i. In simple regression, the slope coefficient β captures the change in the response variable Y associated with a one unit change in the predictor X. In multiple regression, the coefficient β_j of a predictor X_j denotes the change in the response variable Y associated with a one unit change in X_j when none of the other predictors change; i.e., each β coefficient in multiple regression plays the role of a slope, as long as the other predictors are not changing.

Multiple regression can be thought of as the model for the mean of the response Y in a population where the mean depends on the values of the predictors, rather than being constant. For example, consider a setting with two binary predictors such as statin use and sex; the predictors partition the population into four subgroups, and the four predicted values from the model are estimates of the mean in each of the four groups.

GUIDED PRACTICE 7.5

Figure 7.9 shows an estimated regression model for RFFT with predictors Statin and Gender, where Gender is coded 0 for males and 1 for females.[11] Based on the model, what are the estimated mean RFFT scores for the four groups defined by these two categorical predictors?[12]

	Estimate	Std. Error	t value	Pr(>|t|)
(Intercept)	70.4068	1.8477	38.11	0.0000
Statin	-9.9700	2.9011	-3.44	0.0006
Gender	0.6133	2.4461	0.25	0.8021

Figure 7.9: R summary output for the multiple regression model of RFFT versus statin use and sex in prevend.samp.

Datasets for multiple regression have n cases, usually indexed algebraically by i, where i takes on values from 1 to n; 1 denotes the first case in the dataset and n denotes the last case. The dataset prevend.samp contains $n = 500$ observations. Algebraic representations of the data must indicate both the case number and the predictor in the set of p predictors. For case i in the dataset, the variable X_{ij} denotes predictor X_j; the response for case i is simply Y_i, since there can only be one response variable. The dataset prevend.samp has many possible predictors, some of which are examined later in this chapter. The analysis in Section 7.2 used $p = 2$ predictors, Statin and Age.

Just as in Chapter 2, upper case letters are used when thinking of data as a set of random observations subject to sampling from a population, and lower case letters are used for observed values. In a dataset, it is common for each row to contain the information on a single case; the observations in row i of a dataset with p predictors can be written as $(y_i, x_{i1}, x_{i2}, \ldots, x_{ip})$.

For any given set of estimates $b_1, b_2, \ldots, b_p$ and predictors $x_{i1}, x_{i2}, \ldots, x_{ip}$, predicted values of the response can be calculated using

$$\hat{y}_i = b_0 + b_1 x_{i1} + b_2 x_{i2} + \cdots + b_p x_{ip},$$

where $b_0, b_1, \ldots, b_p$ are estimates of the coefficients $\beta_0, \beta_1, \ldots, \beta_p$ obtained using the principle of least squares estimation.

As in simple regression, each prediction has an associated residual, which is the difference between the observed value y_i and the predicted value $\hat{y}_i$, or $e_i = y_i - \hat{y}_i$. The least squares estimate of the model is the set of estimated coefficients $b_0, b_1, \ldots b_p$ that minimizes $e_1^2 + e_2^2 + \cdots e_n^2$. Explicit formulas for the estimates involve advanced matrix theory, but are rarely used in practice. Instead, estimates are calculated using software such as as R, Stata, or Minitab.

---

[11] Until recently, it was common practice to use gender to denote biological sex. Gender is different than biological sex, but this text uses the original names in published datasets.

[12] The prediction equation for the model is $\widehat{RFFT} = 70.41 - 9.97(\text{Statin}) + 0.61(\text{Gender})$. Both Statin and Gender can take on values of either 0 or 1; the four possible subgroups are statin non-user / male (0, 0), statin non-user / female (0, 1), statin user / male (1, 0), statin user / female (1, 1). Predicted RFFT scores for these groups are 70.41, 71.02, 60.44, and 61.05, respectively.

7.4.2   Hypothesis tests and confidence intervals

Using t-tests for individual coefficients

The test of the null hypothesis $H_0 : \beta_k = 0$ is a test of whether the predictor X_k is associated with the response variable. When a coefficient of a predictor equals 0, the predicted value of the response does not change when the predictor changes; i.e., a value of 0 indicates there is no association between the predictor and response. Due to the inherent variability in observed data, an estimated coefficient b_k will almost never be 0 even when the model coefficient β_k is. Hypothesis testing can be used to assess whether the estimated coefficient is significantly different from 0 by examining the ratio of the estimated coefficient to its standard error.

When the assumptions of multiple regression hold, at least approximately, this ratio has a t-distribution with $n - (p + 1) = n - p - 1$ degrees of freedom when the model coefficient is 0. The formula for the degrees of freedom follows a general rule that appears throughout statistics—the degrees of freedom for an estimated model is the number of cases in the dataset minus the number of estimated parameters. There are $p + 1$ parameters in the multiple regression model, one for each of the p predictors and one for the intercept.

SAMPLING DISTRIBUTIONS OF ESTIMATED COEFFICIENTS

Suppose

$$\hat{y} = b_0 + b_1 x_i + b_2 x_i + \cdots + b_p x_i$$

is an estimated multiple regression model from a dataset with n observations on the response and predictor variables, and let b_k be one of the estimated coefficients. Under the hypothesis $H_0 : \beta_k = 0$, the standardized statistic

$$\frac{b_k}{\text{s.e.}(b_k)}$$

has a t-distribution with $n - p - 1$ degrees of freedom.

This sampling distribution can be used to conduct hypothesis tests and construct confidence intervals.

TESTING A HYPOTHESIS ABOUT A REGRESSION COEFFICIENT

A test of the two-sided hypothesis

$$H_0 : \beta_k = 0 \text{ vs. } H_A : \beta_k \neq 0$$

is rejected with significance level α when

$$\frac{|b_k|}{\text{s.e.}(b_k)} > t_{\text{df}}^{\star},$$

where $t_{\text{df}}^{\star}$ is the point on a t-distribution with $n - p - 1$ degrees of freedom and area $(1 - \alpha/2)$ in the left tail.

For one-sided tests, $t_{df}^\star$ is the point on a t-distribution with $n-p-1$ degrees of freedom and area $(1-\alpha)$ in the left tail. A one-sided test of H_0 against $H_A : \beta_k > 0$ rejects when the standardized coefficient is greater than $t_{df}^\star$; a one-sided test of H_0 against $H_A : \beta_k < 0$ rejects when the standardized coefficient is less than $t_{df}^\star$.

CONFIDENCE INTERVALS FOR REGRESSION COEFFICIENT

A two-sided $100(1-\alpha)\%$ confidence interval for the model coefficient β_k is

$$b_k \pm \text{s.e.}(b_k) \times t_{df}^\star.$$

All statistical software packages provide an estimate s of the standard deviation of the residuals ϵ.

The F-statistic for an overall test of the model

When all the model coefficients are 0, the predictors in the model, considered as a group, are not associated with the response; i.e., the response variable is not associated with any linear combination of the predictors. The F-statistic is used to test this null hypothesis of no association, using the following idea.

The variability of the predicted values about the overall mean response can be estimated by

$$\text{MSM} = \frac{\sum_i (\hat{y}_i - \overline{y})^2}{p}.$$

In this expression, p is the number of predictors and is the degrees of freedom of the numerator sum of squares (derivation not given here). The term **MSM** is called the model sum of squares because it reflects the variability of the values predicted by the model ($\hat{y}_i$) about the mean ($\overline{y}$) response.[13] In an extreme case, MSM will have value 0 when all the predicted values coincide with the overall mean; in this scenario, a model would be unnecessary for making predictions, since the average of all observations could be used to make a prediction.

The variability in the residuals can be measured by

$$\text{MSE} = \frac{\sum_i (y_i - \hat{y}_i)^2}{n-p-1}.$$

MSE is called the mean square of the errors since residuals are the observed 'errors', the differences between predicted and observed values.

When MSM is small compared to MSE, the model has captured little of the variability in the data, and the model is of little or no value. The F-statistic is given by

$$F = \frac{\text{MSM}}{\text{MSE}}.$$

The formula is not used for calculation, since the numerical value of the F-statistic is a routine part of the output of regression software.

---

[13]It turns out that $\overline{y}$ is also the mean of the predicted values.

THE *F*-STATISTIC IN REGRESSION

The F-statistic in regression is used to test the null hypothesis

$$H_0 : \beta_1 = \beta_2 = \cdots = \beta_p = 0$$

against the alternative that at least one of the coefficients is not 0.

Under the null hypothesis, the sampling distribution of the F-statistic is an F-distribution with parameters $(p, n - p - 1)$, and the null hypothesis is rejected if the value of the F-statistic is in the right tail of the distribution of the sampling distribution with area α, where α is the significance level of the test.

The F-test is inherently one-sided—deviations from the null hypothesis of any form will push the statistic to the right tail of the F-distribution. The p-value from the right tail of the F-distribution should never be doubled. Students also sometimes make the mistake of assuming that if the null hypothesis of the F-test is rejected, all coefficients must be non-zero, instead of at least one. A significant p-value for the F-statistic suggests that the predictor variables in the model, when considered as a group, are associated with the response variable.

In practice, it is rare for the F-test not to reject the null hypothesis, since most regression models are used in settings where a scientist has prior evidence that at least some of the predictors are useful.

Confidence and Prediction Intervals

The confidence and prediction intervals discussed in Section 6.5 can be extended to multiple regression. Predictions based on specific values of the predictors are made by evaluating the estimated model at those values, and both confidence intervals for the mean and prediction intervals for a new observation are constructed using the corresponding standard errors. The formulas for standard errors in the multiple predictor setting are beyond the scope of this text, and there are no simple approximate formulas that can be calculated by hand. They are always computed in software.

Figure 7.5 shows the estimated regression model used to examine the association of age and statin use with RFFT score in PREVEND. As shown in Example 7.5, the predicted RFFT score for a 67-year-old statin user is 57.6 points. Software can be used to show that a 95% confidence interval for the mean RFFT score for 67-year-old statin users is (49.2, 58.9) points, while a 95% prediction interval for the RFFT score of a particular 67-year statin user is (7.8, 99.4) points. Just as with simple linear regression, the prediction interval is wider than the confidence interval for the mean because it accounts for both variability in the estimated mean and variability in a new observation of the response, RFFT score.

7.5 Categorical predictors with several levels

In the initial model fit with the PREVEND data, the variable Statin is coded 0 if the participant was not using statins, and coded 1 if the participant was a statin user. The category coded 0 is referred to as the reference category; in this model, statin non-users (Statin = 0) are the reference category. The estimated coefficient β_{Statin} is the change in the average response between the reference category and the category Statin = 1.

Since the variable Statin is categorical, the numerical codes 0 and 1 are simply labels for statin non-users and users. The labels can be specified more explicitly in software. For example, in R, categorical variables can be coded as factors; the levels of the variable are displayed as text (such as "NonUser" or "User"), while the data remain stored as integers. The R output with the variable Statin.factor is shown in Figure 7.10, where 0 corresponds to the label "NonUser" and 1 corresponds to "User". The predictor variable is now labeled Statin.factorUser; the estimate -10.05 is the change in mean RFFT from the "NonUser" (reference) category to the "User" category. Note how the reference category is not explicitly labeled; instead, it is contained within the intercept.

	Estimate	Std. Error	t value	Pr(>|t|)
(Intercept)	70.7143	1.3808	51.21	0.0000
Statin.factorUser	-10.0534	2.8792	-3.49	0.0005

Figure 7.10: R summary output for the simple regression model of RFFT versus statin use in prevend.samp, with Statin converted to a factor called Statin.factor that has levels NonUser and User.

For a categorical variable with two levels, estimates from the regression model remain the same regardless of whether the categorical predictor is treated as numerical or not. A "one unit change" in the numerical sense corresponds exactly to the switch between the two categories. However, this is not true for categorical variables with more than two levels.

This idea will be explored with the categorical variable Education, which indicates the highest level of education that an individual completed in the Dutch educational system: primary school, lower secondary school, higher secondary education, or university education. In the PREVEND dataset, educational level is coded as either 0, 1, 2, or 3, where 0 denotes at most a primary school education, 1 a lower secondary school education, 2 a higher secondary education, and 3 a university education. Figure 7.11 shows the distribution of RFFT by education level; RFFT scores tend to increase as education level increases.

In a regression model with a categorical variable with more than two levels, one of the categories is set as the reference category, just as in the setting with two levels for a categorical predictor. The remaining categories each have an estimated coefficient, which corresponds to the estimated change in response relative to the reference category.

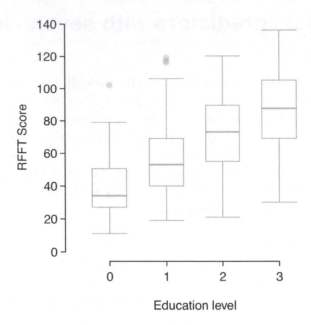

Figure 7.11: Box plots for RFFT score by education level in prevend.samp.

EXAMPLE 7.6

Is RFFT score associated with educational level? Interpret the coefficients from the following model. Figure 7.12 provides the R output for the regression model of RFFT versus educational level in prevend.samp. The variable Education has been converted to Education.factor, which has levels Primary, LowerSecond, HigherSecond, and Univ.

It is clearest to start with writing the model equation:

$$\widehat{\text{RFFT}} = 40.94 + 14.78(\text{EduLowerSecond}) + 32.13(\text{EduHigherSecond}) + 44.96(\text{EduUniv})$$

Each of the predictor levels can be thought of as binary variables that can take on either 0 or 1, where only one level at most can be a 1 and the rest must be 0, with 1 corresponding to the category of interest. For example, the predicted mean RFFT score for individuals in the Lower Secondary group is given by

$$\widehat{\text{RFFT}} = 40.94 + 14.78(1) + 32.13(0) + 44.96(0) = 55.72.$$

The value of the LowerSecond coefficient, 14.78, is the change in predicted mean RFFT score from the reference category Primary to the LowerSecond category.

Participants with a higher secondary education scored approximately 32.1 points higher on the RFFT than individuals with only a primary school education, and have estimated mean RFFT score $40.94 + 32.13 = 73.07$. Those with a university education have estimated mean RFFT score $40.94 + 44.96 = 85.90$.

The intercept value, 40.94, corresponds to the estimated mean RFFT score for individuals who at most completed primary school. From the regression equation,

$$\widehat{\text{RFFT}} = 40.94 + 14.78(0) + 32.13(0) + 44.96(0) = 40.94.$$

The p-values indicate that the change in mean score between participants with only a primary school education and any of the other categories is statistically significant.

	Estimate	Std. Error	t value	Pr(>|t|)
(Intercept)	40.9412	3.2027	12.78	0.0000
Education.factorLowerSecond	14.7786	3.6864	4.01	0.0001
Education.factorHigherSecond	32.1335	3.7631	8.54	0.0000
Education.factorUniv	44.9639	3.6835	12.21	0.0000

Figure 7.12: R summary output for the regression model of RFFT versus educational level in prevend.samp, with Education converted to a factor called Education.factor that has levels Primary, LowerSecond, HigherSecond, and Univ.

EXAMPLE 7.7

Suppose that the model for predicting RFFT score from educational level is fitted with Education, using the original numerical coding with 0, 1, 2, and 3; the R output is shown in Figure 7.13. What does this model imply about the change in mean RFFT between groups? Explain why this model is flawed.

According to this model, the change in mean RFFT between groups increases by 15.158 for any one unit change in Education. For example, the change in means between the groups coded 0 and 1 is necessarily equal to the change in means between the groups coded 2 and 3, since the predictor changes by 1 in both cases.

It is unreasonable to assume that the change in mean RFFT score when comparing the primary school group to the lower secondary group will be equal to the difference in means between the higher secondary group and university group. The numerical codes assigned to the groups are simply short-hand labels, and are assigned arbitrarily. As a consequence, this model would not provide consistent results if the numerical codes were altered; for example, if the primary school group and lower secondary group were relabeled such that the predictor changes by 2, the estimated difference in mean RFFT would change.

	Estimate	Std. Error	t value	Pr(>|t|)
(Intercept)	41.148	2.104	19.55	0.0000
Education	15.158	1.023	14.81	0.0000

Figure 7.13: R summary output for the simple regression model of RFFT versus educational level in prevend.samp, where Education is treated as a numerical variable. Note that it would be incorrect to fit this model; Figure 7.12 shows the results from the correct approach.

Categorical variables can be included in multiple regression models with other predictors, as is shown in the next section. Section 7.9 discusses the connection between ANOVA and regression models with only one categorical predictor.

7.6   Reanalyzing the PREVEND data

The earlier models fit to examine the association between cognitive ability and statin use showed that considering statin use alone could be misleading. While older participants tended to have lower RFFT scores, they were also more likely to be taking statins. Age was found to be a **confounder** in this setting—is it the only confounder?

Potential confounders are best identified by considering the larger scientific context of the analysis. For the PREVEND data, there are two natural candidates for potential confounders: education level and presence of cardiovascular disease. The use of medication is known to vary by education levels, often because individuals with more education tend to have higher incomes and consequently, better access to health care; higher educational levels are associated with higher RFFT scores, as shown by model 7.12. Individuals with cardiovascular disease are often prescribed statins to lower cholesterol; cardiovascular disease can lead to vascular dementia and cognitive decline.

Figure 7.14 contains the result of a regression of RFFT with statin use, adding the possible confounders age, educational level, and presence of cardiovascular disease. The variables Statin, Education and CVD have been converted to factors, and Age is a continuous predictor.

The coefficient for statin use shows the importance of adjusting for confounders. In the initial model for RFFT that only included statin use as a predictor, statin use was significantly associated with decreased RFFT scores. After adjusting for age, statins were no longer significantly associated with RFFT scores, but the model suggested that statin use could be associated with *increased* RFFT scores. This final model suggests that, after adjusting for age, education, and the presence of cardiovascular disease, statin use is associated with an increase in RFFT scores of approximately 4.7 points. The p-value for the slope coefficient for statin use is 0.056, which suggests moderately strong evidence of an association (significant at $\alpha = 0.10$, but not $\alpha = 0.05$).

	Estimate	Std. Error	t value	Pr(>|t|)
(Intercept)	99.0351	6.3301	15.65	0.0000
Statin.factorUser	4.6905	2.4480	1.92	0.0559
Age	-0.9203	0.0904	-10.18	0.0000
Education.factorLowerSecond	10.0883	3.3756	2.99	0.0029
Education.factorHigherSecond	21.3015	3.5777	5.95	0.0000
Education.factorUniv	33.1246	3.5471	9.34	0.0000
CVD.factorPresent	-7.5665	3.6516	-2.07	0.0388

Figure 7.14: R summary output for the multiple regression model of RFFT versus statin use, age, education, and presence of cardiovascular disease in prevend.samp.

The R^2 for the model is 0.4355; a substantial increase from the model with only statin use and age as predictors, which had an R^2 of 0.2852. The adjusted R^2 for the model is 0.4286, close to the R^2 value, which suggests that the additional predictors increase the strength of the model enough to justify the additional complexity.

Figure 7.15 shows a plot of residuals vs predicted RFFT scores from the model in Figure 7.14 and a normal probability plot of the residuals. These plots show that the model fits the data reasonably well. The residuals show a slight increase in variability for larger predicted values, and the normal probability plot shows the residuals depart slightly from normality in the extreme tails. Model assumptions never hold exactly, and the possible violations shown in this figure are not sufficient reasons to discard the model.

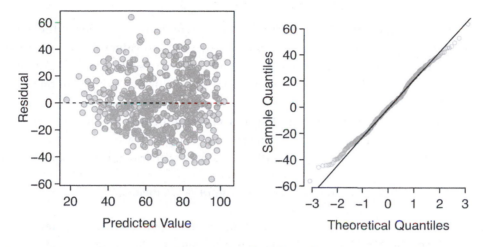

Figure 7.15: A histogram and normal probability plot of the residuals from the linear model for RFFT vs. statin use, age, educational level and presence of cardiovascular disease in the PREVEND data.

It is quite possible that even the model summarized in Figure 7.14 is not the best one to understand the association of cognitive ability with statin use. There be other confounders that are not accounted for. Possible predictors that may be confounders but have not been examined are called **residual confounders**. Residual confounders can be other variables in a dataset that have not been examined, or variables that were not measured in the study. Residual confounders exist in almost all observational studies, and represent one of the main reasons that observational studies should be interpreted with caution. A randomized experiment is the best way to eliminate residual confounders. Randomization ensures that, at least on average, all predictors are not associated with the randomized intervention, which eliminates one of the conditions for confounding. A randomized trial may be possible in some settings; there have been many randomized trials examining the effect of using statins. However, in many other settings, such as a study of the association of marijuana use and later addiction to controlled substances, randomization may not be possible or ethical. In those instances, observational studies may be the best available approach.

7.7   Interaction in regression

An important assumption in the multiple regression model

$$y = \beta_0 + \beta_1 x_1 + \beta_2 x_2 + \dots + \beta_p x_p + \varepsilon$$

is that when one of the predictor variables x_j changes by 1 unit and none of the other variables change, the predicted response changes by β_j, regardless of the values of the other variables. A statistical **interaction** occurs when this assumption is not true, such that the relationship of one explanatory variable x_j with the response depends on the particular value(s) of one or more other explanatory variables.

Interaction is most easily demonstrated in a model with two predictors, where one of the predictors is categorical and the other is numerical.[14] Consider a model that might be used to predict total cholesterol level from age and diabetes status (either diabetic or non-diabetic):

$$E(\text{TotChol}) = \beta_0 + \beta_1(\text{Age}) + \beta_2(\text{Diabetes}). \tag{7.8}$$

Figure 7.16 shows the R output for a regression estimating model 7.8, using data from a sample of 500 adults from the NHANES dataset (nhanes.samp.adult.500). Total cholesterol (TotChol) is measured in mmol/L, Age is recorded in years, and Diabetes is a factor level with the levels No (non-diabetic) and Yes (diabetic) where 0 corresponds to No and 1 corresponds to Yes.

	Estimate	Std. Error	t value	Pr(>|t|)
(Intercept)	4.8000	0.1561	30.75	0.0000
Age	0.0075	0.0030	2.47	0.0137
DiabetesYes	-0.3177	0.1607	-1.98	0.0487

Figure 7.16: Regression of total cholesterol on age and diabetes, using nhanes.samp.adult.500.

---

[14]Interaction effects between numerical variables and between more than two variables can be complicated to interpret. A more complete treatment of interaction is best left to a more advanced course; this text will only examine interaction in the setting of models with one categorical variable and one numerical variable.

EXAMPLE 7.9

Using the output in Figure 7.16, write the model equation and interpret the coefficients for age and diabetes. How does the predicted total cholesterol for a 60-year-old individual compare to that of a 50-year-old individual, if both have diabetes? What if both individuals do not have diabetes?

_____

$$\widehat{TotChol} = 4.80 + 0.0075(Age) - 0.32(DiabetesYes)$$

The coefficient for age indicates that with each increasing year of age, predicted total cholesterol increases by 0.0075 mmol/L. The coefficient for diabetes indicates that diabetics have an average total cholesterol that is 0.32 mmol/L lower than non-diabetic individuals.

If both individuals have diabetes, then the change in predicted total cholesterol level can be determined directly from the coefficient for Age. An increase in one year of age is associated with a 0.0075 increase in total cholesterol; thus, an increase in ten years of age is associated with $10(0.0075) = 0.075$ mmol/L increase in predicted total cholesterol.

The calculation does not differ if both individuals are non-diabetic. According to the model, the relationship between age and total cholesterol remains the same regardless of the values of the other variable in the model.

EXAMPLE 7.10

Using the output in Figure 7.16, write two separate model equations: one for diabetic individuals and one for non-diabetic individuals. Compare the two models.

_____

For non-diabetics (Diabetes = 0), the linear relationship between average cholesterol and age is $\widehat{TotChol} = 4.80 + 0.0075(Age) - 0.32(0) = 4.80 + 0.0075(Age)$.

For diabetics (Diabetes = 1), the linear relationship between average cholesterol and age is $\widehat{TotChol} = 4.80 + 0.0075(Age) - 0.32(1) = 4.48 + 0.0075(Age)$.

The lines predicting average cholesterol as a function of age in diabetics and non-diabetics are parallel, with the same slope and different intercepts. While predicted total cholesterol is higher overall in non-diabetics (as indicated by the higher intercept), the rate of change in predicted average total cholesterol by age is the same for both diabetics and non-diabetics.

This relationship can be expressed directly from the model equation 7.8. For non-diabetics, the population regression line is $E(TotChol) = \beta_0 + \beta_1(Age)$. For diabetics, the line is $E(TotChol) = \beta_0 + \beta_1(Age) + \beta_2 = \beta_0 + \beta_2 + \beta_1(Age)$. The lines have the same slope β_1 but intercepts β_0 and $\beta_0 + \beta_2$.

However, a model that assumes the relationship between cholesterol and age does not depend on diabetes status might be overly simple and potentially misleading. Figure 7.17(b) shows a scatterplot of total cholesterol versus age where the least squares models have been fit separately for non-diabetic and diabetic individuals. The blue line in the plot is estimated using only non-diabetic individuals, while the red line was fit using data from diabetic individuals. The lines are not parallel, and in fact, have slopes with different signs. The plot suggests that among non-diabetics, age is positively associated with total cholesterol. Among diabetics, however, age is negatively associated with total cholesterol.

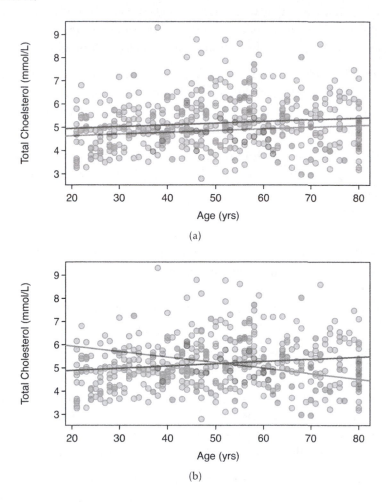

Figure 7.17: Scatterplots of total cholesterol versus age in nhanes.samp.adult.500, where blue represents non-diabetics and red represents diabetics. Plot (a) shows the model equations written out in Example 7.10, estimated from the entire sample of 500 individuals. Plot (b) shows least squares models that are fit separately; coefficients of the blue line are estimated using only data from non-diabetics, while those of the red line are estimated using only data from diabetics.

With the addition of another parameter (commonly referred to as an interaction term), a linear regression model can be extended to allow the relationship of one explanatory variable with the response to vary based on the values of other variables in the model. Consider the model

$$E(\text{TotChol}) = \beta_0 + \beta_1(\text{Age}) + \beta_2(\text{Diabetes}) + \beta_3(\text{Diabetes} \times \text{Age}). \qquad (7.11)$$

The interaction term allows the slope of the association with age to differ by diabetes status. Among non-diabetics (Diabetes = 0), the model reduces to the earlier one,

$$E(\text{TotChol}) = \beta_0 + \beta_1(\text{Age}).$$

Among the diabetic participants, the model becomes

$$E(\text{TotChol}) = \beta_0 + \beta_1(\text{Age}) + \beta_2 + \beta_3(\text{Age})$$
$$= \beta_0 + \beta_2 + (\beta_1 + \beta_3)(\text{Age}).$$

Unlike in the original model, the slopes of the population regression lines for non-diabetics and diabetics are now different: β_1 versus $\beta_1 + \beta_3$.

Figure 7.18 shows the R output for a regression estimating model 7.11. In R, the syntax Age:DiabetesYes represents the (Age × Diabetes) interaction term.

	Estimate	Std. Error	t value	Pr(>|t|)
(Intercept)	4.6957	0.1597	29.40	0.0000
Age	0.0096	0.0031	3.10	0.0020
DiabetesYes	1.7187	0.7639	2.25	0.0249
Age:DiabetesYes	-0.0335	0.0123	-2.73	0.0067

Figure 7.18: Regression of total cholesterol on age and diabetes with an interaction term, using nhanes.samp.adult.500

EXAMPLE 7.12

Using the output in Figure 7.18, write the overall model equation, the model equation for non-diabetics, and the model equation for diabetics.

The overall model equation is

$$\widehat{\text{TotChol}} = 4.70 + 0.0096(\text{Age}) + 1.72(\text{DiabetesYes}) - 0.034(\text{Age} \times \text{DiabetesYes}).$$

For non-diabetics (Diabetes = 0), the linear relationship between average cholesterol and age is

$$\widehat{\text{TotChol}} = 4.70 + 0.0096(\text{Age}) + 1.72(0) - 0.034(\text{Age} \times 0) = 4.70 + 0.0096(\text{Age}).$$

For diabetics (Diabetes = 1), the linear relationship between average cholesterol and age is

$$\widehat{\text{TotChol}} = 4.70 + 0.0096(\text{Age}) + 1.72(1) - 0.034(\text{Age} \times 1) = 6.42 - 0.024(\text{Age}).$$

The estimated equations for non-diabetic and diabetic individuals show the same qualitative behavior seen in Figure 7.17(b), where the slope is positive in non-diabetics and negative in diabetics. However, note that the lines plotted in the figure were estimated from two separate model fits on non-diabetics and diabetics; in contrast, the equations from the interaction model are fit using data from all individuals.

It is more efficient to model the data using a single model with an interaction term than working with subsets of the data.[15] Additionally, using a single model allows for the calculation of a t-statistic and p-value that indicates whether there is statistical evidence of an interaction. The p-value for the Age:Diabetes interaction term is significant at the $\alpha = 0.05$ level. Thus, the estimated model suggests there is strong evidence for an interaction between age and diabetes status when predicting total cholesterol.

Residual plots can be used to assess the quality of the model fit. Figure 7.19 shows that the residuals have roughly constant variance in the region with the majority of the data (predicted values between 4.9 and 5.4 mmol/L). However, there are more large positive residuals than large negative residuals, which suggests that the model tends to underpredict; i.e., predict values of TotChol that are smaller than the observed values.[16] Figure 7.20 shows that the residuals do not fit a normal distribution in the tails. In the right tails, the sample quantiles are larger than the theoretical quantiles, implying that there are too many large residuals. The left tail is a better fit; however, there are too few large negative residuals since the sample quantiles in the left tail are closer to 0 than the theoretical quantiles.

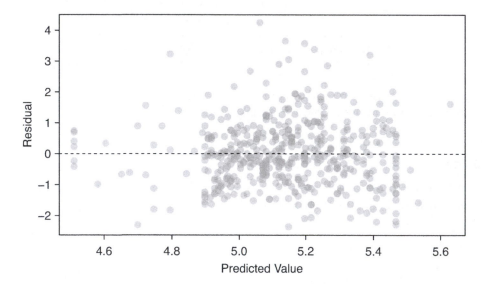

Figure 7.19: A scatterplot of residuals versus predicted values in the model for total cholesterol that includes age, diabetes status, and the interaction of age and diabetes status.

---

[15]In more complex settings, such as those with potential interaction between several variables or between two numerical variables, it may not be clear how to subset the data in a way that reveals interactions. This is another advantage to using an interaction term and single model fit to the entire dataset.

[16]Recall that model residuals are calculated as $y_i - \hat{y}_i$; i.e., $\text{TotChol}_i - \widehat{\text{TotChol}}_i$.

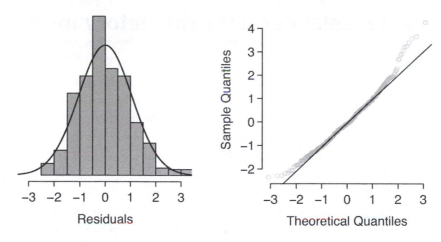

Figure 7.20: A histogram of the residuals and a normal probability plot of the residuals from the linear model for total cholesterol versus age, diabetes status, and the interaction of age and diabetes status.

It is also important to note that the model explains very little of the observed variability in total cholesterol—the multiple R^2 of the model is 0.032. While the model falls well short of perfection, it may be reasonably adequate in applied settings. In the setting of a large study, such as one to examine factors affecting cholesterol levels in adults, a model like the one discussed here is typically a starting point for building a more refined model. Given these results, a research team might proceed by collecting more data. Regression models are commonly used as tools to work towards understanding a phenomenon, and rarely represent a 'final answer'.

There are some important general points that should not be overlooked when interpreting this model. The data cannot be used to infer causality; the data simply show associations between total cholesterol, age, and diabetes status. Each of the NHANES surveys are cross-sectional; they are administered to a sample of US residents with various ages and other demographic features during a relatively short period of time. No single individual has had his or her cholesterol levels measured over a period of many years, so the model slope for diabetes is not indicative of an individual's cholesterol level declining (or increasing) with age.

Finally, the interpretation of a model often requires additional contextual information that is relevant to the study population but not captured in the dataset. What might explain increased age being associated with lower cholesterol for diabetics, but higher cholesterol for non-diabetics? The guidelines for the use of cholesterol-lowering statins suggest that these drugs should be prescribed more often in older individuals, and even more so in diabetic individuals. It is a reasonable speculation that the interaction between age and diabetes status seen in the NHANES data is a result of more frequent statin use in diabetic individuals.

7.8   Model selection for explanatory models

Previously, multiple regression modeling was shown in the context of estimating an association while adjusting for possible confounders. Another application of multiple regression is explanatory modeling, in which the goal is to construct a model that explains the observed variation in the response variable. In this context, there is no pre-specified primary predictor of interest; explanatory modeling is concerned with identifying predictors associated with the response. It is typically desirable to have a small model that avoids including variables which do not contribute much towards the R^2.

The intended use of a regression model influences the way in which a model is selected. Approaches to model selection vary from those based on careful study of a relatively small set of predictors to purely algorithmic methods that screen a large set of predictors and choose a final model by optimizing a numerical criterion. Algorithmic selection methods have gained popularity as researchers have been able to collect larger datasets, but the choice of an algorithm and the optimization criterion require more advanced material and are not covered here. This section illustrates model selection in the context of a small set of potential predictors using only the tools and ideas that have been discussed earlier in this chapter and in Chapter 6.

Generally, model selection for explanatory modeling follows these steps:

1. *Data exploration.* Using numerical and graphical approaches, examine both the distributions of individual variables and the relationships between variables.

2. *Initial model fitting.* Fit an initial model with the predictors that seem most highly associated with the response variable, based on the data exploration.

3. *Model comparison.* Work towards a model that has the highest adjusted R^2.

   – Fit new models without predictors that were either not statistically significant or only marginally so and compare the adjusted R^2 between models; drop variables that decrease the adjusted R^2.

   – If the initial set of variables is relatively small, it is prudent to add variables not in the initial model and check the adjusted R^2; add variables that increase the adjusted R^2.

   – Examine whether interaction terms may improve the adjusted R^2.

4. *Model assessment.* Use residual plots to assess the fit of the final model.

The process behind model selection will be illustrated with a case study in which a regression model is built to examine the association between the abundance of forest birds in a habitat patch and features of a patch.

Abundance of forest birds: introduction

Habitat fragmentation is the process by which a habitat in a large contiguous space is divided into smaller, isolated pieces; human activities such as agricultural development can result in habitat fragmentation. Smaller patches of habitat are only able to support limited populations of organisms, which reduces genetic diversity and overall population fitness. Ecologists study habitat fragmentation to understand its effect on species abundance. The forest.birds dataset in the oibiostat package contains a subset of the variables from a 1987 study analyzing the effect of habitat fragmentation on bird abundance in the Latrobe Valley of southeastern Victoria, Australia.[17]

The dataset consists of the following variables, measured for each of the 57 patches.

- abundance: average number of forest birds observed in the patch, as calculated from several independent 20-minute counting sessions.

- patch.area: patch area, measured in hectares. 1 hectare is 10,000 square meters and approximately 2.47 acres.

- dist.nearest: distance to the nearest patch, measured in kilometers.

- dist.larger: distance to the nearest patch larger than the current patch, measured in kilometers.

- altitude: patch altitude, measured in meters above sea level.

- grazing.intensity: extent of livestock grazing, recorded as either "light", "less than average", "average", "moderately heavy", or "heavy".

- year.of.isolation: year in which the patch became isolated due to habitat fragmentation.

- yrs.isolation: number of years since patch became isolated due to habitat fragmentation.[18]

The following analysis is similar to analyses that appear in Logan (2011)[19] and Quinn & Keough (2002).[20] In the approach here, the grazing intensity variable is treated as a categorical variable; Logan and Quinn & Keough treat grazing intensity as a numerical variable, with values 1-5 corresponding to the categories. The implications of these approaches are discussed at the end of the section.

[17]Loyn, R.H. 1987. "Effects of patch area and habitat on bird abundances, species numbers and tree health in fragmented Victorian forests." Printed in Nature Conservation: The Role of Remnants of Native Vegetation. Saunders DA, Arnold GW, Burbridge AA, and Hopkins AJM eds. Surrey Beatty and Sons, Chipping Norton, NSW, 65-77, 1987.

[18]The Loyn study completed data collection in 1983; yrs.isolation = 1983 − year.of.isolation.

[19]Logan, M., 2011. Biostatistical design and analysis using R: a practical guide. John Wiley & Sons, Ch. 9.

[20]Quinn, G.P. and Keough, M.J., 2002. Experimental design and data analysis for biologists. Cambridge University Press, Ch. 6.

Data exploration

The response variable for the model is abundance. Numerical summaries calculated from software show that abundance ranges from 1.5 to 39.6. Figure 7.21 shows that the distribution of abundance is bimodal, with modes at small values of abundance and at between 25 and 30 birds. The median (21.0) and mean (19.5) are reasonably close, which confirms the distribution is near enough to symmetric to be used in the model without a transformation. The boxplot confirms that the distribution has no outliers.

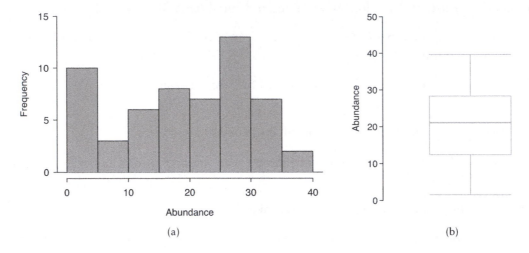

(a)   (b)

Figure 7.21: A histogram (a) and boxplot (b) of abundance in the forest.birds data.

There are six potential predictors in the model; the variable year.of.isolation is only used to calculate the more informative variable yrs.isolation. The plots in Figure 7.22 reveal right-skewing in patch.area, dist.nearest, dist.larger, and yrs.isolation; these might benefit from a log transformation. The variable altitude is reasonably symmetric, and the predictor grazing.factor is categorical and so does not take transformations. Figure 7.23 shows the distributions of log.patch.area, log.dist.nearest, log.dist.larger, and log.yrs.isolation, which were created through a natural log transformation of the original variables. All four are more nearly symmetric. These will be more suitable for inclusion in a model than the untransformed versions.

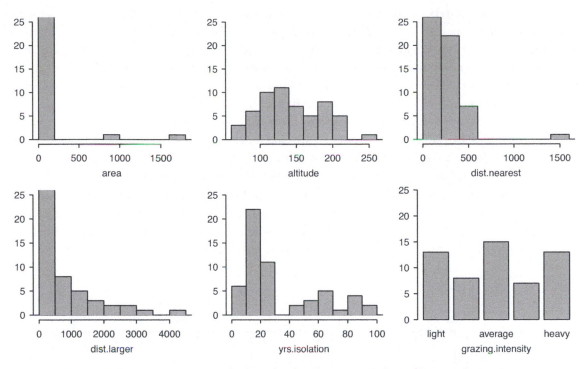

Figure 7.22: Histograms and a barplot for the potential predictors of abundance.

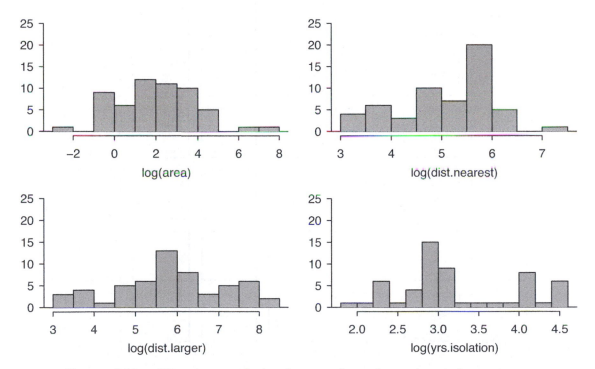

Figure 7.23: Histograms of the log-transformed versions of patch.area, dist.nearest, dist.larger, and yrs.isolation.

A **scatterplot matrix** can be useful for visualizing the relationships between the predictor and response variables, as well as the relationships between predictors. Each subplot in the matrix is a simple scatterplot; all possible plots are shown, except for the plots of a variable versus itself. The variable names are listed along the diagonal of the matrix, and the diagonal divides the matrix into symmetric plots. For instance, the first plot in the first row shows abundance on the vertical axis and log.area on the horizontal axis; the first plot in the first column shows abundance on the horizontal axis and log.area on the vertical axis. Note that for readability, grazing.intensity appears with values 1 - 5, with 1 denoting "light" and 5 denoting "heavy" grazing intensity.

The plots in the first row of Figure 7.24 show the relationships between abundance and the predictors.[21] There is a strong positive association between abundance with log.area, and a strong negative association between abundance and log.yrs.isolation. The variables log.dist.near.patch and log.dist.larger seem weakly positively associated with abundance. There is high variance of abundance and somewhat similar centers for the first four categories, but abundance does clearly tend to be lower in the "high grazing" category versus the others.

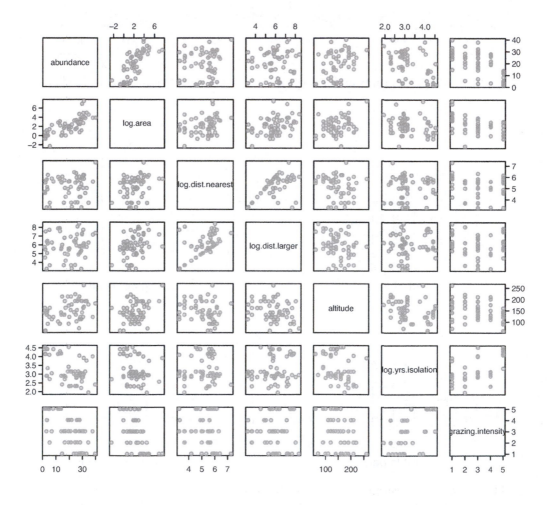

Figure 7.24: Scatterplot matrix of abundance and the possible predictors: log.area, log.dist.near.patch, log.dist.larger.patch, altitude, log.yrs.isolation, and grazing.intensity.

---

[21] Traditionally, the response variable (i.e., the dependent variable) is plotted on the vertical axis; as a result, it seems more natural to look at the first row where abundance is on the y-axis. It is equally valid, however, to assess the association of abundance with the predictors from the plots in the first column.

The variables log.dist.nearest and log.dist.larger appear strongly associated; a model may only need one of the two, as they may be essentially "redundant" in explaining the variability in the response variable.[22] In this case, however, since both are only weakly associated with abundance, both may be unnecessary in a model.

A numerical approach confirms some of the features observable from the scatterplot matrix. Figure 7.25 shows the correlations between pairs of numerical variables in the dataset. Correlations between abundance and log.area and between abundance and log.yrs.isolation are relatively high, at 0.74 and -0.48, respectively. In contrast, the correlation between abundance and the two variables log.dist.nearest and log.dist.larger are much smaller, at 0.13 and 0.12. Additionally, the two potential predictors log.dist.nearest and log.dist.larger have a relatively high correlation of 0.60.

	abundance	log.area	log.dist.nearest	log.dist.larger	altitude	log.yrs.isolation
abundance	1.00	0.74	0.13	0.12	0.39	-0.48
log.area	0.74	1.00	0.30	0.38	0.28	-0.25
log.dist.nearest	0.13	0.30	1.00	0.60	-0.22	0.02
log.dist.larger	0.12	0.38	0.60	1.00	-0.27	0.15
altitude	0.39	0.28	-0.22	-0.27	1.00	-0.29
log.yrs.isolation	-0.48	-0.25	0.02	0.15	-0.29	1.00

Figure 7.25: A correlation matrix for the numerical variables in forest.birds.

Initial model fitting

Based on the data exploration, the initial model should include the variables log.area, altitude, log.yrs.isolation, and grazing.intensity; a summary of this model is shown in Figure 7.26. The R^2 and adjusted R^2 for this model are, respectively, 0.728 and 0.688. The model explains about 73% of the variability in abundance.

	Estimate	Std. Error	t value	Pr(>|t|)
(Intercept)	14.1509	6.3006	2.25	0.0293
log.area	3.1222	0.5648	5.53	0.0000
altitude	0.0080	0.0216	0.37	0.7126
log.yrs.isolation	0.1300	1.9193	0.07	0.9463
grazing.intensityless than average	0.2967	2.9921	0.10	0.9214
grazing.intensityaverage	-0.1617	2.7535	-0.06	0.9534
grazing.intensitymoderately heavy	-1.5936	3.0350	-0.53	0.6019
grazing.intensityheavy	-11.7435	4.3370	-2.71	0.0094

Figure 7.26: Initial model: regression of abundance on log.area, altitude, log.yrs.isolation and grazing.intensity.

Two of the variables in the model are not statistically significant at the $\alpha = 0.05$ level: altitude and log.yrs.isolation. Only one of the categories of grazing.intensity (heavy grazing) is highly significant.

---

[22]Typically, the predictor that is less strongly correlated with the response variable is the one that is "redundant" and will be statistically insignificant when included in a model with the more strongly correlated predictor. This is not always the case, and depends on the other variables in the model.

Model comparison

First, fit models excluding the predictors that were not statistically significant: altitude and
log.yrs.isolation. Models excluding either variable have adjusted R^2 of 0.69, and a model excluding both variables has an adjusted R^2 of 0.70, a small but noticeable increase from the initial
model. This suggests that these two variables can be dropped. At this point, the working model
includes only log.area and grazing.intensity; this model has $R^2 = 0.727$ and is shown in Figure 7.27.

	Estimate	Std. Error	t value	Pr(>|t|)
(Intercept)	15.7164	2.7674	5.68	0.0000
log.area	3.1474	0.5451	5.77	0.0000
grazing.intensityless than average	0.3826	2.9123	0.13	0.8960
grazing.intensityaverage	-0.1893	2.5498	-0.07	0.9411
grazing.intensitymoderately heavy	-1.5916	2.9762	-0.53	0.5952
grazing.intensityheavy	-11.8938	2.9311	-4.06	0.0002

Figure 7.27: Working model: regression of abundance on log.area and
grazing.intensity.

It is prudent to check whether the two distance-related variables that were initially excluded
might increase the adjusted R^2, even though this seems unlikely. When either or both of these
variables are added, the adjusted R^2 decreases from 0.70 to 0.69. Thus, these variables are not
added to the working model.

In this working model, only one of the coefficients associated with grazing intensity is statistically significant; when compared to the baseline grazing category (light grazing), heavy grazing
is associated with a reduced predicted mean abundance of 11.9 birds (assuming that log.area is
held constant). Individual categories of a categorical variable cannot be dropped, so a data analyst has the choice of leaving the variable as is, or collapsing the variable into fewer categories.
For this model, it might be useful to collapse grazing intensity into a two-level variable, with one
category corresponding to the original classification of heavy, and another category corresponding
to the other four categories; i.e., creating a version of grazing intensity that only has the levels
"heavy" and "not heavy". This is supported by the data exploration; a plot of abundance versus
grazing.intensity shows that the centers of the distributions of abundance in the lowest four grazing intensity categories are roughly similar, relative to the center in the heavy grazing category. The
model with the binary version of grazing intensity, grazing.binary, is shown in Figure 7.28. The
model with grazing.binary has adjusted $R^2 = 0.71$, which is slightly larger than 0.70 in the more
complex model with grazing.intensity; the model explains 72% of the variability in abundance
($R^2 = 0.724$).

Incorporating an interaction term did not improve the model; adding a parameter for the
interaction between log.area and grazing.binary decreased the adjusted R^2 to 0.709. Thus, the
model shown in Figure 7.28 is the final model.

	Estimate	Std. Error	t value	Pr(>|t|)
(Intercept)	15.3736	1.4507	10.60	0.0000
log.area	3.1822	0.4523	7.04	0.0000
grazing.binaryheavy	-11.5783	1.9862	-5.83	0.0000

Figure 7.28: Final model: regression of abundance on log.area and
grazing.binary.

Model assessment

The fit of a model can be assessed using various residual plots. Figure 7.29 shows a histogram and normal probability plot of the residuals for the final model. Both show that the residuals follow the shape of a normal density in the middle range (between -10 and 10) but fit less well in the tails. There are too many large positive and large negative values) residuals.

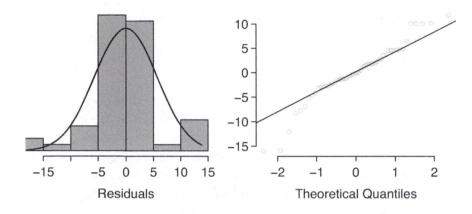

Figure 7.29: Histogram and normal probability plot of residuals in the model for abundance with predictors `log.area` and `grazing.binary`.

Figure 7.30 gives a more detailed look at the residuals, plotting the residuals against predicted values and against the two predictors in the model, `log.area` and `grazing.level`. Recall that residual values closer to 0 are indicative of a more accurate prediction; positive values occur when the predicted value from the model is smaller than the observed value, and vice versa for negative values. Residuals are a measure of the prediction error of a model.

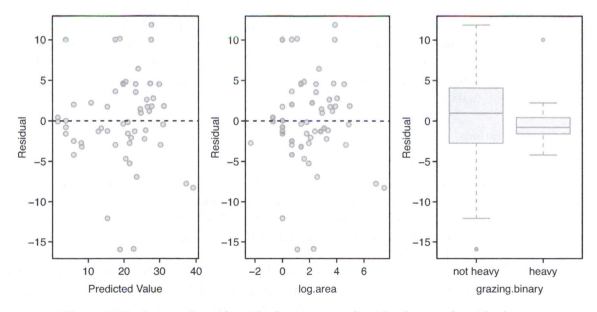

Figure 7.30: Scatterplots of residuals versus predicted values and residuals versus `log.area`, and a side-by-side boxplot of residuals by `grazing.binary`. In the middle plot, red points correspond to values where grazing level is "heavy" and blue points correspond to "not heavy".

In the left plot, the large positive and large negative residuals visible from Figure 7.29 are evident; the large positive residuals occur across the range of predicted values, while the large negative residuals occur around 20 (predicted birds). The middle plot shows that the large positive and negative residuals occur at intermediate values of log.area; i.e., for values of log.area between 0 and 4, or equivalently for values of area between $\exp(0) = 1$ and $\exp(4) = 54.5$ hectares. In the same range, there are also relatively accurate predictions; most residuals are between -5 and 5. Both the middle plot and the right plot show that the prediction error is smaller for patches with heavy grazing than for patches where grazing intensity was between "light" and "moderately heavy". Patches with heavy grazing are represented with red points; note how the red points mostly cluster around the $y = 0$ line, with the exception of one outlier with a residual value of about 10.

Conclusions

The relatively large R^2 for the final model (0.72) suggests that patch area and extent of grazing (either heavy or not) explain a large amount of the observed variability in bird abundance. Of the features measured in the study, these two are the most highly associated with bird abundance. Larger area is associated with an increase in abundance; when grazing intensity does not change, the model predicts an increase in average abundance by 3.18 birds for every one unit increase in log area (or equivalently, when area is increased by a factor of $\exp(1) = 2.7$). A patch with heavy grazing is estimated to have a mean abundance of about 11.58 birds lower than a patch that has not been heavily grazed.

The residual plots imply that the final model may not be particularly accurate. For most observations, the predictions are accurate between ± 5 birds, but there are several instances of over-predictions as high as around 10 and under-predictions of about 15. Additionally, the accurate and inaccurate predictions occur at similar ranges of of log.area; if the model only tended to be inaccurate at a specific range, such as for patches with low area, it would be possible to provide clearer advice about when the model is unreliable. The residuals plots do suggest that the model is more reliable for patches with heavy grazing, although there is a slight tendency towards over-prediction.

Based on these results, the ecologists might decide to proceed by collecting more data. Currently, the model seems to adequately explain the variability in bird abundance for patches that have been heavily grazed, but perhaps there are additional variables that are associated with bird abundance, especially in patches that are not heavily grazed. Adding these variables might improve model residuals, in addition to raising R^2.

Final considerations

Might a model including all the predictor variables be better than the final model with only log.area and grazing.binary? The model is shown in Figure 7.31. The R^2 for this model is 0.729 and the adjusted R^2 is 0.676. While the R^2 is essentially the same as for the final model, the adjusted R^2 is noticeably lower. The residual plots in Figure 7.32 do not indicate that this model is an especially better fit, although the residuals are slightly closer to normality. There would be little gained from using the larger model.

In fact, there is an additional reason to avoid the larger model. When building regression models, it is important to consider that the complexity of a model is limited by sample size (i.e., the number of observations in the data). Attempting to estimate too many parameters from a small dataset can produce a model with unreliable estimates; the model may be 'overfit', in the sense that it fits the data used to build it particularly well, but will fail to generalize to a new set of data. Methods for exploring these issues are covered in more advanced regression courses.

	Estimate	Std. Error	t value	Pr(>|t|)
(Intercept)	10.8120	9.9985	1.08	0.2852
log.area	2.9720	0.6587	4.51	0.0000
log.dist.near.patch	0.1390	1.1937	0.12	0.9078
log.dist.larger.patch	0.3496	0.9301	0.38	0.7087
altitude	0.0117	0.0233	0.50	0.6169
log.yrs.isolation	0.2155	1.9635	0.11	0.9131
grazing.intensityless than average	0.5163	3.2631	0.16	0.8750
grazing.intensityaverage	0.1344	2.9870	0.04	0.9643
grazing.intensitymoderately heavy	-1.2535	3.2000	-0.39	0.6971
grazing.intensityheavy	-12.0642	4.5657	-2.64	0.0112

Figure 7.31: Full model: regression of abundance on all 6 predictors in forest.birds.

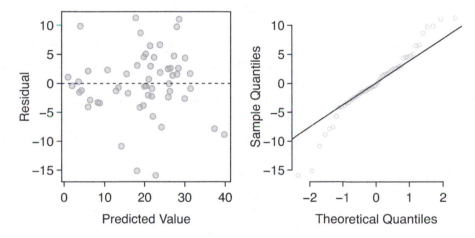

Figure 7.32: Residual plots for the full model of abundance that includes all predictors.

A general rule of thumb is to avoid fitting a model where there are fewer than 10 observations per parameter; e.g., to fit a model with 3 parameters, there should be at least 30 observations in the data. In a regression context, all of the following are considered parameters: an intercept term, a slope term for a numerical predictor, a slope term for each level of a categorical predictor, and an interaction term. In forest.birds, there are 56 cases, but fitting the full model involves estimating 10 parameters. The rule of thumb suggests that for these data, a model can safely support at most 5 parameters.

As mentioned earlier, other analyses of forest.birds have treated grazing.intensity as a numerical variable with five values. One advantage to doing so is to produce a more stable model; only one slope parameter needs to be estimated, rather than four. However, treating grazing.intensity as a numerical variable requires assuming that any one unit change is associated with the same change in population mean abundance; under this assumption, a change between "light" and "less than average" (codes 1 to 2) is associated with the same change in population mean abundance as between "moderately heavy" to "heavy" (codes 4 to 5) grazing. Previous model fitting has shown that this assumption is not supported by the data, and that changes in mean abundance between adjacent levels in grazing intensity are not constant. In this text, it is our recommendation that categorical variables should not be treated as numerical variables.

7.9   The connection between ANOVA and regression

Regression with categorical variables and ANOVA are essentially the same method, but with some important differences in the information provided by the analysis. Earlier in this chapter, the strength of the association between RFFT scores and educational level was assessed with regression. Figure 7.33 shows the results of an ANOVA to analyze the difference in RFFT scores between education groups.

	Df	Sum Sq	Mean Sq	F value	Pr(>F)
as.factor(Education)	3	115040.88	38346.96	73.30	0.0000
Residuals	496	259469.32	523.12		

Figure 7.33: Summary of ANOVA of RFFT by Education Levels

In this setting, the F-statistic is used to test the null hypothesis of no difference in mean RFFT score by educational level against the alternative that at least two of the means are different. The F-statistic is 73.3 and highly significant.

The F-statistic can also be calculated for regression models, although it has not been shown in the regression model summaries in this chapter. In regression, the F-statistic tests the null hypothesis that all regression coefficients are equal to 0 against the alternative that least one of the coefficients is not equal to 0.

Although the phrasing of the hypotheses in ANOVA versus regression may seem different initially, they are equivalent. Consider the regression model for predicting RFFT from educational level—each of the coefficients in the model is an estimate of the difference in mean RFFT for a particular education level versus the baseline category of Education = 0. A significant F-statistic indicates that at least one of the coefficients is not zero; i.e., that at least one of the mean levels of RFFT differs from the baseline category. If all the coefficients were to equal zero, then the differences between the means would be zero, implying all the mean RFFT levels are equal. It is reasonable, then, that the F-statistic associated with the RFFT versus Education regression model is also 73.3.

The assumptions behind the two approaches are identical. Both ANOVA and linear regression assume that the groups are independent, that the observations within each group are independent, that the response variable is approximately normally distributed, and that the standard deviations of the response are the same across the groups.

The regression approach provides estimates of the mean at the baseline category (the intercept) and the differences of the means between each category and the baseline, along with a t-statistic and p-value for each comparison. From regression output, it is easy to calculate all the estimated means; to do the same with ANOVA requires calculating summary statistics for each group. Additionally, diagnostic plots to check model assumptions are generally easily accessible in most computing software.

Why use ANOVA at all if fitting a linear regression model seems to provide more information? A case can be be made that the most important first step in analyzing the association between a response and a categorical variable is to compute and examine the F-statistic for evidence of any effect, and that only when the F-statistic is significant does it become appropriate to proceed to examine the nature of the differences. ANOVA displays the F-statistic prominently, emphasizing its importance. It is available in regression output, but may not always be easy to locate; the focus of regression is on the significance of the individual coefficients. ANOVA has traditionally been used in carefully designed experiments. There are complex versions of ANOVA that are appropriate for experiments in which several different factors are set at a range of levels. More complex versions of ANOVA are beyond the scope of this text and are covered in more advanced books.

Section 5.5 discussed the use of Bonferroni corrections when testing hypotheses about pairwise differences among the group means when conducting ANOVA. In principle, Bonferroni corrections can be applied in regression with categorical variables, but that is not often done. In designed experiments in which ANOVA has historically been used, the goal was typically to show definitively that a categorical predictor, often a treatment or intervention, was associated with a response variable so that the treatment could be adopted for clinical use. In experiments where the predictor can be manipulated by a scientist and cases are randomized to one of several levels of a predictor, the association can be interpreted as causal. It can be particularly important to control Type I error probabilities in those settings. Regression is often thought of as an exploratory technique, used in observational studies to discover associations that can be explored in further studies. Strict control of Type I error probabilities may be less critical in such settings.

At the introductory level, ANOVA is useful in that it provides more direct access to Type I error control and pairwise comparisons with t-tests. In practice, with the use of techniques not covered in this text, any analysis done via the ANOVA approach can also be approached with regression modeling.

7.10 Notes

This chapter and the previous chapter cover only the basic principles behind linear regression, and are meant to provide useful tools for getting started with data analysis. This section summarizes the most important ideas in the chapter and makes reference to some related topics that have not been discussed in detail.

Important ideas

Keep a clear view of the purpose. Is the goal of constructing the model to understand the relationship between the response and a particular predictor after adjusting for confounders? Or is the goal to understand the joint association between a response and a set of predictors?

Avoid rushing into model fitting. Before fitting models, examine the data. Assess whether the response variable has an approximate normal distribution, or at least a symmetric distribution; a log transformation will often produce approximate normality. Examine the relationships between the response and predictors, as well as the relationships between predictors; check for nonlinear trends or outliers.

Remember the context of the problem. Context is important at each stage of a regression analysis. The best approach for constructing a model from a small number of potential predictors is based on considering the context of the problem and including predictors that have either been shown in the past to be associated with the response or for which there is a plausible working hypothesis about association with the response. When interpreting coefficients, consider whether the model results cohere with the underlying biological or medical context.

Critically examine residual plots. All models are approximations, so it is not necessary to be concerned about relatively minor violations of assumptions; residual plots are seldom as well behaved as those for the PREVEND data. In some cases, like with the California DDS data, residual plots show obvious major violations. With intermediate cases such as in the forest.birds plots, examine the plots closely and provide a detailed assessment of where the model seems less reliable.

Related topics

Stepwise model selection. Many introductory texts recommend using "stepwise" regression. Forward stepwise regression adds predictors one by one according to a set criterion (usually by smallest p-value). Backward stepwise regression eliminates variables one by one from a larger model until a criterion is met. Stepwise methods can be useful, and are usually automated in statistical software. However, there are weaknesses—the final models are data-dependent and chance alone can lead to spurious variables being included. In very large datasets, stepwise regression can lead to substantially incorrect models.

Prediction models. An application of regression not discussed in this chapter is predictive modeling, in which the goal is to construct a model that best predicts outcomes. The focus is on overall predictive accuracy; significance of individual coefficients is less important. Evaluating a model's predictive accuracy involves advanced methods such as cross-validation, in which the original data sample is divided into a training set and a test set, similar to the approach used with the Golub leukemia data in Chapter 1. Prediction models are typically built from large datasets, using automated model selection procedures like stepwise regression.

Prediction intervals. Predicted values from regression have an inherent uncertainty because model parameters are only estimates. There are two types of interval estimates used with prediction: confidence intervals for a predicted mean response from a set of values for the predictors, and prediction intervals that show the variability in the predicted value for a new response (i.e., for a case not in the dataset) given a set of values for the predictor variables. Prediction intervals are wider than confidence intervals for a predicted mean because prediction intervals are subject to both the variability in a predicted mean response and the variability of an individual observation about its mean.

Controlling Type I error in regression. Control of Type I error probabilities becomes more critical in regression models with very large numbers of potential predictors. Datasets containing measurements on genetic data often contain large numbers of potential predictors for a response for many cases; a stricter significance level is used to maintain an overall error rate of $\alpha = 0.05$. For example, in genome-wide association studies, the accepted "genome-wide significance rate" for an individual marker to be considered significantly associated with an outcome is 5×10^{-8}.

Because there are so many tools available in multiple regression, this chapter has a larger collection of labs than most other chapters. Lab 1 introduces the multiple regression model, illustrating one its most common uses—estimating an association between a response variable and predictor of interest while adjusting for possible confounding. Lab 2 discusses the residual plots used to check assumptions for multiple regression and introduces adjusted R^2 using the California DDS dataset initially introduced in Chapter 1.

Lab 3 explores how the association between a response variable and categorical predictors with more than two levels can be be estimated using multiple regression. This topic extends the earlier material in Chapter 6, Lab 4. Lab 4 introduces the concept of a statistical interaction using the NHANES dataset, examining whether the association between BMI and age among women is different than that among men.

Multiple regression is often used to examine associations between response variables and a small set of pre-specified predictors. It can also be used to explore and select models between a response variable and a set of candidate predictors. Lab 5 discusses explanatory modeling, in which the goal is to construct a model that effectively explains the observed variation in the response variable.

7.11   Exercises

7.11.1   Introduction to multiple linear regression

There are not currently exercises available for this section.

7.11.2   Simple versus multiple regression

7.1   PREVEND, Part I. The summary table below shows the results of a multiple regression model of RFFT score versus statin use and age.

	Estimate	Std. Error	t value	Pr(>|t|)
(Intercept)	137.8822	5.1221	26.92	0.0000
Statin	0.8509	2.5957	0.33	0.7432
Age	-1.2710	0.0943	-13.48	0.0000

In a clinical setting, the interpretive focus lies on reporting the nature of the association between the primary predictor and the response, while specifying which potential confounders have been adjusted for. Briefly respond to a clinician who is concerned about a possible association between statin use and decreased cognitive function, based on the above analysis.

7.2   PREVEND, Part II. Can the results of the analysis in Exercise 7.1 be used to conclude that as one ages, one's cognitive function (as measured by RFFT score) declines? Explain your answer.

7.3   Baby weights, Part I. The Child Health and Development Studies investigate a range of topics. One study considered all pregnancies between 1960 and 1967 among women in the Kaiser Foundation Health Plan in the San Francisco East Bay area. The variable smoke is coded 1 if the mother is a smoker, and 0 if not. The variable parity is 1 if the child is the first born, and 0 otherwise. The summary table below shows the results of a linear regression model for predicting the average birth weight of babies, measured in ounces, based on the smoking status of the mother and whether the child is the first born.[23]

	Estimate	Std. Error	t value	Pr(>|t|)
(Intercept)	123.57	0.72	172.75	0.0000
smoke	-8.96	1.03	-8.68	0.0000
parity	-1.98	1.15	-1.72	0.0859

(a) Write the equation of the regression model.

(b) Interpret the model slopes in the context of the data.

(c) Calculate the estimated difference in mean birth weight for two infants born to non-smoking mothers, if one is first born and the other is not.

(d) Calculate the estimated difference in mean birth weight for two infants born to mothers who are smokers, if one is first born and the other is not.

(e) Calculate the predicted mean birth weight for a first born baby born to a mother who is not a smoker.

---

[23]Child Health and Development Studies, Baby weights data set.

7.4 Wolbachia, Part I. *Wolbachia* is a microbial symbiont estimated to be hosted by about 40% of all arthropod species, transmitted primarily from females to their offspring through the eggs. Researchers conducted a study on a wasp species to understand the effect of *Wolbachia* on the lifetime reproductive success of an insect host. They estimated the realized lifetime reproductive success of female wasps by collecting them soon after they die naturally in the field, counting the number of eggs remaining in their ovaries and quantifying *Wolbachia* density in their body.

In the first stage of the experiment, researchers estimated potential reproductive success by collecting female wasps as they emerged from eggs then dissecting them to count the number of eggs in their ovaries. These data were used to create a predictive model for initial number of eggs based on tibia length (an indicator of body size) and *Wolbachia* density. Tibia length was measured in μm, and *Wolbachia* density in units of -ddCt.

	Estimate	Std. Error	t value	Pr(>|t|)
(Intercept)	-18.82	27.26	-0.69	0.497
wolbachia	1.77	1.07	-1.65	0.111
tibia	0.357	0.15	2.38	0.0258

(a) Write the model equation.

(b) Interpret the model coefficients in the context of the data.

(c) Predict mean initial egg count for a wasp with tibia length of 171.4286 μm and *Wolbachia* density of -3.435 -ddCt.

7.11.3 Evaluating the fit of a multiple regression model

7.5 Baby weights, Part III. We considered the variables smoke and parity, one at a time, in modeling birth weights of babies in Exercise 7.3. A more realistic approach to modeling infant weights is to consider all possibly related variables at once. Other variables of interest include length of pregnancy in days (gestation), mother's age in years (age), mother's height in inches (height), and mother's pregnancy weight in pounds (weight). Below are three observations from this data set.

	bwt	gestation	parity	age	height	weight	smoke
1	120	284	0	27	62	100	0
2	113	282	0	33	64	135	0
⋮	⋮	⋮	⋮	⋮	⋮	⋮	⋮
1236	117	297	0	38	65	129	0

The summary table below shows the results of a regression model for predicting the average birth weight of babies based on all of the variables included in the data set.

	Estimate	Std. Error	t value	Pr(>|t|)
(Intercept)	-80.41	14.35	-5.60	0.0000
gestation	0.44	0.03	15.26	0.0000
parity	-3.33	1.13	-2.95	0.0033
age	-0.01	0.09	-0.10	0.9170
height	1.15	0.21	5.63	0.0000
weight	0.05	0.03	1.99	0.0471
smoke	-8.40	0.95	-8.81	0.0000

(a) Write the equation of the regression model that includes all of the variables.

(b) Interpret the slopes of gestation and age in this context.

(c) The coefficient for parity is different than in the linear model shown in Exercise 7.3. Why might there be a difference?

(d) Calculate the residual for the first observation in the data set.

(e) The variance of the residuals is 249.28, and the variance of the birth weights of all babies in the data set is 332.57. Calculate the R^2 and the adjusted R^2. Note that there are 1,236 observations in the data set.

7.6 Absenteeism, Part I. Researchers interested in the relationship between absenteeism from school and certain demographic characteristics of children collected data from 146 randomly sampled students in rural New South Wales, Australia, in a particular school year. Below are three observations from this data set.

	eth	sex	lrn	days
1	0	1	1	2
2	0	1	1	11
⋮	⋮	⋮	⋮	⋮
146	1	0	0	37

The summary table below shows the results of a linear regression model for predicting the average number of days absent based on ethnic background (eth: 0 - aboriginal, 1 - not aboriginal), sex (sex: 0 - female, 1 - male), and learner status (lrn: 0 - average learner, 1 - slow learner).[24]

	Estimate	Std. Error	t value	Pr(>|t|)
(Intercept)	18.93	2.57	7.37	0.0000
eth	-9.11	2.60	-3.51	0.0000
sex	3.10	2.64	1.18	0.2411
lrn	2.15	2.65	0.81	0.4177

(a) Write the equation of the regression model.

(b) Interpret each one of the slopes in this context.

(c) Calculate the residual for the first observation in the data set: a student who is aboriginal, male, a slow learner, and missed 2 days of school.

(d) The variance of the residuals is 240.57, and the variance of the number of absent days for all students in the data set is 264.17. Calculate the R^2 and the adjusted R^2. Note that there are 146 observations in the data set.

[24]W. N. Venables and B. D. Ripley. *Modern Applied Statistics with S*. Fourth Edition. Data can also be found in the R MASS package. New York: Springer, 2002.

7.7 Baby weights, Part VI. Exercise 7.5 presents a regression model for predicting the average birth weight of babies based on length of gestation, parity, height, weight, and smoking status of the mother. Use the following plots to assess whether the assumptions for linear regression are reasonably met. Discuss your reasoning.

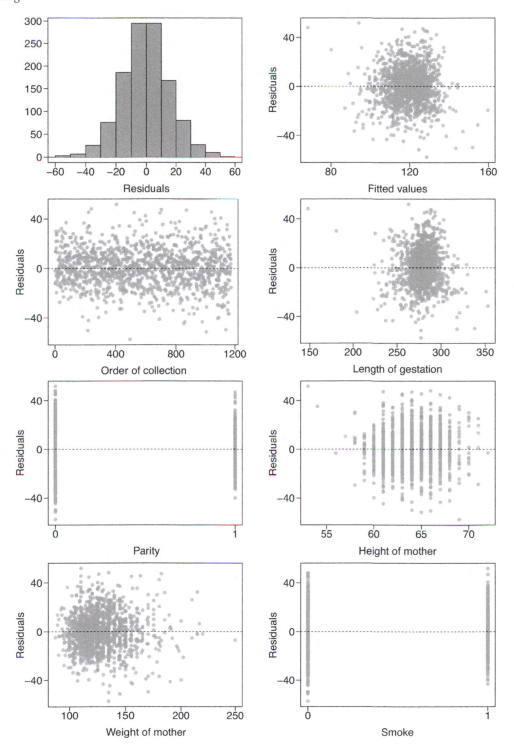

7.8 Toxemia and birth weight. A model was fit for a random sample of 100 low birth weight infants born in two teaching hospitals in Boston, Massachusetts, regressing birthweight on the predictors gestational age and toxemia status. The condition toxemia, also known as preeclampsia, is characterized by high blood pressure and protein in urine by the 20th week of pregnancy; left untreated, toxemia can be life-threatening. Birth weight was measured in grams and gestational age measured in weeks.

	Estimate	Std. Error	t value	Pr(>|t|)
(Intercept)	-1286.200	234.918	-5.475	0.0000
toxemiaYes	-206.591	51.078	-4.045	0.0001
gestage	84.048	8.251	10.188	0.0000

(a) Write the model equation.

(b) Interpret the coefficients of the model, and comment on whether the intercept has a meaningful interpretation.

(c) Predict the average birth weight for an infant born to a mother diagnosed with toxemia with gestational age 31 weeks.

(d) Evaluate whether the assumptions for linear regression are reasonably satisfied.

(e) A simple regression model with only toxemia status as a predictor had $R^2 = 0.0001$ and $R^2_{adj} = 0.010$; in this model, the slope estimate for toxemia status is 7.785, with $p = 0.907$. The simple regression model and multiple regression model disagree regarding the nature of the association between birth weight and toxemia. Briefly explain a potential reason behind the discrepancy. Which model do you prefer for understanding the relationship between birth weight and toxemia, and why?

7.9 Multiple regression fact checking. Determine which of the following statements are true and false. For each statement that is false, explain why it is false.

(a) Suppose a numerical variable x has a coefficient of $b_1 = 2.5$ in the multiple regression model. Suppose also that the first observation has $x_1 = 7.2$, the second observation has a value of $x_1 = 8.2$, and these two observations have the same values for all other predictors. Then the predicted value of the second observation will be 2.5 higher than the prediction of the first observation based on the multiple regression model.

(b) If a regression model's first variable has a coefficient of $b_1 = 5.7$, then if we are able to influence the data so that an observation will have its x_1 be 1 larger than it would otherwise, the value y_1 for this observation would increase by 5.7.

(c) Suppose we fit a multiple regression model based on a data set of 472 observations. We also notice that the distribution of the residuals includes some skew but does not include any particularly extreme outliers. Because the residuals are not nearly normal, we should not use this model and require more advanced methods to model these data.

7.11.4 The general multiple linear regression model

7.10 Cherry trees. Timber yield is approximately equal to the volume of a tree, however, this value is difficult to measure without first cutting the tree down. Instead, other variables, such as height and diameter, may be used to predict a tree's volume and yield. Researchers wanting to understand the relationship between these variables for black cherry trees collected data from 31 such trees in the Allegheny National Forest, Pennsylvania. Height is measured in feet, diameter in inches (at 54 inches above ground), and volume in cubic feet.[25]

	Estimate	Std. Error	t value	Pr(>|t|)
(Intercept)	-57.99	8.64	-6.71	0.00
height	0.34	0.13	2.61	0.01
diameter	4.71	0.26	17.82	0.00

(a) Calculate a 95% confidence interval for the coefficient of height, and interpret it in the context of the data.

(b) One tree in this sample is 79 feet tall, has a diameter of 11.3 inches, and is 24.2 cubic feet in volume. Determine if the model overestimates or underestimates the volume of this tree, and by how much.

7.11 GPA. A survey of 55 Duke University students asked about their GPA, number of hours they study at night, number of nights they go out, and their gender. Summary output of the regression model is shown below. Note that male is coded as 1.

	Estimate	Std. Error	t value	Pr(>|t|)
(Intercept)	3.45	0.35	9.85	0.00
studyweek	0.00	0.00	0.27	0.79
sleepnight	0.01	0.05	0.11	0.91
outnight	0.05	0.05	1.01	0.32
gender	-0.08	0.12	-0.68	0.50

(a) Calculate a 95% confidence interval for the coefficient of gender in the model, and interpret it in the context of the data.

(b) Would you expect a 95% confidence interval for the slope of the remaining variables to include 0? Explain

7.12 Trait inheritance of high blood pressure. One research question of public health interest is to determine the extent to which high blood pressure is a genetic phenomenon. In 20 families, the systolic blood pressure of the mother, father, and first-born child in the family were measured (in units of mm Hg). A multiple linear regression model using Y = child's blood pressure, X_1 = mother's blood pressure, and X_2 = father's blood pressure led to the following estimate of a least squares line: $E(Y) = -15.69 + 0.415X_1 + 0.423X_2$. The standard errors associated with b_0, b_1, and b_2, respectively, are 23.65, 0.125, and 0.119. The least squares fit produced $R^2 = 0.597$ and $MSE = 113.8$.

(a) What proportion of the variability of a child's systolic blood pressure is explained by this model?

(b) Does the least squares line indicate statistically significant associations between each of the parent's systolic blood pressures and that of the child? Explain your answer.

(c) What is the predicted systolic blood pressure for a child whose mother's and father's systolic blood pressure is 125 mm Hg and 140 mm Hg, respectively?

(d) A colleague tells you that something must be wrong with your model because your fitted intercept is negative, but blood pressures are never negative. How do you respond?

(e) Briefly describe three different plots for assessing the appropriateness or fit of the above regression model.

---

[25]D.J. Hand. *A handbook of small data sets*. Chapman & Hall/CRC, 1994.

7.13 Wolbachia, Part II. Exercise 7.4 introduced a study about Wolbachia and reproductive success in a wasp host. The following table shows the model coefficients for a model predicting the number of eggs laid over a lifetime from the predictor variables wolbachia density and tibia length. A higher number of eggs laid over a lifetime is indicative of greater reproductive success. The model has $R^2 = 0.314$ and degrees of freedom 34. The F-statistic is 7.782, with p-value 0.0016.

	Estimate	Std. Error	t value	Pr(>|t|)
(Intercept)	-17.88	28.63	-0.62	0.537
wolbachia	4.28	1.25	3.42	0.002
tibia	0.272	0.16	1.69	0.010

(a) Write the model equation.

(b) Interpret the slope coefficient of wolbachia.

(c) Assess the evidence for whether *Wolbachia* is beneficial for its host in nature, based on these data.

(d) Compute and interpret a 95% confidence interval for the population slope of wolbachia.

(e) Interpret the significance of the F-statistic.

7.14 Difficult encounters, Part I. A study was conducted at a university outpatient primary care clinic in Switzerland to identify factors associated with difficult doctor-patient encounters. The data consist of 527 patient encounters, conducted by the 27 medical residents employed at the clinic. After each encounter, the attending physician completed two questionnaires: the Difficult Doctor Patient Relationship Questionnaire (DDPRQ-10) and the patient's vulnerability grid (PVG).

A higher score on the DDPRQ-10 indicates a more difficult encounter. The maximum possible score is 60 and encounters with score 30 and higher are considered difficult.

A model was fit for the association of DDPRQ-10 score with features of the attending physician: age, sex, and years of training. The model has F-statistic of 0.23 on 3 and 286 degrees of freedom, with p-value 0.876.

	Estimate	Std. Error	t value	Pr(>|t|)
(Intercept)	30.594	2.886	10.601	0.0000
age	-0.016	0.104	-0.157	0.876
sexM	-0.535	0.781	-0.686	0.494
yrs.train	0.096	0.215	0.445	0.656

(a) As a group, are these physician features useful for predicting DDPRQ-10 score?

(b) Is there evidence of a significant association between DDPRQ-10 score and any of the physician features?

7.11.5 Categorical predictors with several levels

7.15 Prison isolation experiment, Part III.
Exercises 5.35 and 5.47 introduced an experiment conducted with the goal of identifying a treatment that reduces subjects' psychopathic deviant T scores on the MMPI test. Exercise 5.35 evaluated the success of each individual treatment, and in exercise 5.47, ANOVA was used to compare the success of the three treatments. This exercise uses multiple regression to examine the intervention effect.

For this problem, a treatment variable (labeled treatment) has been constructed with three levels:

(1) Therapeutic for sensory restriction plus the 15 minute "therapeutic" tape advising that professional help is available.

(2) Neutral for sensory restriction plus a 15 minute "emotionally neutral" tap on training hunting dogs.

(3) Absent for sensory restriction but no taped message.

Forty-two subjects were randomly assigned to these treatment groups, and an MMPI test was administered before and after the treatment. Investigators hoped that the interventions would lower MMPI scores. The table below shows the result of a multiple regression in R where the response variable trt.effect is the change in MMPI score (pre-intervention - post-intervention) and the predictor variable is treatment.

	Estimate	Std. Error	t value	Pr(>|t|)
(Intercept)	-3.2143	2.6174	-1.23	0.2268
treatmentNeutral	6.0714	3.7015	1.64	0.1090
treatmentTherapeutic	9.4286	3.7015	2.55	0.0149

In this model, the residual standard error is 9.79, the F-statistic is 3.33 with 2 and 39 degrees of freedom; $P(F_{2,39} > 3.33) = 0.0461$.

(a) Interpret the meaning of a positive value for trt.effect versus a negative value.

(b) Write the estimated model equation.

(c) Calculate the predicted value for trt.effect for a patient in the neutral tape group.

(d) Does the intercept have a meaningful interpretation in this model?

(e) What is the interpretation of the two slope coefficients in the regression model?

(f) Describe the tested hypotheses that correspond to each of the p-values in the last column of the table.

7.16 Poverty and educational level. This question uses data from 500 randomly selected adults in the larger NHANES dataset. Poverty is measured as a ratio of family income to poverty guidelines. Smaller numbers indicate more poverty, and ratios of 5 or larger were recorded as 5. The Education variable indicates the highest level of education achieved: either 8th grade, 9 - 11th grade, high school, some college, or college grad.

	Estimate	Std. Error	t value	Pr(>|t|)
(Intercept)	1.4555	0.2703	5.38	0.0000
Education9 - 11th Grade	0.9931	0.3302	3.01	0.0028
EducationHigh School	1.0900	0.3113	3.50	0.0005
EducationSome College	1.4943	0.2976	5.02	0.0000
EducationCollege Grad	2.4948	0.2958	8.43	0.0000

In this model, the residual standard error is 1.46, the F-statistic is 28.09 with 4 and 456 degrees of freedom; $P(F_{4,456} > 28.09) < 0.0001$.

(a) Write the estimated model equation.

(b) Calculate the predicted poverty ratio for an individual who at most completed high school.

(c) Interpret the estimated intercept value.

(d) Interpret the slope coefficient for EducationCollege Grad, and describe the tested hypotheses that correspond to the p-value for this slope coefficient.

(e) Assess whether educational level, overall, is associated with poverty. Be sure to include any relevant numerical evidence as part of your answer.

7.17 Prison isolation experiment, Part IV. Exercise 7.15 used regression to examine the effect of three interventions on prisoner MMPI scores. The response variable in the regression was trt.effect, the change in MMPI score (pre-intervention - post-intervention).

Instead of estimating the intervention effect through the change in scores, suppose one is interested in predicting a post-intervention score based on the pre-intervention score for an individual and a particular intervention.

(a) The table below shows an alternative regression model that can be fit to the data. In this model, the response variable is the post-intervention MMPI value (post, not shown explicitly in the table) and the predictors are the pre-intervention score (pre) and the treatment, coded as in problem 7.15.

Write the estimated equation for this model.

	Estimate	Std. Error	t value	Pr(>|t|)
(Intercept)	28.4053	12.2949	2.31	0.0264
pre	0.6593	0.1628	4.05	0.0002
treatmentNeutral	-5.7307	3.5545	-1.61	0.1152
treatmentTherapeutic	-9.7450	3.5540	-2.74	0.0093

(b) In this model, describe in general terms the association of the pre-intervention and post-intervention scores.

(c) Does the pre-intervention score appear to be an important predictor of a post intervention score?

(d) What is the predicted post-intervention score for an individual with a pre-intervention score of 73 and receiving no tape after the isolation?

(e) Explain the interpretation of the coefficients for coefficient of treatmentNeutral. Is there strong statistical evidence that it is an important predictor?

7.18 Resilience, Part I. The American Psychological Association defines resilience as "the process of adapting well in the face of adversity, trauma, tragedy, threats, or even significant sources of stress". Studies have suggested that resilience is an important factor in contributing to how medical students perceive their quality of life and educational environment.

Survey data were collected from 1,350 students across 25 medical schools. At each school, 54 students were randomly selected to participate in the study. Participants completed questionnaires measuring resilience, quality of life, perception of educational environment, depression symptoms, and anxiety symptoms.

The following regression model was fit to analyze the relationship between resilience and depressive symptoms. Resilience was categorized as: very low, low, moderately low, moderately high, high, and very high. Depressive symptoms were measured on a scale of 0 to 63 points, with higher scores indicating either more numerous or more severe depressive symptoms; this questionnaire is called the Beck Depression Inventory (BDI).

In this model, the residual standard error is 5.867, the F-statistic is 118.1 with 5 and 1344 degrees of freedom; $P(F_{5,1344} > 118.1) < 0.0001$.

	Estimate	Std. Error	t value	Pr(>|t|)
(Intercept)	4.9754	0.4118	12.08	0.0000
resHigh	2.2936	0.4987	4.60	0.0000
resModHigh	4.3005	0.5181	8.30	0.0000
resModLow	6.7108	0.5938	11.30	0.0000
resLow	9.6538	0.7458	12.94	0.0000
resVeryLow	15.6453	0.7518	20.81	0.0000

(a) Describe the overall trend in language accessible to someone who has not taken a statistics course.

(b) Does the intercept have a meaningful interpretation? Explain your answer.

(c) Compare the predicted mean BDI score for someone with low resilience to that of someone with very low resilience.

(d) + (e) **Continue to the next page for parts (d) and (e).**

(d) Assess whether level of resilience, overall, is associated with depressive symptoms as measured by BDI score. Be sure to include any relevant numerical evidence as part of your answer.

(e) A model was fitted predicting BDI score from resilience, with the categories numerically coded from 1 to 6, with 1 being very high resilience and 6 being very low resilience. This model has a single slope estimate of 2.76 with p-value < 0.0001.

    i. Using this model, compare the predicted mean BDI score for someone with low resilience to that of someone with very low resilience. Compare this answer to the one from part (c).

    ii. What does this model imply about the change in mean BDI score between groups?

    iii. Explain why this model is flawed.

7.11.6 Reanalyzing the PREVEND data

There are not currently exercises available for this section.

7.11.7 Interaction in regression

7.19 Prison isolation experiment, Part V. Exercise 7.17 used regression to predict a post-intervention score based on pre-intervention score and a particular intervention.

The following table shows a model incorporating interaction between pre-intervention score and intervention.

	Estimate	Std. Error	t value	Pr(>|t|)
(Intercept)	-17.5790	17.7090	-0.99	0.3275
pre	1.2813	0.2376	5.39	0.0000
treatmentNeutral	67.7518	30.6168	2.21	0.0333
treatmentTherapeutic	64.4183	24.2124	2.66	0.0116
pre:treatmentNeutral	-0.9890	0.4082	-2.42	0.0206
pre:treatmentTherapeutic	-1.0080	0.3266	-3.09	0.0039

(a) Write the model equation.

(b) Interpret the model coefficients.

(c) Write a separate model equation for each intervention group.

(d) Do these data suggest that there is a statistically significant difference in association between pre- and post-intervention scores by treatment group? Explain your answer.

7.20 Vitamin D. A study was conducted to evaluate Vitamin D status among schoolchildren in Thailand. Exposure to sunlight allows the body to produce serum $25(OH)D$, which is a marker of Vitamin D status; serum level is measured in units of nmol/L and having serum level below 50 nmol/L is indicative of Vitamin D deficiency. The following model was fit to predict serum $25(OH)D$ level from age, sex, and their interaction.

	Estimate	Std. Error	t value	Pr(>|t|)
(Intercept)	97.7709	4.7732	20.48	0.0000
age	-3.0156	0.4774	-6.32	0.0000
sexM	-16.2848	7.0740	-2.30	0.0217
age:sexM	2.9369	0.7054	4.16	0.0000

(a) Write the model equation.

(b) Interpret the model coefficients.

(c) Is there statistically significant evidence that the association between serum $25(OH)D$ level and age differs by sex? Explain your answer.

7.21 PREVEND, Part III. Exercise 7.1 showed a multiple regression model predicting RFFT score from statin use and age. For this problem, an interaction term is added between statin use and age.

	Estimate	Std. Error	t value	Pr(>|t|)
(Intercept)	140.2031	5.6209	24.94	0.0000
Statin	-13.9720	15.0113	-0.93	0.3524
Age	-1.3149	0.1040	-12.65	0.0000
Statin:Age	0.2474	0.2468	1.00	0.3166

(a) Write the model equation.

(b) Interpret the model coefficients.

(c) Is there statistically significant evidence that the association between RFFT score and age differs by whether someone is a statin user? Explain your answer.

7.22 Antibiotic consumption, Part I. Antibiotic resistance represents a major public health challenge. Overuse of antibiotics in clinical settings is thought to be a major contributor to increased antibiotic resistance. A study was conducted across several regions in China to investigate the impact of a 2011 law prohibiting over-the-counter (OTC) sales of antibiotics in private pharmacies. The study team collected data on average monthly antibiotic consumption in 621 counties, in addition to information on socioeconomic determinants such as percentage of population illiterate.

The following model was fit to investigate whether the relationship between monthly antibiotic consumption and percentage of population (over 25 years of age) with an advanced degree differs between counties that are located in a metropolitan area and those that are not.

	Estimate	Std. Error	t value	Pr(>|t|)
(Intercept)	0.8482	0.3311	2.56	0.0107
metroYes	2.3035	0.9612	2.40	0.0169
edu	0.5711	0.0319	17.90	0.0000
metroYes:edu	-0.1838	0.0752	-2.44	0.0148

(a) Interpret the model coefficients, including any relevant inferential results.

(b) Make a prediction of average monthly antibiotic consumption for a county in a metropolitan area where 10% of the population over 25 years old has an advanced degree.

7.11.8 Model selection for explanatory variables

7.23 Baby weights, Part VII. Suppose the starting point for model selection for the birth weight data were the full model, with all variables. The table below shows the adjusted R^2 for the full model as well as the adjusted R^2 values for all models with one fewer predictor variable. Based on examining the table from Exercise 7.5 and the following table, identify which variable, if any, should be removed from the model first. Explain your answer.

	Model	Adjusted R^2
1	Full model	0.2541
2	No gestation	0.1031
3	No parity	0.2492
4	No age	0.2547
5	No height	0.2311
6	No weight	0.2536
7	No smoking status	0.2072

7.24 Absenteeism, Part II. Suppose the starting point for model selection for the absenteeism data were the full model, with all variables. The table below shows the adjusted R^2 for the full model as well as the adjusted R^2 values for all models with one fewer predictor variable. Based on examining the table from Exercise 7.6 and the following table, identify which variable, if any, should be removed from the model first. Explain your answer.

	Model	Adjusted R^2
1	Full model	0.0701
2	No ethnicity	-0.0033
3	No sex	0.0676
4	No learner status	0.0723

7.25 Baby weights, Part VIII. Exercise 7.5 shows a regression model for predicting the average birth weight of babies based on all variables included in the dataset: length of pregnancy in days (gestation), mother's age in years (age), mother's height in inches (height), and mother's pregnancy weight in pounds (weight).

The following plots show the relationships between the response and the numerical predictor variables, in addition to the relationships between the response and two categorical predictor variables.

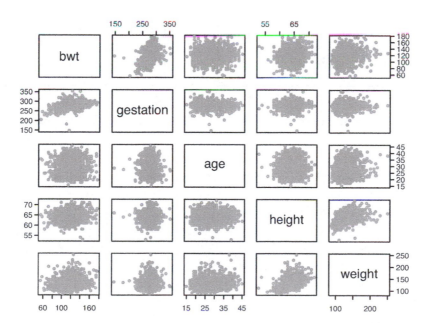

(a) Examine the relationship between the response variable and the predictor variables. Describe what you see. Which predictor variables seem like they would be useful to include in an initial model?

(b) Identify any predictors that seem related to each other.

7.26 Antibiotic consumption, Part II. Exercise 7.22 introduced a study conducted about antibiotic consumption in China. One aim of the study was to develop a prediction model for predicting monthly average antibiotic consumption based on county-level data. The following plots show the association between monthly average antibiotic consumption and four potential predictor variables: proportion female inhabitants (female), average life expectancy in years (lifeexp), proportion of population illiterate (illiterate), and population density in 1,000 people / km^2 (popdensity).

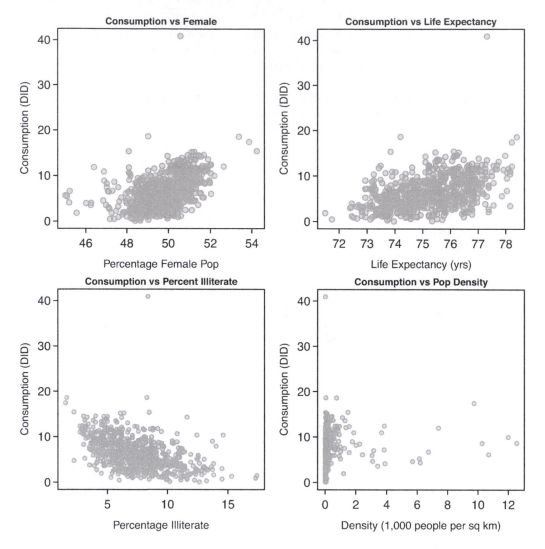

(a) Summarize what you see.

(b) Identify any predictor variables that might benefit from a natural log transformation and briefly justify your choices.

7.11.9 The connection between ANOVA and regression

7.27 Prison isolation experiment, Part VI. Problem 7.15 used a regression model to examine the effect of the interventions on possibly reducing psychopathic deviant T scores on prisoners. The regression model is shown in the problem statement.

(a) The value of the F-statistic is 3.33 with 2 and 39 degrees of freedom and $P(F_{2,39} > 3.33) = 0.0461$. In terms of the variables in the regression model, state the null hypothesis that corresponds to the F-statistic.

(b) Describe the relationship between the coefficients from the linear model and the usual summary statistics for the three sets of difference scores.

(c) Explain why the null hypothesis in this regression model is equivalent to the null hypothesis when these data were analyzed using ANOVA in Problem 5.47.

(d) Explain whether the assumptions for this regression model differ from those used in ANOVA.

7.28 Resilience, Part II. Exercise 7.18 shows a regression model for the association of BDI score with resilience level.

(a) In terms of the variables in the regression model, state the null hypothesis that corresponds to the F-statistic.

(b) Describe the relationship between the coefficients from the linear model and the usual summary statistics for the six sets of BDI scores.

(c) Explain why the null hypothesis used in this regression model is equivalent to the null hypothesis that would be used if these data were analyzed with an ANOVA approach.

Chapter 8

Inference for
categorical data

Previous chapters discussed methods of inference for numerical data; in this chapter, those methods are extended to categorical data, such as binomial proportions or data in two-way tables. While various details of the methods may change, such as the calculations for a test statistic or the distributions used to find a p-value, the core ideas and principles behind inference remain the same.

Categorical data arise frequently in medical research because disease outcomes and patient characteristics are often recorded in natural categories such as types of treatment received, whether or not disease advanced to a later stage, or whether or not a patient responded initially to a treatment. In the simplest settings, a binary outcome (yes/no, success/failure, etc) is recorded for a single group of participants, in hopes of learning more about the population from which the participants were drawn. The binomial distribution is often used for the statistical model in this setting, and inference about the binomial probability of success provides information about a population proportion p. In more complex settings, participant characteristics are recorded in a categorical variable with two or more levels, and the outcome or response variable itself has two or more levels. In these instances, data are usually summarized in two-way tables with two or more rows and two or more columns.

As with all methods of inference, it is important to understand how the data were collected and whether the data may be viewed as a random sample from a well-identified population, at least approximately. This issue is at least as important as the formulas for test statistics and confidence intervals, and is often overlooked.

Be careful about the notation in this chapter—since p is the standard notation for a population proportion and for a probability, p does double duty in this chapter as a population parameter and significance level.

For labs, slides, and other resources, please visit
www.openintro.org/book/biostat

8.1   Inference for a single proportion

Advanced melanoma is an aggressive form of skin cancer that until recently was almost uniformly fatal. In rare instances, a patient's melanoma stopped progressing or disappeared altogether when the patient's immune system successfully mounted a response to the cancer. Those observations led to research into therapies that might trigger an immune response in cancer. Some of the most notable successes have been in melanoma, particularly with two new therapies, nivolumab and ipilimumab.[1]

A 2013 report in the New England Journal of Medicine by Wolchok et al. reported the results of a study in which patients were treated with both nivolumab and ipilimumab.[2] Fifty-three patients were given the new regimens concurrently, and the response to therapy could be evaluated in 52 of the 53. Of the 52 evaluable patients, 21 (40%) experienced a response according to commonly accepted criteria. In previous studies, the proportion of patients responding to one of these agents was 30% or less. How might one compare the new data to past results?

The data from this study are binomial data, with success defined as a response to therapy. Suppose the number of patients who respond in a study like this is represented by the random variable X, where X is binomial with parameters n (the number of trials, where each trial is represented by a patient) and p (the unknown population proportion of response). From formulas discussed in Chapter 3, the mean of X is np and the standard deviation of X is $\sqrt{np(1-p)}$.

Inference about p is based on the sample proportion $\hat{p}$, where $\hat{p} = X/n$. In this case, $\hat{p} = 21/52 = 0.404$. If the sample proportion is nearly normally distributed, the normal approximation to the binomial distribution can be used to conduct inference; this method is commonly used. When X does not have an approximately normal distribution, exact inference can based on the binomial distribution for X. Both the normal approximation and exact methods are covered in this chapter.

8.1.1   Inference using the normal approximation

A **sample proportion** can be described as a sample mean. If each success in the melanoma data is represented as a 1 and each failure as a 0, then the sample proportion is the mean of the 52 numerical outcomes:

$$\hat{p} = \frac{0 + 1 + 1 + \cdots + 0}{52} = 0.404.$$

The distribution of $\hat{p}$ is nearly normal when the distribution of successes and failures is not too strongly skewed.

---

[1] The -mab suffix in these therapies stands for monoclonal antibody, a therapeutic agent made by identical immune cells that are all clones of a unique parent cell from a patient.

[2] N Engl J Med 2013;369:122-33. DOI: 10.1056/NEJMoa1302369

CONDITIONS FOR THE SAMPLING DISTRIBUTION OF $\hat{p}$ BEING NEARLY NORMAL

The sampling distribution for $\hat{p}$, calculated from a sample of size n from a population with a success proportion p, is nearly normal when

1. the sample observations are independent and

2. at least 10 successes and 10 failures are expected in the sample, i.e. $np \geq 10$ and $n(1-p) \geq 10$. This is called the **success-failure condition**.

If these conditions are met, then the sampling distribution of $\hat{p}$ is approximately normal with mean p and standard error

$$SE_{\hat{p}} = \sqrt{\frac{p(1-p)}{n}}. \tag{8.1}$$

$\hat{p}$
sample
proportion

p
population
proportion

When conducting inference, the population proportion p is unknown. Thus, to construct a confidence interval, the sample proportion $\hat{p}$ can be substituted for p to check the success-failure condition and compute the standard error. In a hypothesis test, p_0 is substituted for p.

Confidence intervals for a proportion

When using the normal approximation to the sampling distribution of $\hat{p}$, a confidence interval for a proportion has the same structure as a confidence interval for a mean; it is centered at the point estimate, with a margin of error calculated from the standard error and appropriate $z^\star$ value. The formula for a 95% confidence interval is

$$\hat{p} \pm 1.96\sqrt{\frac{\hat{p}(1-\hat{p})}{n}}.$$

EXAMPLE 8.2

Using the normal approximation, construct an approximate 95% confidence interval for the response probability for patients with advanced melanoma who were administered the combination of nivolumab and ipilimumab.

The independence and success-failure assumptions should be checked first. Since the outcome of one patient is unlikely to influence that of other patients, the observations are independent. The success-failure condition is satisfied since $n\hat{p} = (52)(.404) = 21 > 10$ and $n\hat{p}(1 - \hat{p}) = (52)(.596) = 31 > 10$.

The point estimate for the response probability, based on a sample of size $n = 52$, is $\hat{p} = 0.404$. For a 95% confidence interval, $z^\star = 1.96$. The standard error is estimated as: $\sqrt{\frac{\hat{p}(1-\hat{p})}{n}} = \sqrt{\frac{(0.404)(1-0.404)}{52}} = 0.068$. The confidence interval is

$$0.404 \pm 1.96(0.068) \rightarrow (0.27, 0.54)$$

The approximate 95% confidence interval for p, the population response probability of melanoma patients to the combination of these new drugs, is (0.27, 0.54) or (27%, 54%).

GUIDED PRACTICE 8.3

In New York City on October 23rd, 2014, a doctor who had recently been treating Ebola patients in Guinea went to the hospital with a slight fever and was subsequently diagnosed with Ebola. Soon after, a survey conducted by the Marist Poll, an organization with a carefully designed methodology for drawing random samples from identified populations, found that 82% of New Yorkers favored a "mandatory 21-day quarantine for anyone who has come in contact with an Ebola patient."[3] a) Verify that the sampling distribution of $\hat{p}$ is nearly normal. b) Construct a 95% confidence interval for p, the proportion of New York adults who supported a quarantine for anyone who has come into contact with an Ebola patient.[4]

Did the participants in the melanoma trial constitute a random sample? Patients who participate in clinical trials are unlikely to be a random sample of patients with the disease under study since the patients or their physicians must be aware of the trial, and patients must be well enough to travel to a major medical center and be willing to receive an experimental therapy that may have serious side effects.

Investigators in the melanoma trial were aware that the observed proportion of patients responding in a clinical trial may be different than the hypothetical response probability in the population of patients with advanced melanoma. Study teams try to minimize these systematic differences by following strict specifications for deciding whether patients are eligible for a study. However, there is no guarantee that the results observed in a sample will be replicated in the general population.

Small, initial studies in which there is no control group, like the one described here, are early steps in exploring the value of a new therapy and are used to justify further study of a treatment when the results are substantially different than expected. The largest observed response rate in previous trials of 30% was close to the lower bound of the confidence interval from the study (27%, 54%), so the results were considered adequate justification for continued research on this treatment.

Hypothesis testing for a proportion

Just as with inference for population means, confidence intervals for population proportions can be used when deciding whether to reject a null hypothesis. It is useful in most settings, however, to calculate the p-value for a test as a measure of the strength of the evidence contradicting the null hypothesis.

When using the normal approximation for the distribution of $\hat{p}$ to conduct a hypothesis test, one should always verify that $\hat{p}$ is nearly normal under H_0 by checking the independence and success-failure conditions. Since a hypothesis test is based on the distribution of the test statistic under the null hypothesis, the success-failure condition is checked using the null proportion p_0, not the estimate $\hat{p}$.

According to the normal approximation to the binomial distribution, the number of successes in n trials is normally distributed with mean np_0 and standard deviation $\sqrt{np(1-p_0)}$. This approximation is valid when np_0 and $n(1-p_0)$ are both at least 10.[5]

---

[3]Poll ID NY141026 on maristpoll.marist.edu.

[4]a) The poll is based on a simple random sample and consists of fewer than 10% of the adult population of New York, which makes independence a reasonable assumption. The success-failure condition is satisfied since, $1042(0.82) > 5$ and $1042(1 - 0.82) > 5$. b) $0.82 \pm 1.96\sqrt{\frac{0.82(1-0.82)}{1042}} \rightarrow (0.796, 0.844)$.

[5]The normal approximation to the binomial distribution was discussed in Section 3.2 of Chapter 3.

Under the null hypothesis, the sample proportion $\hat{p} = X/n$ is approximately distributed as

$$N\left(p_0, \sqrt{\frac{p_0(1-p_0)}{n}}\right).$$

The test statistic z for the null hypothesis $H_0 : p = p_0$ based on a sample of size n is

$$z = \frac{\text{point estimate - null value}}{SE}$$

$$= \frac{\hat{p} - p_0}{\sqrt{\frac{(p_0)(1-p_0)}{n}}}.$$

EXAMPLE 8.4

Suppose that out of a cohort of 120 patients with stage 1 lung cancer at the Dana-Farber Cancer Institute (DFCI) treated with a new surgical approach, 80 of the patients survive at least 5 years, and suppose that National Cancer Institute statistics indicate that the 5-year survival probability for stage 1 lung cancer patients nationally is 0.60. Do the data collected from 120 patients support the claim that the DFCI population treated with this new form of surgery has a different 5-year survival probability than the national population? Let $\alpha = 0.10$, since this is an early study of the new surgery.

Test the hypothesis $H_0 : p = 0.60$ versus the alternative, $H_A : p \neq 0.60$, using $\alpha = 0.10$. If we assume that the outcome of one patient at DFCI does not influence the outcome of other patients, the independence condition is met, and the success-failure condition is satisfied since $(120)(0.60) = 80 > 5$ and $(120)(1 - 0.60) = 40 > 5$. The test statistic is the z-score of the point estimate:

$$z = \frac{\text{point estimate - null value}}{SE} = \frac{0.67 - 0.60}{\sqrt{\frac{(0.60)(1-0.60)}{120}}} = 1.57.$$

The p-value is the probability that a standard normal variable is larger than 1.57 or smaller than -1.57, $P(|Z| > 1.57) = 0.12$; since the p-value is greater than 0.10, there is insufficient evidence to reject H_0 in favor of H_A. There is not convincing evidence that the survival probability at DFCI differs from the national survival probability. Had a more traditional 0.05 significance level been used, the data would be even less convincing.

EXAMPLE 8.5

Using the data from the study in advanced melanoma, use the normal approximation to the sampling distribution of $\hat{p}$ to test the null hypothesis that the response probability to the novel combined therapy is 30% against a one-sided alternative that the response proportion is greater than 30%. Let $\alpha = 0.10$.

The test statistic has value

$$z = (0.404 - 0.30)/\sqrt{(0.30)(0.70)/52} = 1.64.$$

The one-sided p-value is $P(Z \geq 1.64) = 0.05$; there is sufficient evidence to reject the null hypothesis at $\alpha = 0.10$. This is an example of where a two-sided test and a one-sided test yield different conclusions.

GUIDED PRACTICE 8.6

One of the questions on the National Health and Nutrition Examination Survey (introduced in Chapter 5) asked participants whether they participated in moderate or vigorous intensity sports, fitness, or recreational activities. In a random sample of 135 adults, 76 answered "Yes" to the question. Based on this evidence, are a majority of American adults physically active?[6]

8.1.2   Inference using exact methods

When the normal approximation to the distribution of $\hat{p}$ may not be accurate, inference is based on exact binomial probabilities. Calculating confidence intervals and p-values based on the binomial distribution can be done by hand, with tables of the binomial distribution, or (more easily and accurately) with statistical software. The logic behind computing a p-value is discussed here, but the formulas for a confidence interval are complicated and are not shown.

The p-value for a hypothesis test corresponds to the sum of the probabilities of all events that are as or more extreme than the sample result. Let X be a binomial random variable with parameters n and p_0, where $\hat{p} = x/n$ and x is the observed number of events. If $\hat{p} \leq p_0$, then the one-tail probability equals $P(X \leq x)$; if $\hat{p} > p_0$, then the one-tail probability equals $P(X \geq x)$. These probabilities are calculated using the approaches from Chapter 3. Two-tailed probabilities are calculated by doubling the appropriate one-tailed value.

EXAMPLE 8.7

In 2009, the FDA Oncology Drug Advisory Committee (ODAC) recommended that the drug Avastin be approved for use in glioblastoma, a form of brain cancer. Tumor shrinkage after taking a drug is called a response; out of 85 patients, 24 exhibited a response. Historically, response probabilities for brain cancer drugs were approximately 0.05, or about 5%. Assess whether there is evidence that the response probability for Avastin is different from previous drugs.

$H_0 : p = 0.05$; $H_A : p \neq 0.05$. Let $\alpha = 0.05$.

The independence condition is satisfied, but the success-failure condition is not, since $np_0 = (85)(0.05) = 4.25 < 5$, so this is a setting where exact binomial probabilities should be used to calculate a p-value.

The sample proportion $\hat{p}$ equals $x/n = 24/85 = 0.28$. Since $\hat{p} > p_0$, calculate the two-sided p-value from $2 \times P(X \geq 24)$, where $X \sim \text{Binom}(85, 0.05)$.

Calculating the p-value is best done in software; the R command pbinom returns a value of 5.3486×10^{-12}.[7]

The p-value is highly significant and suggests that the response probability for Avastin is higher than for previous brain cancer drugs. The FDA staff considered this evidence sufficiently strong to justify approval for the use of the drug, even though the FDA normally requires evidence from two independently conducted randomized trials.

---

[6]The observations are independent. Check success-failure: $np_0 = n(1 - p_0) = 135(0.5) > 10$. $H_0 : p = 0.5$; $H_A : p > 0.5$. Calculate the z-score: $z = \frac{0.56 - 0.50}{\sqrt{\frac{0.5(1-0.5)}{135}}} = 1.39$. The p-value is 0.08. Since the p-value is larger than 0.05, there is insufficient evidence to reject H_0; there is not convincing evidence that a majority of Americans are physically active, although the data suggest that may be the case.

[7]2*pbinom(q = 23, size = 85, p = 0.05, lower.tail = FALSE)

GUIDED PRACTICE 8.8

Medical consultants assist patients with all aspects of an organ donation surgery, with the goal of reducing the possibility of complications during the medical procedure and recovery. To attract customers, one consultant noted that while the usual proportion of complications in liver donation surgeries in the United States is about 10%, only 3 out of her 62 clients experienced complications with liver donor surgeries. Is there evidence to suggest that the proportion of complications in her patients is lower than the national average?[8]

8.1.3 Choosing a sample size when estimating a proportion

Whenever possible, a sample size for a study should be estimated before data collection begins. Section 5.4 explored the calculation of sample sizes that allow a hypothesis test comparing two groups to have adequate power. When estimating a proportion, preliminary sample size calculations are often done to estimate a sample size large enough to make the **margin of error** m in a confidence interval sufficiently small for the interval to be useful. Recall that the margin of error m is the term that is added to and subtracted from the point estimate. Statistically, this means estimating a sample size n so that the sample proportion is within some margin of error m of the actual proportion with a certain level of confidence. When the normal approximation is used for a binomial proportion, a sample size sufficiently large to have a margin of error of m will satisfy

$$m = (z^\star)(\text{s.e.}(\hat{p})) = z^\star \sqrt{\frac{(p)(1-p)}{n}}.$$

Algebra can be used to show that the above equation implies

$$n = \frac{(z^\star)^2(p)(1-p)}{m^2}.$$

In some settings a preliminary estimate for p can be used to calculate n. When no estimate is available, calculus can be used to show that $p(1-p)$ has its largest value when $p = 0.50$, and that conservative value for p is often used to ensure that n is sufficiently large regardless of the value of the unknown population proportion p. In that case, n satisfies

$$n \geq \frac{(z^\star)^2(0.50)(1-0.50)}{m^2} = \frac{(z^\star)^2}{4m^2}.$$

---

[8]Assume that the 62 patients in her dataset may be viewed as a random sample from patients receiving a donated liver. The sample proportion $\hat{p} = 3/62 = 0.048$. Under the null hypothesis, the expected number of complications is $62(0.10) = 6.2$, so the normal approximation may not be accurate and it is best to use exact binomial probabilities. Since $\hat{p} \leq p_0$, find the p-value by calculating $P(X \leq 3)$ when X has a binomial distribution with parameters $n = 62$, $p = 0.10$: $P(X \leq 3) = 0.121$. There is not sufficient evidence to suggest that the proportion of complications among her patients is lower than the national average.

EXAMPLE 8.9

Donor organs for organ transplant are scarce. Studies are conducted to explore whether the population of eligible organs can be expanded. Suppose a research team is studying the possibility of transplanting lungs from hepatitis C positive individuals; recipients can be treated with one of the new drugs that cures hepatitis C. Preliminary studies in organ transplant are often designed to estimate the probability of a successful organ graft 6 months after the transplant. How large should a study be so that the 95% confidence interval for the probability of a successful graft at 6 months is no wider than 20%?

A confidence interval no wider than 20% has a margin of error of 10%, or 0.10. Using the conservative value p = 0.50,

$$n = \frac{(1.96)^2}{(4)(0.10^2)} = 96.04.$$

Sample sizes are always rounded up, so the study should have 97 patients.

Since the study will likely yield a value $\hat{p}$ different from 0.50, the final margin of error will be smaller than ±0.10.

When the confidence coefficient is 95%, 1.96 can replaced by 2 and the sample size formula reduces to

$$n = 1/m^2.$$

This remarkably simple formula is often used by practitioners for a quick estimate of sample size.

GUIDED PRACTICE 8.10

A 2015 estimate of Congress' approval rating was 19%.[9] Using this estimate, how large should an additional survey be to produce a margin of error of 0.04 with 95% confidence?[10]

---

[9]www.gallup.com/poll/183128/five-months-gop-congress-approval-remains-low.aspx
[10]Apply the formula

$$1.96 \times \sqrt{\frac{p(1-p)}{n}} \approx 1.96 \times \sqrt{\frac{0.19(1-0.19)}{n}} \leq 0.04 \quad \rightarrow \quad n \geq 369.5.$$

A sample size of 370 or more would be reasonable.

8.2 Inference for the difference of two proportions

Just as inference can be done for the difference of two population means, conclusions can also be drawn about the difference of two population proportions: $p_1 - p_2$.

8.2.1 Sampling distribution of the difference of two proportions

The normal model can be applied to $\hat{p}_1 - \hat{p}_2$ if the sampling distribution for each sample proportion is nearly normal and if the samples are independent random samples from the relevant populations.

CONDITIONS FOR THE SAMPLING DISTRIBUTION OF $\hat{P}_1 - \hat{P}_2$ TO BE APPROXIMATELY NORMAL

The difference $\hat{p}_1 - \hat{p}_2$ tends to follow a normal model when

- each of the two samples are random samples from a population,
- the two samples are independent of each other, and
- each sample proportion follows (approximately) a normal model. This condition is satisfied when $n_1 p_1, n_1(1 - p_1), n_2 p_2$ and $n_2(1 - p_2)$ are all ≥ 10.

The standard error of the difference in sample proportions is

$$SE_{\hat{p}_1 - \hat{p}_2} = \sqrt{SE_{\hat{p}_1}^2 + SE_{\hat{p}_2}^2} = \sqrt{\frac{p_1(1 - p_1)}{n_1} + \frac{p_2(1 - p_2)}{n_2}}, \qquad (8.11)$$

where p_1 and p_2 are the population proportions, and n_1 and n_2 are the two sample sizes.

8.2.2 Confidence intervals for $p_1 - p_2$

When calculating confidence intervals for a difference of two proportions using the normal approximation to the binomial, the two sample proportions are used to verify the success-failure condition and to compute the standard error.

EXAMPLE 8.12

The way a question is phrased can influence a person's response. For example, Pew Research Center conducted a survey with the following question:[11]

> As you may know, by 2014 nearly all Americans will be required to have health insurance. [People who do not buy insurance will pay a penalty] while [People who cannot afford it will receive financial help from the government]. Do you approve or disapprove of this policy?

For each randomly sampled respondent, the statements in brackets were randomized: either they were kept in the order given above, or the order of the two statements was reversed. Figure 8.1 shows the results of this experiment. Calculate and interpret a 90% confidence interval of the difference in the probability of approval of the policy.

First the conditions for the use of a normal model must be verified. The Pew Research Center uses sampling methods that produce random samples of the US population (at least approximately) and because each group was a simple random sample from less than 10% of the population, the observations are independent, both within the samples and between the samples. The success-failure condition also holds for each sample, so the normal model can be used for confidence intervals for the difference in approval proportions. The point estimate of the difference in support, where $\hat{p}_1$ corresponds to the original ordering and $\hat{p}_2$ to the reversed ordering:

$$\hat{p}_1 - \hat{p}_2 = 0.47 - 0.34 = 0.13.$$

The standard error can be computed from Equation (8.11) using the sample proportions:

$$SE \approx \sqrt{\frac{0.47(1 - 0.47)}{771} + \frac{0.34(1 - 0.34)}{732}} = 0.025.$$

For a 90% confidence interval, $z^\star = 1.65$:

$$\text{point estimate} \pm z^\star \times SE \quad \rightarrow \quad 0.13 \pm 1.65 \times 0.025 \quad \rightarrow \quad (0.09, 0.17).$$

With 90% confidence, the proportion approving the 2010 health care law ranged between 9% and 17% depending on the phrasing of the question. The Pew Research Center interpreted this modestly large difference as an indication that for most of the public, opinions were still fluid on the health insurance mandate. The law eventually passed as the Affordable Health Care Act (ACA).

	Sample size (n_i)	Approve (%)	Disapprove (%)	Other
Original ordering	771	47	49	3
Reversed ordering	732	34	63	3

Figure 8.1: Results for a Pew Research Center poll where the ordering of two statements in a question regarding healthcare were randomized.

[11] www.people-press.org/2012/03/26/public-remains-split-on-health-care-bill-opposed-to-mandate. Sample sizes for each polling group are approximate.

8.2.3 Hypothesis testing for $p_1 - p_2$

Hypothesis tests for $p_1 - p_2$ are usually testing the null hypothesis of no difference between p_1 and p_2; i.e. $H_0 : p_1 - p_2 = 0$. Under the null hypothesis, $\hat{p}_1 - \hat{p}_2$ is normally distributed with mean 0 and standard deviation $\sqrt{p(1-p)(\frac{1}{n_1} + \frac{1}{n_2})}$, where under the null hypothesis $p = p_1 = p_2$.

Since p is unknown, an estimate is used to compute the standard error of $\hat{p}_1 - \hat{p}_2$; p can be estimated by $\hat{p}$, the weighted average of the sample proportions $\hat{p}_1$ and $\hat{p}_2$:

$$\hat{p} = \frac{n_1 \hat{p}_1 + n_2 \hat{p}_2}{n_1 + n_2} = \frac{x_1 + x_2}{n_1 + n_2},$$

where x_1 is the number of observed events in the first sample and x_2 is the number of observed events in the second sample. This pooled proportion $\hat{p}$ is also used to check the success-failure condition.

The test statistic z for testing $H_0 : p_1 = p_2$ versus $H_A : p_1 \neq p_2$ equals:

$$z = \frac{\hat{p}_1 - \hat{p}_2}{\sqrt{\hat{p}(1-\hat{p})\left(\frac{1}{n_1} + \frac{1}{n_2}\right)}}.$$

EXAMPLE 8.13

The use of screening mammograms for breast cancer has been controversial for decades because the overall benefit on breast cancer mortality is uncertain. Several large randomized studies have been conducted in an attempt to estimate the effect of mammogram screening. A 30-year study to investigate the effectiveness of mammograms versus a standard non-mammogram breast cancer exam was conducted in Canada with 89,835 female participants.[12] During a 5-year screening period, each woman was randomized to either receive annual mammograms or standard physical exams for breast cancer. During the 25 years following the screening period, each woman was screened for breast cancer according to the standard of care at her health care center.

At the end of the 25 year follow-up period, 1,005 women died from breast cancer. The results by intervention are summarized in Figure 8.2.

Assess whether the normal model can be used to analyze the study results.

Since the participants were randomly assigned to each group, the groups can be treated as independent, and it is reasonable to assume independence of patients within each group. Participants in randomized studies are rarely random samples from a population, but the investigators in the Canadian trial recruited participants using a general publicity campaign, by sending personal invitation letters to women identified from general population lists, and through contacting family doctors. In this study, the participants can reasonably be thought of as a random sample.

The pooled proportion $\hat{p}$ is

$$\hat{p} = \frac{x_1 + x_2}{n_1 + n_2} = \frac{500 + 505}{500 + 44,425 + 505 + 44,405} = 0.0112.$$

Checking the success-failure condition for each group:

$$\hat{p} \times n_{mgm} = 0.0112 \times 44,925 = 503 \qquad (1 - \hat{p}) \times n_{mgm} = 0.9888 \times 44,925 = 44,422$$
$$\hat{p} \times n_{ctrl} = 0.0112 \times 44,910 = 503 \qquad (1 - \hat{p}) \times n_{ctrl} = 0.9888 \times 44,910 = 44,407$$

All values are at least 10.

The normal model can be used to analyze the study results.

	Death from breast cancer?	
	Yes	No
Mammogram	500	44,425
Control	505	44,405

Figure 8.2: Summary results for the mammogram study.

---

[12]Miller AB. 2014. *Twenty five year follow-up for breast cancer incidence and mortality of the Canadian National Breast Screening Study: randomised screening trial.* BMJ 2014;348:g366 doi: 10.1136/bmj.g366

EXAMPLE 8.14

Do the results from the study provide convincing evidence of a difference in the proportion of breast cancer deaths between women who had annual mammograms during the screening period versus women who received annual screening with physical exams?

The null hypothesis is that the probability of a breast cancer death is the same for the women in the two groups. If group 1 represents the mammogram group and group 2 the control group, $H_0 : p_1 = p_2$ and $H_A : p_1 \neq p_2$. Let $\alpha = 0.05$.

Calculate the test statistic z:

$$z = \frac{0.01113 - 0.01125}{\sqrt{(0.0112)(1 - 0.0112)\left(\frac{1}{44,925} + \frac{1}{44,910}\right)}} = -0.17.$$

The two-sided p-value is $P|Z| \geq 0.17 = 0.8650$, which is greater than 0.05. There is insufficient evidence to reject the null hypothesis; the observed difference in breast cancer death rates is reasonably explained by chance.

Evaluating medical treatments typically requires accounting for additional evidence that cannot be evaluated from a statistical test. For example, if mammograms are much more expensive than a standard screening and do not offer clear benefits, there is reason to recommend standard screenings over mammograms. This study also found that a higher proportion of diagnosed breast cancer cases in the mammogram screening arm (3250 in the mammogram group vs 3133 in the physical exam group), despite the nearly equal number of breast cancer deaths. The investigators inferred that mammograms may cause over-diagnosis of breast cancer, a phenomenon in which a breast cancer diagnosed with mammogram and subsequent biopsy may never become symptomatic. The possibility of over-diagnosis is one of the reasons mammogram screening remains controversial.

EXAMPLE 8.15

Calculate a 95% confidence interval for the difference in proportions of deaths from breast cancer from the Canadian study.

———————

The independence and random sampling conditions have already been discussed. The success failure condition should be checked for each sample, since this is not a hypothesis testing context (i.e., there is no null hypothesis). For the mammogram group, $\hat{p}_1 = 0.01113$; $n_1\hat{p}_1 = (0.1113)(44,925) = 500$ and $n_1(1 - \hat{p}_1) = 39,925$. It is easy to show that the success failure condition is holds for the control group as well.

The point estimate for the difference in the probability of death is

$$\hat{p}_1 - \hat{p}_2 = 0.01113 - 0.01125 = -0.00012,$$

or 0.012%.

The standard error for the estimated difference uses the individual estimates of the probability of a death:

$$SE \approx \sqrt{\frac{0.01113(1 - 0.01113)}{44,925} + \frac{0.01125(1 - 0.01125)}{44,910}} = 0.0007.$$

The 95% confidence interval is given by

$$-0.00012 \pm (1.96)(0.0007) = (-0.0015, 0.0013).$$

With 95% confidence, the difference in the probability of death is between -0.15% and 0.13%. As expected from the large p-value, the confidence interval contains the null value 0.

8.3 Inference for two or more groups

The comparison of the proportion of breast cancer deaths between the two groups can also be approached using a two-way contingency table, which contains counts for combinations of outcomes for two variables. The results for the mammogram study in this format are shown in Figure 8.3.

Previously, the main question of interest was stated as, "Is there evidence of a difference in the proportion of breast cancer deaths between the two screening groups?" If the probability of a death from breast cancer does not depend the method of screening, then screening method and outcome are independent. Thus, the question can be re-phrased: "Is there evidence that screening method is associated with outcome?"

Hypothesis testing in a two-way table assesses whether the two variables of interest are associated (i.e., not independent). The approach can be applied to settings with two or more groups and for responses that have two or more categories. The observed number of counts in each table cell are compared to the number of **expected counts**, where the expected counts are calculated under the assumption that the null hypothesis of no association is true. A χ^2 test of significance is based on the differences between observed and expected values in the cells.

Death from BC	Yes	No	Total
Mammogram	500	44,425	44,925
Control	505	44,405	44,910
Total	1,005	88,830	89,835

Figure 8.3: Results of the mammogram study, as a contingency table with marginal totals.

GUIDED PRACTICE 8.16

Formulate hypotheses for a contingency-table approach to analyzing the mammogram data.[13]

8.3.1 Expected counts

If type of breast cancer screening had no effect on outcome in the mammogram data, what would the expected results be?

Recall that if two events A and B are independent, then $P(A \cap B) = P(A)P(B)$. Let A represent assignment to the mammogram group and B the event of death from breast cancer. Under independence, the number of individuals out of 89,835 that are expected to be in the mammogram screening group and die from breast cancer equals:

$$(89,835)P(A)P(B) = (89,835)\left(\frac{44,925}{89,835}\right)\left(\frac{1,005}{89,835}\right) = 502.6.$$

Note that the quantities 44,925 and 1,005 are the row and column totals corresponding to the upper left cell of Figure 8.3, and 89,835 is the total number n of observations in the table. A general formula for computing expected counts for any cell can be written from the marginal totals and the total number of observations.

---

[13]H_0: There is no association between type of breast cancer screening and death from breast cancer. H_A: There is an association between type of breast cancer screening and death from breast cancer.

COMPUTING EXPECTED COUNTS IN A TWO-WAY TABLE

To calculate the expected count for the ith row and j^{th} column, compute

$$\text{Expected Count}_{\text{row }i,\text{ col }j} = \frac{(\text{row }i\text{ total}) \times (\text{column }j\text{ total})}{\text{table total}}.$$

EXAMPLE 8.17

Calculate expected counts for the data in Figure 8.3.

$$E_{1,1} = \frac{44,925 \times 1,005}{89,835} = 502.6 \qquad E_{1,2} = \frac{44,925 \times 88,830}{89,835} = 44,422.4$$

$$E_{2,1} = \frac{2,922 \times 1,005}{89,835} = 502.4 \qquad E_{2,2} = \frac{7,078 \times 88,830}{89,835} = 44,407.6$$

Death from BC	Yes	No	Total
Mammogram	500 (**502.6**)	44,425 (**44,422.4**)	44,925
Control	505 (**502.4**)	44,405 (**44,407.6**)	44,910
Total	1,005	88,830	89,835

Figure 8.4: Results of the mammogram study, with (**expected counts**). The expected counts should also sum to the row and column totals; this can be a useful check for accuracy.

EXAMPLE 8.18

If a newborn is HIV$^+$, should he or she be treated with nevirapine (NVP) or a more expensive drug, lopinarvir (LPV)? In this setting, success means preventing virologic failure; i.e., growth of the virus. A randomized study was conducted to assess whether there is an association between treatment and outcome.[14] Of the 147 children administered NVP, about 41% experienced virologic failure; of the 140 children administered LPV, about 19% experienced virologic failure. Construct a table of observed counts and a table of expected counts.

Convert the proportions to count data: 41% of 147 is approximately 60, and 19% of 140 is approximately 27. The observed results are given in Figure 8.5.

Calculate the expected counts for each cell:

$$E_{1,1} = \frac{87 \times 147}{287} = 44.6 \qquad E_{1,2} = \frac{87 \times 140}{287} = 42.4$$

$$E_{2,1} = \frac{200 \times 147}{287} = 102.4 \qquad E_{2,2} = \frac{200 \times 140}{287} = 97.6$$

The expected counts are summarized in Figure 8.6.

---

[14]Violari A, et al. N Engl J Med 2012; 366:2380-2389 DOI: 10.1056/NEJMoa1113249

	NVP	LPV	Total
Virologic Failure	60	27	87
Stable Disease	87	113	200
Total	147	140	287

Figure 8.5: Observed counts for the HIV study.

	NVP	LPV	Total
Virologic Failure	44.6	42.4	87
Stable Disease	102.4	97.6	200
Total	147	140	287

Figure 8.6: Expected counts for the HIV study.

8.3.2 The χ^2 test statistic

Previously, test statistics have been constructed by calculating the difference between a point estimate and a null value, then dividing by the standard error of the point estimate to standardize the difference. The χ^2 statistic is based on a different idea. In each cell of a table, the difference *observed - expected* is a measure of the discrepancy between what was observed in the data and what should have been observed under the null hypothesis of no association. If the row and column variables are highly associated, that difference will be large. Two adjustments are made to the differences before the final statistic is calculated. First, since both positive and negative differences suggest a lack of independence, the differences are squared to remove the effect of the sign. Second, cells with larger counts may have larger discrepancies by chance alone, so the squared differences in each cell are scaled by the number expected in the cell under the hypothesis of independence. The final χ^2 statistic is the sum of these standardized squared differences, where the sum has one term for each cell in the table.

The χ^2 test statistic is calculated as:

$$\chi^2 = \sum_{\text{all cells}} \frac{(\text{observed} - \text{expected})^2}{\text{expected}}.$$

χ^2
chi-square
test statistic

The theory behind the χ^2 test and its sampling distribution relies on the same normal approximation to the binomial distribution that was introduced earlier. The cases in the dataset must be independent and each expected cell count should be at least 10. The second condition can be relaxed in tables with more than 4 cells.

CONDITIONS FOR THE χ^2 TEST

Two conditions that must be checked before performing a χ^2 test:

Independence. Each case that contributes a count to the table must be independent of all the other cases in the table.

Sample size. Each expected cell count must be greater than or equal to 10. For tables larger than 2×2, it is appropriate to use the test if no more than 1/5 of the expected counts are less than 5, and all expected counts are greater than 1.

EXAMPLE 8.19

For the mammogram data, check the conditions for the χ^2 test and calculate the χ^2 test statistic.

Independence is a reasonable assumption, since individuals have been randomized to either the treatment or control group. Each expected cell count is greater than 10.

$$\chi^2 = \sum_{\text{all cells}} \frac{(\text{observed} - \text{expected})^2}{\text{expected}}$$

$$= \frac{(500 - 502.6)^2}{502.6} + \frac{(44,425 - 44,422.4)^2}{44,422.4} + \frac{(505 - 502.4)^2}{502.4} + \frac{(44,405 - 44,407.6)^2}{44,407.6}$$

$$= 0.02.$$

GUIDED PRACTICE 8.20

For the HIV data, check the conditions for the χ^2 test and calculate the χ^2 test statistic.[15]

8.3.3   Calculating p-values for a χ^2 distribution

The **chi-square distribution** is often used with data and statistics that are positive and right-skewed. The distribution is characterized by a single parameter, the degrees of freedom. Figure 8.7 demonstrates three general properties of chi-square distributions as the degrees of freedom increases: the distribution becomes more symmetric, the center moves to the right, and the variability increases.

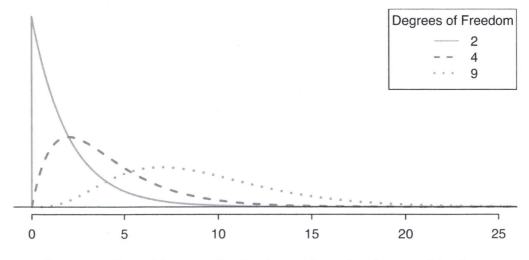

Figure 8.7: Three chi-square distributions with varying degrees of freedom.

---

[15]Independence holds, since this is a randomized study. The expected counts are greater than 10. $\chi^2 = \frac{(60-44.6)^2}{44.6} + \frac{(27-42.4)^2}{42.4} + \frac{(87-102.4)^2}{102.4} + \frac{(113-97.6)^2}{97.6} = 14.7.$

The χ^2 statistic from a contingency table has a sampling distribution that approximately follows a χ^2 distribution with degrees of freedom $df = (r-1)(c-1)$, where r is the number of rows and c is the number of columns. Either statistical software or a table can be used to calculate p-values from the χ^2 distribution. The **chi-square table** is partially shown in Figure 8.8, and a more complete table is presented in Appendix B.3 on page 468. This table is very similar to the t-table: each row provides values for distributions with different degrees of freedom, and a cut-off value is provided for specified tail areas. One important difference from the t-table is that the χ^2 table only provides upper tail values.

Upper tail		0.3	0.2	0.1	0.05	0.02	0.01	0.005	0.001
df	1	1.07	1.64	2.71	3.84	5.41	6.63	7.88	10.83
	2	2.41	3.22	4.61	5.99	7.82	9.21	10.60	13.82
	3	3.66	4.64	6.25	7.81	9.84	11.34	12.84	16.27
	4	4.88	5.99	7.78	9.49	11.67	13.28	14.86	18.47
	5	6.06	7.29	9.24	11.07	13.39	15.09	16.75	20.52
	6	7.23	8.56	10.64	12.59	15.03	16.81	18.55	22.46
	7	8.38	9.80	12.02	14.07	16.62	18.48	20.28	24.32

Figure 8.8: A section of the chi-square table. A complete table is in Appendix B.3 on page 468.

EXAMPLE 8.21

Calculate an approximate p-value for the mammogram data, given that the χ^2 statistic equals 0.02. Assess whether the data provides convincing evidence of an association between screening group and breast cancer death.

The degrees of freedom in a 2×2 table is 1, so refer to the values in the first column of the probability table. The value 0.02 is less than 1.07, so the p-value is greater than 0.3. The data do not provide convincing evidence of an association between screening group and breast cancer death. This supports the conclusions from Example 8.14, where the p-value was calculated to be 0.8650 and is visualized in Figure 8.9.

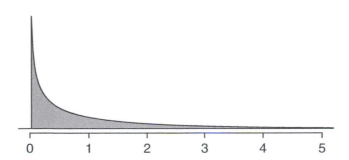

Figure 8.9: The p-value for the mammogram data is shaded on the χ^2 distribution with $df = 1$. The shaded area is to the right of x = 0.02.

GUIDED PRACTICE 8.22

Calculate an approximate p-value for the HIV data. Assess whether the data provides convincing evidence of an association between treatment and outcome at the $\alpha = 0.01$ significance level.[16]

---

[16]The χ^2 statistic is 14.7. For degrees of freedom 1, the tail area beyond 14.7 is smaller than 0.001. There is evidence to suggest that treatment is not independent of outcome.

8.3.4   Interpreting the results of a χ^2 test

If the p-value from a χ^2 test is small enough to provide evidence to reject the null hypothesis of no association, it is important to explore the results further to understand direction of the observed association. This is done by examining the residuals, the standardized differences of the *observed - expected*, for each cell. Instead of using squared differences, the residuals are based on the differences themselves, and the standardizing or scaling factor is $\sqrt{\text{expected}}$. Calculating residuals can be particularly helpful for understanding the results from large tables.

For each cell in a table, the residual equals:

$$\frac{\text{observed} - \text{expected}}{\sqrt{\text{expected}}}.$$

Residuals with a large magnitude contribute the most to the χ^2 statistic. If a residual is positive, the observed value is greater than the expected value, and vice versa for a negative residual.

EXAMPLE 8.23

In the FAMuSS study introduced in Chapter 1, researchers measured a variety of demographic and genetic characteristics for about 1,300 participants, including data on race and genotype at a specific locus on the ACTN3 gene (Figure 8.10). Is there evidence of an association between genotype and race?

First, check the assumptions for applying a χ^2 test. It is reasonable to assume independence, since it is unlikely that any participants were related to each other. None of the expected counts, as shown in Figure 8.11, are less than 5.

H_0: Race and genotype are independent.

H_A: Race and genotype are not independent.

Let $\alpha = 0.05$.

Calculate the χ^2 statistic:

$$\chi^2 = \sum_{\text{all cells}} \frac{(\text{observed} - \text{expected})^2}{\text{expected}}$$

$$= \frac{(16 - 7.85)^2}{7.85} + \frac{(6 - 11.84)^2}{11.84} + \dots + \frac{(5 - 6.22)^2}{6.22}$$

$$= 19.4.$$

Calculate the p-value: for a table with 3 rows and 5 columns, the χ^2 statistic is distributed with $(3 - 1)(5 - 1) = 8$ degrees of freedom. From the table, a χ^2 value of 19.4 corresponds to a tail area between 0.01 and 0.02. Thus, there is sufficient evidence to reject the null hypothesis of independence between race and genotype.

The p-value can be obtained using the R function pchisq (pchisq(19.4, df = 8, lower.tail = FALSE)), which returns a value of 0.012861.

To further explore the differences in genotype distribution between races, calculate residuals for each cell (Figure 8.12). The largest residuals are in the first row; there are many more African Americans with the CC genotype than expected under independence, and fewer with the CT genotype than expected. The residuals in the second row indicate a similar trend for Asians, but with a less pronounced difference. These results suggest further directions for research; a future study could enroll a larger number of African American and Asian participants to examine whether the observed trend holds with a more representative sample. Geneticists might also be interested in exploring whether this genetic difference between populations has an observable phenotypic effect.

	CC	CT	TT	Sum
African American	16	6	5	27
Asian	21	18	16	55
Caucasian	125	216	126	467
Hispanic	4	10	9	23
Other	7	11	5	23
Sum	173	261	161	595

Figure 8.10: Observed counts for race and genotype data from the FAMuSS study.

	CC	CT	TT	Sum
African Am	7.85	11.84	7.31	27.00
Asian	15.99	24.13	14.88	55.00
Caucasian	135.78	204.85	126.36	467.00
Hispanic	6.69	10.09	6.22	23.00
Other	6.69	10.09	6.22	23.00
Sum	173.00	261.00	161.00	595.00

Figure 8.11: Expected counts for race and genotype data from the FAMuSS study.

	CC	CT	TT	Sum
African Am	**2.91**	**-1.70**	-0.85	0.00
Asian	**1.25**	**-1.25**	0.29	0.00
Caucasian	-0.93	0.78	-0.03	0.00
Hispanic	-1.04	-0.03	1.11	0.00
Other	0.12	0.29	-0.49	0.00
Sum	0.00	0.00	0.00	0.00

Figure 8.12: Residuals for race and genotype data from the FAMuSS study.

EXAMPLE 8.24

In Guided Practice 8.22, the p-value was found to be smaller than 0.001, suggesting that treatment is not independent of outcome. Does the evidence suggest that infants should be given nevirapine or lopinarvir?

In a 2×2 table, it is relatively easy to directly compare observed and expected counts. For nevirapine, more infants than expected experienced virologic failure ($60 > 44.6$), while fewer than expected reached a stable disease state ($87 < 102.4$). For lopinarvir, fewer infants than expected experienced virologic failure ($27 < 42.4$), and more infants than expected reached a stable disease state ($113 > 97.6$) (Figure 8.13). The outcomes for infants on lopinarvir are better than for those on nevirapine; combined with the results of the significance test, the data suggest that lopinarvir is associated with better treatment outcomes.

	NVP	LPV	Total
Virologic Failure	60 **44.6**	27 **42.4**	87
Stable Disease	87 **102.4**	113 **97.6**	200
Total	147	140	287

Figure 8.13: Observed and **(expected)** counts for the HIV study.

GUIDED PRACTICE 8.25

Confirm the conclusions reached in Example 8.24 by analyzing the residuals.[17]

---

[17] $R_{1,1} = \frac{(44.6-60)}{\sqrt{44.6}} = 2.31$; $R_{1,2} = \frac{(42.4-27)}{\sqrt{27}} = -2.37$; $R_{2,1} = \frac{(87-102.4)}{\sqrt{102.4}} = -1.53$; $R_{2,2} = \frac{(113-97.6)}{\sqrt{97.6}} = 1.56$. The positive residuals for the upper left and lower right cells indicate that more infants than expected experienced virologic failure on NVP and stable disease on LPV; vice versa for the upper right and lower left cells. The larger magnitude of the residuals for the two NVP cells indicates that most of the discrepancy between observed and expected counts is for outcomes related to NVP.

GUIDED PRACTICE 8.26

Chapter 1 started with the discussion of a study examining whether exposure to peanut products reduce the rate of a child developing peanut allergies. Children were randomized either to the peanut avoidance or the peanut consumption group; at 5 years of age, each child was tested for peanut allergy using an oral food challenge (OFC). The results of the OFC are reproduced in Figure 8.14; failing the food challenge indicates an allergic reaction. Assess whether there is evidence for exposure to peanut allergy reducing the chance of developing peanut allergies.[18]

	FAIL OFC	PASS OFC	Sum
Peanut Avoidance	36	227	263
Peanut Consumption	5	262	267
Sum	41	489	530

Figure 8.14: LEAP Study Results.

8.3.5 Fisher's exact test

If sample sizes are too small, the χ^2 distribution does not yield accurate p-values for assessing independence of the row and column variables in a table. When expected counts in a table are less than 10, **Fisher's exact test** is often used to calculate exact levels of significance. This test is usually applied to 2×2 tables. It can be applied to larger tables, but the logic behind the test is complex and the calculations involved are computationally intensive, so this section covers only 2×2 tables.

Clostridium difficile is a bacterium that causes inflammation of the colon. Antibiotic treatment is typically not effective, particularly for patients who experience multiple recurrences of infection. Infusion of feces from healthy donors has been reported as an effective treatment for recurrent infection. A randomized trial was conducted to compare the efficacy of donor-feces infusion versus vancomycin, the antibiotic typically prescribed to treat *C. difficile* infection. The results of the trial are shown in Figure 8.15.[19] A brief calculation shows that all of the expected cell counts are less than 10, so the χ^2 test should not be used as a test for association.

Under the null hypothesis, the probabilities of cure in the fecal infusion and vancomycin groups are equal; i.e., individuals in one group are just as likely to be cured as individuals in the other group. Suppose the probability that an individual is cured, given that he or she was assigned to the fecal infusion group, is p_1 and the probability an individual is cured in the vancomycin group is p_2. Researchers were interested in testing the null hypothesis $H_0: p_1 = p_2$.

	Cured	Uncured	Sum
Fecal Infusion	13	3	16
Vancomycin	4	9	13
Sum	17	12	29

Figure 8.15: Fecal Infusion Study Results.

[18]The assumptions for conducting a χ^2 test are satisfied. Calculate a χ^2 test statistic: 24.29. The associated p-value is 8.3×10^{-7}. There is evidence to suggest that treatment group is not independent of outcome. Specifically, a residual analysis shows that in the peanut avoidance group, more children than expected failed the OFC; in the peanut consumption group, more children than expected passed the OFC.

[19]These results correspond to the number of patients cured after the first infusion of donor feces and the number of patients cured in the vancomycin-alone group.

The p-value is the probability of observing results as or more extreme than those observed in the study under the assumption that the null hypothesis is true. Previously discussed methods for significance testing have relied on calculating a test statistic associated with a defined sampling distribution, then obtaining p-values from tail areas on the distribution. Fisher's exact test uses a similar approach, but introduces a new sampling distribution.

The p-value for Fisher's exact test is calculated by adding together the individual conditional probabilities of obtaining each table that is as or more extreme than the one observed, under the null hypothesis and given that the marginal totals are considered fixed.

- When the row and column totals are held constant, the value of any one cell in the table determines the rest of the entries. For example, if the marginal sums in Figure 8.15 are known, along with the value in one cell (e.g., the upper right equals 3), it is possible to calculate the values in the other three cells. Thus, when marginal totals are considered fixed, each table represents a unique set of results.

- Extreme tables are those which contradict the null hypothesis of $p_1 = p_2$. In the fecal infusion group, under the null hypothesis of no difference in the population proportion cured, one would expect $\frac{16 \times 17}{29} = 9.38$ cured individuals. The 13 observed cured individuals is extreme in the direction of more being cured than expected under the null hypothesis. An extreme result in the other direction would be, for instance, 1 cured patient in the fecal infusion group and 16 in the vancomycin group.

EXAMPLE 8.27

Of the 17 patients cured, 13 were in the fecal infusion group and 4 were in the vancomycin group. Assume that the marginal totals are fixed (i.e., 17 patients were cured, 12 were uncured, and 16 patients were in the fecal infusion group, while 13 were in the vancomycin group). Enumerate all possible sets of results that are more extreme than what was observed, in the same direction.

―――――

The observed results show a case of $\hat{p}_1 > \hat{p}_2$; results that are more extreme consist of cases where more than 13 cured patients were in the fecal infusion group. Under the assumption that the total number of cured patients is constant at 17 and that only 16 patients were assigned to the fecal infusion group (out of 29 patients total), more extreme results are represented by cases where 14, 15, or 16 cured patients were in the fecal infusion group. The following tables illustrate the unique combinations of values for the 4 table cells corresponding to those extreme results.

	Cured	Uncured	Sum
Fecal Infusion	14	2	16
Vancomycin	3	10	13
Sum	17	12	29

	Cured	Uncured	Sum
Fecal Infusion	15	1	16
Vancomycin	2	11	13
Sum	17	12	29

	Cured	Uncured	Sum
Fecal Infusion	16	0	16
Vancomycin	1	12	13
Sum	17	12	29

Calculating a one-sided p-value

Suppose that researchers were interested in testing the null hypothesis against the one-sided alternative, $H_A : p_1 > p_2$. To calculate the one-sided p-value, sum the probabilities of each table representing results as or more extreme than those observed; specifically, sum the probabilities of observing Figure 8.15 and the tables in Example 8.27.

	Cured	Uncured	Sum
Fecal Infusion	a	b	$a+b$
Vancomycin	c	d	$c+d$
Sum	$a+c$	$b+d$	n

Figure 8.16: General Layout of Data in Fecal Infusion Study.

The probability of observing a table with cells a, b, c, d given fixed marginal totals $a+b$, $c+d$, $a+c$, and $b+d$ follows the hypergeometric distribution. The hypergeometric distribution was introduced in Section 3.5.3.

$$P(a,b,c,d) = \text{HGeom}(a+b, c+d, a+c) = \frac{\binom{a+b}{a}\binom{c+d}{c}}{\binom{n}{a+c}} = \frac{(a+b)!\,(c+d)!\,(a+c)!\,(b+d)!}{a!\,b!\,c!\,d!\,n!}.$$

EXAMPLE 8.28

Calculate the probability of observing Figure 8.15, assuming the margin totals are fixed.

$$P(13,3,4,9) = \frac{\binom{16}{13}\binom{13}{4}}{\binom{29}{17}} = \frac{16!\,13!\,17!\,12!}{13!\,3!\,4!\,9!\,29!} = 7.71 \times 10^{-3}.$$

The value 0.0077 represents the probability of observing 13 cured patients out of 16 individuals in the fecal infusion group and 1 cured in the vancomycin group, given that there are a total of 29 patients and 17 were cured overall.

EXAMPLE 8.29

Evaluate the statistical significance of the observed data in Figure 8.15 using the one-sided alternative $H_A : p_1 > p_2$.

Calculate the probability of the tables from Example 8.27. Generally, the formula for these tables is

$$P(a, b, c, d) = \frac{\binom{a+b}{a}\binom{c+d}{c}}{\binom{n}{a+c}} = \frac{\binom{16}{a}\binom{13}{c}}{\binom{29}{17}},$$

since the marginal totals from Figure 8.15 are fixed. The value a ranges from 14, 15, 16, while c ranges from 3, 2, 1.

$$P(14, 2, 3, 10) = \frac{\binom{16}{14}\binom{13}{3}}{\binom{29}{17}} = 6.61 \times 10^{-4}$$

$$P(15, 1, 2, 11) = \frac{\binom{16}{15}\binom{13}{2}}{\binom{29}{17}} = 2.40 \times 10^{-5}$$

$$P(16, 0, 1, 12) = \frac{\binom{16}{16}\binom{13}{1}}{\binom{29}{17}} = 2.51 \times 10^{-7}$$

The probability of the observed table is 7.71×10^{-3}, as calculated in the previous example.

The one-sided p-value is the sum of these table probabilities: $(7.71 \times 10^{-3}) + (6.61 \times 10^{-4}) + (2.40 \times 10^{-5}) + (2.51 \times 10^{-7}) = 0.0084$.

The results are significant at the $\alpha = 0.05$ significance level. There is evidence to support the one-sided alternative that the proportion of cured patients in the fecal infusion group is higher than the proportion of cured patients in the vancomycin group. However, it is important to note that two-sided alternatives are the standard in medical literature. Conducting a two-sided test would be especially desirable when evaluating a treatment which lacks randomized trials supporting its efficacy, such as donor-feces infusion.

Calculating a two-sided p-value

There are various methods for calculating a two-sided p-value in the Fisher's exact test setting. When the test is calculated by hand, the most common way to calculate a two-sided p-value is to double the smaller of the one-sided p-values. One other common method used by various statistical computing packages such as R is to classify "more extreme" tables as all tables with probabilities less than that of the observed table, in both directions. The two-sided p-value is the sum of probabilities for the qualifying tables. That approach is illustrated in the next example.

EXAMPLE 8.30

Evaluate the statistical significance of the observed data in Figure 8.15 using the two-sided alternative $H_A : p_1 \neq p_2$.

Identify tables that are more extreme in the other direction of the observed result, i.e. where the proportion of cured patients in the vancomycin group are higher than in the fecal infusion group. Start with the most extreme cases and calculate probabilities until a table has a p-value higher than 7.71×10^{-3}, the probability of the observed table.

The most extreme result in the $\hat{p}_1 < \hat{p}_2$ direction would be if all patients in the vancomycin group were cured; then 13 of the cured patients would be in the vancomycin group and 4 would be in the fecal transplant group. This table has probability 3.5×10^{-5}.

	Cured	Uncured	Sum
Fecal Infusion	4	12	16
Vancomycin	13	0	13
Sum	17	12	29

Continue enumerating tables by decreasing the number of cured patients in the vancomycin group. The table with 5 cured patients in the fecal infusion group has probability 1.09×10^{-3}.

	Cured	Uncured	Sum
Fecal Infusion	5	11	16
Vancomycin	12	1	13
Sum	17	12	29

The table with 6 cured patients in the fecal infusion group has probability 0.012. This value is greater than 7.71×10^{-3}, so it will not be part of the sum to calculate the two-sided p-value.

	Cured	Uncured	Sum
Fecal Infusion	6	10	16
Vancomycin	11	2	13
Sum	17	12	29

As calculated in the previous example, the one-sided p-value is 0.0084. Thus, the two-sided p-value for these data equals $0.0084 + (3.5 \times 10^{-5}) + (1.09 \times 10^{-3}) = 0.0095$. The results are significant at the $\alpha = 0.01$ significance level, and there is evidence to support the efficacy of donor-feces infusion as a treatment for recurrent C. *difficile* infection.

8.4   Chi-square tests for the fit of a distribution

The χ^2 test can also be used to examine the appropriateness of hypothesized distribution for a dataset, most commonly when a set of observations falls naturally into categories as in the examples discussed in this section. As with testing in the two-way table setting, expected counts are calculated based on the assumption that the hypothesized distribution is correct, and the statistic is based on the discrepancies between observed and expected counts. The χ^2 sampling distribution for the test statistic is reasonably accurate when each expected count is at least 5 and follows a χ^2 distribution with $k - 1$ degrees of freedom, where k is the number of categories. Some guidelines recommend that no more than 1/5 of the cells have expected counts less than 5, but the stricter requirement that all cells have expected counts greater than 5 is safer.

When used in this setting, the χ^2 test is often called a **'goodness-of-fit' test**, a term that is often misunderstood. Small p-values of the test suggest evidence that a hypothesized distribution is not a good model, but non-significant p-values do not imply that the hypothesized distribution is the best model for the data, or even a good one. In the logic of hypothesis testing, failure to reject a null hypothesis cannot be viewed as evidence that the null hypothesis is true.

EXAMPLE 8.31

The participants in the FAMuSS study were volunteers at a university, and so did not come from a random sample of the US population. The participants may not be representative of the general United States population. The χ^2 test can be used to test the null hypothesis that the participants are racially representative of the general population. Figure 8.17 shows the number observed by racial category in FAMuSS and the proportions of the US population in each of those categories.[20]

Under the null hypothesis, the sample proportions should equal the population proportions. For example, since African Americans are 0.128 of the general proportion, $(0.128)(595) = 76.16$ African Americans would be expected in the sample. The rest of the expected counts are shown in Figure 8.18.

Since each expected count is greater than or equal to 5, the χ^2 distribution can be used to calculate a p-value for the test.

$$\chi^2 = \sum_{\text{all cells}} \frac{(\text{observed} - \text{expected})^2}{\text{expected}}$$

$$= \frac{(27 - 76.16)^2}{76.16} + \frac{(55 - 5.95)^2}{5.95} + \frac{(467 - 478.38)^2}{478.38} + \frac{(46 - 34.51)^2}{34.51}$$

$$= 440.18.$$

There are 3 degrees of freedom, since $k = 4$. The χ^2 statistic is extremely large, and the associated tail area is smaller than 0.001. There is more than sufficient evidence to reject the null hypothesis that the sample is representative of the general population. A comparison of the observed and expected values (or the residuals) indicates that the largest discrepancy is with the over-representation of Asian participants.

---

[20]The US Census Bureau considers Hispanic as a classification separate from race, on the basis that Hispanic individuals can be any race. In order to facilitate the comparison with the FAMuSS data, participants identified as "Hispanic" have been merged with the "Other" category.

Race	African American	Asian	Caucasian	Other	Total
FAMuSS	27	55	467	46	595
US Census	0.128	0.01	0.804	0.058	1.00

Figure 8.17: Representation by race in the FAMuSS study versus the general population.

Race	African American	Asian	Caucasian	Other	Total
Observed	27	55	467	46	595
Expected	76.16	5.95	478.38	34.51	595

Figure 8.18: Actual and expected counts in the FAMuSS data.

EXAMPLE 8.32

According to Mendelian genetics, alleles segregate independently; if an individual is heterozygous for a gene and has alleles A and B, then the alleles have an equal chance of being passed to an offspring. Under this framework, if two individuals with genotype AB mate, then their offspring are expected to exhibit a 1:2:1 genotypic ratio; 25% of the offspring will be AA, 50% will be AB, and 50% will be BB. The term "segregation distortion" refers to a deviation from expected Mendelian frequencies.

At a specific gene locus in the plant *Arabidopsis thaliana*, researchers have observed 84 AA individuals, 233 AB individuals, and 134 BB individuals. Is there evidence of segregation disorder at this locus? Conduct the test at $\alpha = 0.0001$ to account for multiple testing, since the original study examined approximately 250 locations across the genome.

The Mendelian proportions are 25%, 50%, and 25%. Thus, the expected counts in a group of 451 individuals are: 112.75 AA, 225.50 AB, and 112.75 BB. No expected count is less than 5.

$$\chi^2 = \sum_{\text{all cells}} \frac{(\text{observed} - \text{expected})^2}{\text{expected}}$$

$$= \frac{(84 - 112.75)^2}{112.75} + \frac{(233 - 225.50)^2}{225.50} + \frac{(134 - 112.75)^2}{112.75}$$

$$= 11.59.$$

There are 2 degrees of freedom, since $k = 3$. The p-value is between 0.005 and 0.001, which is greater than $\alpha = 0.0001$. There is insufficient evidence to reject the null hypothesis that the offspring ratios correspond to expected Mendelian frequencies; i.e., there is not evidence of segregation distortion at this locus.

8.5 Outcome-based sampling: case-control studies

8.5.1 Introduction

The techniques so far in this chapter have often relied on the assumption that the data were collected using random sampling from a population. When cases come from a random sample, the sample proportion of observations with a particular outcome should accurately estimate the population proportion, given that the sample size is large enough. When studying rare outcomes, however, moderate sized samples may contain few or none of the outcomes. Persistent pulmonary hypertension of the newborn (PPHN) is a dangerous condition in which the blood vessels in the lungs of a newborn do not relax immediately after birth, leading to inadequate oxygenation. The condition is rare, occurring in about 1.9 per 1,000 live births, so it is difficult to study using random sampling. In the early 2000s, anecdotal evidence began to accumulate that the risk of the condition might be increased if the mother of the newborn had been taking a particular medication for depression, a selective serotonin reuptake inhibitor (SSRI) during the third trimester of pregnancy or even as early as during week 20 of the pregnancy.

One design for studying the issue would enroll two cohorts of women, one in which women were taking SSRIs for depression and one in which they were not. However, if the chance of PPHN was 1.9/1,000 in newborns of a control cohort of 1,000 women, then the probability of observing no cases of PPHN is about 0.15. If the probability of PPHN is elevated among infants born to women taking SSRIs, such as to 3.0/1,000, the chance of observing no cases among 1,000 women is approximately 0.05. Precise measures of the probability of PPHN occurring would require very large cohorts.

An alternative design for studies like this reverses the sampling scheme so that the two cohorts are determined by outcome, rather than exposure; a cohort with the condition and a cohort without the condition are sampled, then exposure to a possible cause is recorded. To apply this design for studying PPHN, a registry of live births could be used to sort births by presence or absence of PPHN. The number in each group in which the mother had been taking SSRIs could then be recorded (based on medical records). Such a design would have the advantage of sufficient numbers of cases with and without PPHN, but it has other limitations which will be discussed later in this section. Traditionally, these studies have been called case-control studies because of the original sampling of individuals with and without a condition. More generally, it is an example of outcome-dependent sampling.

8.5.2 χ^2 tests of association in case-control studies

In 2006, Chambers, et. al reported a case-control study examining the association of SSRI use and persistent pulmonary hypertension in newborns.[21] The study team enrolled 337 women whose infants suffered from PPHN and 836 women with similar characteristics but whose infants did not have PPHN. Among the women whose infants had PPHN, 14 had taken an SSRI after week 20 of the pregnancy. In the cohort of women whose infants did not have PPHN, 6 had been taking the medication after week 20. In the subset of women who had been taking an SSRI, the infants are considered 'exposed' to the medication. The data from the study are summarized in Figure 8.19.

PPHN present	Yes	No	Total
SSRI exposed	14	6	20
SSRI unexposed	323	830	1153
Total	337	836	1173

Figure 8.19: SSRI exposure vs observed number of PPHN cases in newborns.

The sample of women participating in the study are clearly not a random sample drawn from women who had recently given birth; they were identified according to the disease status of their infants. In this sample, the proportion of newborns with PPHN ($337/1173 = 28.7\%$) is much higher than the disease prevalence in the general population.

Even so, the concept of independence between rows and columns under a null hypothesis of no association still holds. If SSRI use had no effect on the occurrence of PPHN, then the proportions of mothers taking SSRIs among the PPHN and non-PPHN infants should be about the same. In other words, the null hypothesis of equal SSRI use among mothers with/without PPHN affected infants is the hypothesis of no association between SSRI use and PPHN. The test of independence can be conducted using the approach introduced earlier in the chapter.

The expected counts shown in Figure 8.20 suggest that the p-value from a χ^2 test may not be accurate; under the null hypothesis, the expected number of PPHN cases in the SSRI exposed group is less than 10.

PPHN present	Yes	No	Total
SSRI exposed	5.80	14.20	20
SSRI unexposed	331.20	811.80	1153
Total	337	836	1173

Figure 8.20: SSRI exposure vs expected number of PPHN cases in newborn.

The p-value from Fisher's exact test is < 0.001 (0.00014, to be precise), so the evidence is strong that SSRI exposure and PPHN are associated. Fisher's exact test is often used in studies of rare conditions or exposures since one or more expected cell counts are typically less than 10.

---

[21]N Engl J Med 2006;354:579-87.

8.5.3 Estimates of association in case-control studies

For data in a 2×2 table, correct point estimates of association depend on the mechanism used to gather the data. In the example of a clinical trial of nevirapine versus lopinarvir discussed in Section 8.3.1, the population proportion of children who would experience virologic failure after treatment with one of the drugs can be estimated by the observed proportion of virologic failures while on that drug. For nevirapine, the proportion of children with virologic failure is $60/147 = 0.41$, while for lopinarvir the proportion is $27/140 = 0.19$. The difference in outcome between the two groups can be summarized by the difference in these proportions. The proportion experiencing virologic failure when treated with nevirapine was 0.12 larger in nevirapine (0.41 - 0.29), so if the two drugs were to be used in a large population, approximately 12% more children treated with nevirapine would experience virologic failure as compared to lopinarvir. The confidence intervals discussed in Section 8.2.2 can be used to express the uncertainty in this estimate.

Since the proportion of virologic failures can be estimated from the trial data, the relative risk of virologic failure can also be used to estimate the association between treatment and virologic failure. Relative risk is the ratio of two proportions, and was introduced in Section 1.6.2. The relative risk of virologic failure with nevirapine versus lopinarvir is $0.41/0.19 = 2.16$. Children treated with nevirapine are estimated to be more than twice as likely to experience virologic failure.

Statistically, the population parameter for the relative risk in the study of HIV$^+$ is a ratio of conditional probabilities:

$$\frac{P(\text{virologic failure}|\text{treatment with nevirapine})}{P(\text{virologic failure}|\text{treatment with lopinarvir})}.$$

In a study like the PPHN case-control study, the natural population parameter of interest would be the relative risk of PPHN for infants exposed to an SSRI during after week 20 of gestation compared to those who were not exposed. However, in the design of this study, participating mothers were sampled and grouped according to whether their infants did or did not suffer from PPHN, rather than assigned to either SSRI exposure or non-exposure. Relative risk of PPHN from exposure to SSRI cannot be estimated from the data because it is not possible to estimate $P(\text{PPHN}|\text{SSRI exposure})$ and $P(\text{PPHN}|\text{no SSRI exposure})$. In case-control studies, association is estimated using **odds** and **odds ratios** rather than relative risk.

The **odds** of SSRI exposure among the cases are given by the fraction

$$\text{odds}_{\text{cases}} = \frac{P(\text{SSRI exposure}|\text{PPHN})}{P(\text{no SSRI exposure}|\text{PPHN})} = \frac{14/337}{323/337} = \frac{14}{323}.$$

The odds of SSRI exposure among the controls are given by the fraction

$$\text{odds}_{\text{controls}} = \frac{P(\text{SSRI exposure}|\text{no PPHN})}{P(\text{no SSRI exposure}|\text{no PPHN})} = \frac{6/836}{830/836} = \frac{6}{830}.$$

The ratio of the odds, the odds ratio, compares the odds of exposure among the cases to the odds of exposure among the controls:

$$OR_{\text{exposure, cases vs. controls}} = \frac{\text{odds}_{\text{cases}}}{\text{odds}_{\text{controls}}} = \frac{14/323}{6/830} = \frac{(14)(830)}{(323)(6)} = 6.00.$$

A population odds ratio of, for example, 1.5, implies that the odds of exposure in cases are 50% larger than the odds of exposure in controls. For this study, the odds ratio of 6.00 implies that the odds of SSRI exposure in infants with PPHN are 6 times as large as the odds of exposure in infants without PPHN. Epidemiologists describe this odds ratio as the odds of exposure given presence of PPHN compared to the odds of exposure given absence of PPHN. An OR greater than 1 suggests that the exposure may be a risk factor for the disease or condition under study. Epidemiologists also use the term **relative odds** as a synonym for odds ratio.

Surprisingly, the odds ratio of exposure comparing cases to controls is equivalent to the odds ratio of disease comparing exposed to unexposed.[22] With a specific example, it is easy to see how the fraction for the odds ratios are numerically equivalent:

$$OR_{\text{disease, exposed versus unexposed}} = \frac{\text{odds}_{\text{exposed}}}{\text{odds}_{\text{unexposed}}} = \frac{14/6}{323/830} = \frac{(14)(830)}{(6)(323)} = 6.00.$$

Despite the apparently restrictive nature of the case-control sampling design, the odds ratio of interest, the odds ratio for disease given exposure, can be estimated from case-control data.

Epidemiologists rely on one additional result, called the rare disease assumption. When a disease is rare, the odds ratio for the disease given exposure is approximately equal to the relative risk of the disease given exposure. These identities are the reason case-control studies are widely used in settings in which a disease is rare: it allows for the relative risk of disease given exposure to be estimated, even if the study design is based on sampling cases and controls then measuring exposure.

In a general 2×2 table of exposure versus disease status (Figure 8.21) the odds ratio for disease given exposure status is the ad/bc.

Disease Status	Present	Absent	Total
Exposed	a	b	$a+b$
Unexposed	c	d	$c+d$
Total	$a+c$	$b+d$	n

Figure 8.21: Exposure vs Disease Status.

In the PPHN case-control data, the odds ratio for PPHN given SSRI exposure status is $(14)(830)/(6)(323) = 6.00$. Because PPHN is a rare condition, the risk of PPHN among infants exposed to an SSRI is estimated to be approximately 6 times that of the risk among unexposed infants. Infants exposed to an SSRI are 600% more likely to suffer from PPHN.

It can be shown that the p-value used in a test of no association (between exposure and disease) is also the p-value for a test of the null hypothesis that the odds ratio is 1.

---

[22]This result can be shown through Bayes' rule.

8.6 Notes

Two-way tables are often used to summarize data from medical research studies, and entire texts have been written about methods of analysis for these tables. This chapter covers only the most basic of those methods.

Until recently, Fisher's exact test could only be calculated for 2×2 tables with small cell counts. Research has produced faster algorithms for enumerating tables and calculating p-values, and the computational power of recent desktop and laptop computers now make it possible to calculate the Fisher test on nearly any 2×2 table. There are also versions of the test that can be calculated on tables with more than 2 rows and/or columns. The practical result for data analysts is that the sample size condition for the validity of the χ^2 test can be made more restrictive. This chapter recommends using the χ^2 test only when cell counts in a 2×2 table are greater than 10; some approaches recommend cell counts larger than 10.

For many years, introductory textbooks recommended using a modified version of the χ^2 test, called the Fisher-Yates test, which adjusted the value of the statistic in small sample sizes to increase the accuracy of the χ^2 sampling distribution in calculating p-values. The Fisher-Yates version of the test is no longer used as often because of the widespread availability of the Fisher test.

The Fisher test is not without controversy, at least in the theoretical literature. Conditioning on the row and column totals allows the calculation of a p-value from the hypergeometric distribution, but in principle restricts inference to the set of tables with the same row and column values. In practice, this is less serious than it may seem. For tables of moderate size, the p-values from the χ^2 and Fisher tests are nearly identical and for tables with small counts, the Fisher test guarantees that the Type I error will be no larger than the specified value of α. In small sample sizes, some statisticians argue that the Fisher-Yates correction is preferable to the Fisher test because of the discrete nature of the hypergeometric distribution. In small tables, for example, an observed p-value of 0.04 may be the largest value that is less than 0.05, such that the Type I error of the test in that situation is 0.04, not 0.05.

Section 8.5.3 does not show the derivation that the odds ratio estimated from a case-control is the same as that from a cohort study. It is long and algebraically more complex than other derivations shown in the text, but it is a direct application of Bayes' rule, applied to each term in the fraction that defines population odds ratio.

The two labs for this chapter examine methods of inference for the success probability in binomial data then generalizes inference for binomial proportions to two-way contingency tables. Lab 2 also discusses measures of association in two-by-two tables. The datasets in the labs are similar to datasets that arise frequently in medical statistics. Lab 1 assesses the evidence for a treatment effect in a single uncontrolled trial of a new drug for melanoma and whether outcomes in stage 1 lung cancer are different among patients treated at Dana-Farber Cancer Institute compared to population based statistics. In Lab 2, students analyze a dataset from a published clinical trial examining the benefit of using a more expensive but potentially more effective drug to treat HIV-positive infants.

8.7 Exercises

8.7.1 Inference for a single proportion

8.1 Vegetarian college students. Suppose that 8% of college students are vegetarians. Determine if the following statements are true or false, and explain your reasoning.

(a) The distribution of the sample proportions of vegetarians in random samples of size 60 is approximately normal since $n \geq 30$.

(b) The distribution of the sample proportions of vegetarian college students in random samples of size 50 is right skewed.

(c) A random sample of 125 college students where 12% are vegetarians would be considered unusual.

(d) A random sample of 250 college students where 12% are vegetarians would be considered unusual.

(e) The standard error would be reduced by one-half if we increased the sample size from 125 to 250.

8.2 Young Americans, Part I. About 77% of young adults think they can achieve the American dream. Determine if the following statements are true or false, and explain your reasoning.[23]

(a) The distribution of sample proportions of young Americans who think they can achieve the American dream in samples of size 20 is left skewed.

(b) The distribution of sample proportions of young Americans who think they can achieve the American dream in random samples of size 40 is approximately normal since $n \geq 30$.

(c) A random sample of 60 young Americans where 85% think they can achieve the American dream would be considered unusual.

(d) A random sample of 120 young Americans where 85% think they can achieve the American dream would be considered unusual.

8.3 Gender equality. The General Social Survey asked a random sample of 1,390 Americans the following question: "On the whole, do you think it should or should not be the government's responsibility to promote equality between men and women?" 82% of the respondents said it "should be". At a 95% confidence level, this sample has 2% margin of error. Based on this information, determine if the following statements are true or false, and explain your reasoning.[24]

(a) We are 95% confident that between 80% and 84% of Americans in this sample think it's the government's responsibility to promote equality between men and women.

(b) We are 95% confident that between 80% and 84% of all Americans think it's the government's responsibility to promote equality between men and women.

(c) If we considered many random samples of 1,390 Americans, and we calculated 95% confidence intervals for each, 95% of these intervals would include the true population proportion of Americans who think it's the government's responsibility to promote equality between men and women.

(d) In order to decrease the margin of error to 1%, we would need to quadruple (multiply by 4) the sample size.

(e) Based on this confidence interval, there is sufficient evidence to conclude that a majority of Americans think it's the government's responsibility to promote equality between men and women.

---

[23]A. Vaughn. "Poll finds young adults optimistic, but not about money". In: *Los Angeles Times* (2011).
[24]National Opinion Research Center, General Social Survey, 2018.

8.4 Elderly drivers. The Marist Poll published a report stating that 66% of adults nationally think licensed drivers should be required to retake their road test once they reach 65 years of age. It was also reported that interviews were conducted on 1,018 American adults, and that the margin of error was 3% using a 95% confidence level.[25]

(a) Verify the margin of error reported by The Marist Poll.

(b) Based on a 95% confidence interval, does the poll provide convincing evidence that *more than* 70% of the population think that licensed drivers should be required to retake their road test once they turn 65?

8.5 Fireworks on July 4th. A local news outlet reported that 56% of 600 randomly sampled Kansas residents planned to set off fireworks on July 4th. Determine the margin of error for the 56% point estimate using a 95% confidence level.[26]

8.6 Life rating in Greece. Greece has faced a severe economic crisis since the end of 2009. A Gallup poll surveyed 1,000 randomly sampled Greeks in 2011 and found that 25% of them said they would rate their lives poorly enough to be considered "suffering".[27]

(a) Describe the population parameter of interest. What is the value of the point estimate of this parameter?

(b) Check if the conditions required for constructing a confidence interval based on these data are met.

(c) Construct a 95% confidence interval for the proportion of Greeks who are "suffering".

(d) Without doing any calculations, describe what would happen to the confidence interval if we decided to use a higher confidence level.

(e) Without doing any calculations, describe what would happen to the confidence interval if we used a larger sample.

8.7 Study abroad. A survey on 1,509 high school seniors who took the SAT and who completed an optional web survey shows that 55% of high school seniors are fairly certain that they will participate in a study abroad program in college.[28]

(a) Is this sample a representative sample from the population of all high school seniors in the US? Explain your reasoning.

(b) Let's suppose the conditions for inference are met. Even if your answer to part (a) indicated that this approach would not be reliable, this analysis may still be interesting to carry out (though not report). Construct a 90% confidence interval for the proportion of high school seniors (of those who took the SAT) who are fairly certain they will participate in a study abroad program in college, and interpret this interval in context.

(c) What does "90% confidence" mean?

(d) Based on this interval, would it be appropriate to claim that the majority of high school seniors are fairly certain that they will participate in a study abroad program in college?

---

[25]Marist Poll, Road Rules: Re-Testing Drivers at Age 65?, March 4, 2011.

[26]Survey USA, News Poll #19333, data collected on June 27, 2012.

[27]Gallup World, More Than One in 10 "Suffering" Worldwide, data collected throughout 2011.

[28]studentPOLL, College-Bound Students' Interests in Study Abroad and Other International Learning Activities, January 2008.

8.8 Legalization of marijuana, Part I. The General Social Survey asked 1,578 US residents: "Do you think the use of marijuana should be made legal, or not?" 61% of the respondents said it should be made legal.[29]

(a) Is 61% a sample statistic or a population parameter? Explain.

(b) Construct a 95% confidence interval for the proportion of US residents who think marijuana should be made legal, and interpret it in the context of the data.

(c) A critic points out that this 95% confidence interval is only accurate if the statistic follows a normal distribution, or if the normal model is a good approximation. Is this true for these data? Explain.

(d) A news piece on this survey's findings states, "Majority of Americans think marijuana should be legalized." Based on your confidence interval, is this news piece's statement justified?

8.9 National Health Plan, Part I. A *Kaiser Family Foundation* poll for US adults in 2019 found that 79% of Democrats, 55% of Independents, and 24% of Republicans supported a generic "National Health Plan". There were 347 Democrats, 298 Republicans, and 617 Independents surveyed.[30]

(a) A political pundit on TV claims that a majority of Independents support a National Health Plan. Do these data provide strong evidence to support this type of statement?

(b) Would you expect a confidence interval for the proportion of Independents who oppose the public option plan to include 0.5? Explain.

8.10 Legalize Marijuana, Part II. As discussed in Exercise 8.8, the General Social Survey reported a sample where about 61% of US residents thought marijuana should be made legal. If we wanted to limit the margin of error of a 95% confidence interval to 2%, about how many Americans would we need to survey?

8.11 National Health Plan, Part II. Exercise 8.9 presents the results of a poll evaluating support for a generic "National Health Plan" in the US in 2019, reporting that 55% of Independents are supportive. If we wanted to estimate this number to within 1% with 90% confidence, what would be an appropriate sample size?

8.12 Acetaminophen and liver damage. It is believed that large doses of acetaminophen (the active ingredient in over the counter pain relievers like Tylenol) may cause damage to the liver. A researcher wants to conduct a study to estimate the proportion of acetaminophen users who have liver damage. For participating in this study, he will pay each subject $20 and provide a free medical consultation if the patient has liver damage.

(a) If he wants to limit the margin of error of his 98% confidence interval to 2%, what is the minimum amount of money he needs to set aside to pay his subjects?

(b) The amount you calculated in part (a) is substantially over his budget so he decides to use fewer subjects. How will this affect the width of his confidence interval?

8.13 College smokers. We are interested in estimating the proportion of students at a university who smoke. Out of a random sample of 200 students from this university, 40 students smoke.

(a) Calculate a 95% confidence interval for the proportion of students at this university who smoke, and interpret this interval in context. (Reminder: Check conditions.)

(b) If we wanted the margin of error to be no larger than 2% at a 95% confidence level for the proportion of students who smoke, how big of a sample would we need?

---

[29]National Opinion Research Center, General Social Survey, 2018.

[30]Kaiser Family Foundation, The Public On Next Steps For The ACA And Proposals To Expand Coverage, data collected between Jan 9-14, 2019.

8.14 2010 Healthcare Law. On June 28, 2012 the U.S. Supreme Court upheld the much debated 2010 healthcare law, declaring it constitutional. A Gallup poll released the day after this decision indicates that 46% of 1,012 Americans agree with this decision. At a 95% confidence level, this sample has a 3% margin of error. Based on this information, determine if the following statements are true or false, and explain your reasoning.[31]

(a) We are 95% confident that between 43% and 49% of Americans in this sample support the decision of the U.S. Supreme Court on the 2010 healthcare law.

(b) We are 95% confident that between 43% and 49% of Americans support the decision of the U.S. Supreme Court on the 2010 healthcare law.

(c) If we considered many random samples of 1,012 Americans, and we calculated the sample proportions of those who support the decision of the U.S. Supreme Court, 95% of those sample proportions will be between 43% and 49%.

(d) The margin of error at a 90% confidence level would be higher than 3%.

8.15 Oral contraceptive use, Part I. In a study of 100 randomly sampled 18 year-old women in an inner city neighborhood, 15 reported that they were taking birth control pills.

(a) Can the normal approximation to the binomial distribution be used to calculate a confidence interval for proportion of women using birth control pills in this neighborhood? Explain your answer.

(b) Compute an approximate 95% confidence interval for the population proportion of women age 18 in this neighborhood taking birth control pills.

(c) Does the interval from part (b) support the claim that, for the young women in this neighborhood, the percentage who use birth control is not significantly different from the national average of 5%? Justify your answer.

8.16 Oral contraceptive use, Part II. Suppose that the study were repeated in a different inner city neighborhood and that out of 50 randomly sampled 18-year-old women, 6 reported that they were taking birth control pills. The researchers would like to assess the evidence that the proportion of 18-year-old women using birth control pills in this neighborhood is greater than the national average of 5%.

(a) Can the normal approximation to the binomial distribution be used to conduct a hypothesis test of the null hypothesis that the proportion of women using birth control pills in this neighborhood is equal to 0.05? Explain your answer.

(b) State the hypotheses for the analysis of interest and compute the p-value.

(c) Interpret the results from part (b) in the context of the data.

---

[31]Gallup, Americans Issue Split Decision on Healthcare Ruling, data collected June 28, 2012.

8.7.2 Inference for the difference of two proportions

8.17 Social experiment, Part I. A "social experiment" conducted by a TV program questioned what people do when they see a very obviously bruised woman getting picked on by her boyfriend. On two different occasions at the same restaurant, the same couple was depicted. In one scenario the woman was dressed "provocatively" and in the other scenario the woman was dressed "conservatively". The table below shows how many restaurant diners were present under each scenario, and whether or not they intervened.

		Scenario		
		Provocative	Conservative	Total
Intervene	Yes	5	15	20
	No	15	10	25
	Total	20	25	45

Explain why the sampling distribution of the difference between the proportions of interventions under provocative and conservative scenarios does not follow an approximately normal distribution.

8.18 Heart transplant success. The Stanford University Heart Transplant Study was conducted to determine whether an experimental heart transplant program increased lifespan. Each patient entering the program was officially designated a heart transplant candidate, meaning that he was gravely ill and might benefit from a new heart. Patients were randomly assigned into treatment and control groups. Patients in the treatment group received a transplant, and those in the control group did not. The table below displays how many patients survived and died in each group.[32]

	control	treatment
alive	4	24
dead	30	45

Suppose we are interested in estimating the difference in survival rate between the control and treatment groups using a confidence interval. Explain why we cannot construct such an interval using the normal approximation. What might go wrong if we constructed the confidence interval despite this problem?

8.19 National Health Plan, Part III. Exercise 8.9 presents the results of a poll evaluating support for a generically branded "National Health Plan" in the United States. 79% of 347 Democrats and 55% of 617 Independents support a National Health Plan.

(a) Calculate a 95% confidence interval for the difference between the proportion of Democrats and Independents who support a National Health Plan ($p_D - p_I$), and interpret it in this context. We have already checked conditions for you.

(b) True or false: If we had picked a random Democrat and a random Independent at the time of this poll, it is more likely that the Democrat would support the National Health Plan than the Independent.

8.20 Sleep deprivation, CA vs. OR, Part I. According to a report on sleep deprivation by the Centers for Disease Control and Prevention, the proportion of California residents who reported insufficient rest or sleep during each of the preceding 30 days is 8.0%, while this proportion is 8.8% for Oregon residents. These data are based on simple random samples of 11,545 California and 4,691 Oregon residents. Calculate a 95% confidence interval for the difference between the proportions of Californians and Oregonians who are sleep deprived and interpret it in context of the data.[33]

---

[32]B. Turnbull et al. "Survivorship of Heart Transplant Data". In: *Journal of the American Statistical Association* 69 (1974), pp. 74–80.

[33]CDC, Perceived Insufficient Rest or Sleep Among Adults — United States, 2008.

8.21 Remdesivir in COVID-19. Remdesivir is an antiviral drug previously tested in animal models infected with coronaviruses like SARS and MERS. As of May 2020, remdesivir has temporary approval from the FDA for use in severely ill COVID-10 patients. A randomized controlled trial conducted in China enrolled 236 patients with severe COVID-19; 158 were assigned to receive remdesivir and 78 to receive a placebo. In the remdesivir group, 103 patients showed clinical improvement; in the placebo group, 45 patients showed clinical improvement.

(a) Conduct a formal comparison of the clinical improvement rates and summarize your findings.

(b) Report and interpret an appropriate confidence interval.

8.22 Sleep deprivation, CA vs. OR, Part II. Exercise 8.20 provides data on sleep deprivation rates of Californians and Oregonians. The proportion of California residents who reported insufficient rest or sleep during each of the preceding 30 days is 8.0%, while this proportion is 8.8% for Oregon residents. These data are based on simple random samples of 11,545 California and 4,691 Oregon residents.

(a) Conduct a hypothesis test to determine if these data provide strong evidence the rate of sleep deprivation is different for the two states. (Reminder: Check conditions)

(b) It is possible the conclusion of the test in part (a) is incorrect. If this is the case, what type of error was made?

8.23 Gender and color preference. A study asked 1,924 male and 3,666 female undergraduate college students their favorite color. A 95% confidence interval for the difference between the proportions of males and females whose favorite color is black ($p_{male} - p_{female}$) was calculated to be (0.02, 0.06). Based on this information, determine if the following statements are true or false, and explain your reasoning for each statement you identify as false.[34]

(a) We are 95% confident that the true proportion of males whose favorite color is black is 2% lower to 6% higher than the true proportion of females whose favorite color is black.

(b) We are 95% confident that the true proportion of males whose favorite color is black is 2% to 6% higher than the true proportion of females whose favorite color is black.

(c) 95% of random samples will produce 95% confidence intervals that include the true difference between the population proportions of males and females whose favorite color is black.

(d) We can conclude that there is a significant difference between the proportions of males and females whose favorite color is black and that the difference between the two sample proportions is too large to plausibly be due to chance.

(e) The 95% confidence interval for ($p_{female} - p_{male}$) cannot be calculated with only the information given in this exercise.

---

[34]L Ellis and C Ficek. "Color preferences according to gender and sexual orientation". In: *Personality and Individual Differences* 31.8 (2001), pp. 1375–1379.

8.24 Prenatal vitamins and Autism. Researchers studying the link between prenatal vitamin use and autism surveyed the mothers of a random sample of children aged 24 - 60 months with autism and conducted another separate random sample for children with typical development. The table below shows the number of mothers in each group who did and did not use prenatal vitamins during the three months before pregnancy (periconceptional period).[35]

		Autism		
		Autism	Typical development	Total
Periconceptional	No vitamin	111	70	181
prenatal vitamin	Vitamin	143	159	302
	Total	254	229	483

(a) State appropriate hypotheses to test for independence of use of prenatal vitamins during the three months before pregnancy and autism.

(b) Complete the hypothesis test and state an appropriate conclusion. (Reminder: Verify any necessary conditions for the test.)

(c) A New York Times article reporting on this study was titled "Prenatal Vitamins May Ward Off Autism". Do you find the title of this article to be appropriate? Explain your answer. Additionally, propose an alternative title.[36]

8.25 Sleep deprived transportation workers. The National Sleep Foundation conducted a survey on the sleep habits of randomly sampled transportation workers and a control sample of non-transportation workers. The results of the survey are shown below.[37]

		Transportation Professionals		
	Control	Pilots	Truck Drivers	Train Operators	Bus/Taxi/Limo Drivers
Less than 6 hours of sleep	35	19	35	29	21
6 to 8 hours of sleep	193	132	117	119	131
More than 8 hours	64	51	51	32	58
Total	292	202	203	180	210

Conduct a hypothesis test to evaluate if these data provide evidence of a difference between the proportions of truck drivers and non-transportation workers (the control group) who get less than 6 hours of sleep per day, i.e. are considered sleep deprived.

8.26 An apple a day keeps the doctor away. A physical education teacher at a high school wanting to increase awareness on issues of nutrition and health asked her students at the beginning of the semester whether they believed the expression "an apple a day keeps the doctor away", and 40% of the students responded yes. Throughout the semester she started each class with a brief discussion of a study highlighting positive effects of eating more fruits and vegetables. She conducted the same apple-a-day survey at the end of the semester, and this time 60% of the students responded yes. Can she used a two-proportion method from this section for this analysis? Explain your reasoning.

---

[35]R.J. Schmidt et al. "Prenatal vitamins, one-carbon metabolism gene variants, and risk for autism". In: *Epidemiology* 22.4 (2011), p. 476.

[36]R.C. Rabin. "Patterns: Prenatal Vitamins May Ward Off Autism". In: *New York Times* (2011).

[37]National Sleep Foundation, 2012 Sleep in America Poll: Transportation Workers' Sleep, 2012.

8.7.3   Inference for two or more groups

8.27   True or false, Part I. Determine if the statements below are true or false. For each false statement, suggest an alternative wording to make it a true statement.

(a) The chi-square distribution, just like the normal distribution, has two parameters, mean and standard deviation.

(b) The chi-square distribution is always right skewed, regardless of the value of the degrees of freedom parameter.

(c) The chi-square statistic is always positive.

(d) As the degrees of freedom increases, the shape of the chi-square distribution becomes more skewed.

8.28   True or false, Part II. Determine if the statements below are true or false. For each false statement, suggest an alternative wording to make it a true statement.

(a) As the degrees of freedom increases, the mean of the chi-square distribution increases.

(b) If you found $\chi^2 = 10$ with $df = 5$ you would fail to reject H_0 at the 5% significance level.

(c) When finding the p-value of a chi-square test, we always shade the tail areas in both tails.

(d) As the degrees of freedom increases, the variability of the chi-square distribution decreases.

8.29   Quitters. Does being part of a support group affect the ability of people to quit smoking? A county health department enrolled 300 smokers in a randomized experiment. 150 participants were assigned to a group that used a nicotine patch and met weekly with a support group; the other 150 received the patch and did not meet with a support group. At the end of the study, 40 of the participants in the patch plus support group had quit smoking while only 30 smokers had quit in the other group.

(a) Create a two-way table presenting the results of this study.

(b) Answer each of the following questions under the null hypothesis that being part of a support group does not affect the ability of people to quit smoking, and indicate whether the expected values are higher or lower than the observed values.

  i. How many subjects in the "patch + support" group would you expect to quit?

  ii. How many subjects in the "patch only" group would you expect to not quit?

8.30   Parasitic worm. Lymphatic filariasis is a disease caused by a parasitic worm. Complications of the disease can lead to extreme swelling and other complications. Here we consider results from a randomized experiment that compared three different drug treatment options to clear people of the this parasite, which people are working to eliminate entirely. The results for the second year of the study are given below:[38]

	Clear at Year 2	Not Clear at Year 2
Three drugs	52	2
Two drugs	31	24
Two drugs annually	42	14

(a) Set up hypotheses for evaluating whether there is any difference in the performance of the treatments, and also check conditions.

(b) Statistical software was used to run a chi-square test, which output:

$$X^2 = 23.7 \qquad\qquad df = 2 \qquad\qquad \text{p-value} = 7.2\text{e-}6$$

Use these results to evaluate the hypotheses from part (a), and provide a conclusion in the context of the problem.

---

[38]Christopher King et al. "A Trial of a Triple-Drug Treatment for Lymphatic Filariasis". In: *New England Journal of Medicine* 379 (2018), pp. 1801–1810.

8.31 PREVEND, Part IV. In the PREVEND data, researchers measured various features of study participants, including data on statin use and highest level of education attained. A two-way table of education level and statin use is shown below.

	Primary	LowerSec	UpperSec	Univ	Sum
NonUser	31	111	107	136	385
User	20	46	27	22	115
Sum	51	157	134	158	500

(a) Set up hypotheses for evaluating whether there is an association between statin use and educational level.

(b) Check assumptions required for an analysis of these data.

(c) Statistical software was used to conduct a χ^2 test: the test statistic is 19.054, with p-value 0.0027. Summarize the conclusions in context of the data, and be sure to comment on the direction of association.

8.32 Diabetes and unemployment. A Gallup poll surveyed Americans about their employment status and whether or not they have diabetes. The survey results indicate that 1.5% of the 47,774 employed (full or part time) and 2.5% of the 5,855 unemployed 18-29 year olds have diabetes.[39]

(a) Create a two-way table presenting the results of this study.

(b) State appropriate hypotheses to test for difference in proportions of diabetes between employed and unemployed Americans.

(c) The sample difference is about 1%. If we completed the hypothesis test, we would find that the p-value is very small (about 0), meaning the difference is statistically significant. Use this result to explain the difference between statistically significant and practically significant findings.

8.33 TB Treatment. Tuberculosis (TB) is an infectious disease caused by the *Mycobacterium tuberculosis* bacteria. Active TB can be cured by adhering to a treatment regimen of several drugs for 6-9 months. A major barrier to eliminating TB worldwide is failure to adhere to treatment; this is known as defaulting from treatment. A study was conducted in Thailand to identify factors associated with default from treatment. The study results indicate that out of 54 diabetic participants, 0 defaulted from treatment; out of 1,180 non-diabetic participants, 54 defaulted from treatment. Participants were recruited at health centers upon diagnosis of TB.

(a) Create a two-way table presenting the results of this study.

(b) State appropriate hypotheses to test for difference in proportions of treatment default between diabetics and non-diabetics.

(c) Check assumptions. You may use a less stringent version of the success-failure condition: the expected number of successes per group should be greater than or equal to 5 (rather than 10).

(d) Formally test whether the proportion of patients who default from treatment differs between diabetics and non-diabetics. Summarize your findings.

---

[39] Gallup Wellbeing, Employed Americans in Better Health Than the Unemployed, data collected Jan. 2, 2011 - May 21, 2012.

8.34 Coffee and Depression. Researchers conducted a study investigating the relationship between caffeinated coffee consumption and risk of depression in women. They collected data on 50,739 women free of depression symptoms at the start of the study in the year 1996, and these women were followed through 2006. The researchers used questionnaires to collect data on caffeinated coffee consumption, asked each individual about physician- diagnosed depression, and also asked about the use of antidepressants. The table below shows the distribution of incidences of depression by amount of caffeinated coffee consumption.[40]

		Caffeinated coffee consumption					
		≤ 1 cup/week	2-6 cups/week	1 cup/day	2-3 cups/day	≥ 4 cups/day	Total
Clinical	Yes	670	373	905	564	95	2,607
depression	No	11,545	6,244	16,329	11,726	2,288	48,132
	Total	12,215	6,617	17,234	12,290	2,383	50,739

(a) What type of test is appropriate for evaluating if there is an association between coffee intake and depression?

(b) Write the hypotheses for the test you identified in part (a).

(c) Calculate the overall proportion of women who do and do not suffer from depression.

(d) Identify the expected count for the highlighted cell, and calculate the contribution of this cell to the test statistic.

(e) The test statistic is $\chi^2 = 20.93$. What is the p-value?

(f) What is the conclusion of the hypothesis test?

(g) One of the authors of this study was quoted on the NYTimes as saying it was "too early to recommend that women load up on extra coffee" based on just this study.[41] Do you agree with this statement? Explain your reasoning.

8.35 Mosquito nets and malaria. This problem examines a hypothetical prospective study about an important problem in the developing world: the use of mosquito nets to prevent malaria in children. The nets are typically used to protect children from mosquitoes while sleeping.

Suppose that in a large region of an African country, 100 households with one child are randomized to receive free mosquito nets for the child in the household and 100 households with one child are randomized to a control group where families do not receive the nets.

You are given the following information:

– In the 100 households receiving the nets, 22 children became infected with malaria.

– In the 100 households without the nets, 30 children became infected with malaria.

– The 200 families selected to participate in the study may be regarded as a random sample from the families in the region, so the 100 families in each group may be regarded as random samples from the population.

– Malaria among children is common in this region, with a prevalence of approximately 25%.

(a) Write down the 2 × 2 contingency table that corresponds to the data from the trial, labeling the table clearly and including the row and column totals.

(b) Under the hypothesis of no association between use of a mosquito net and malaria infection, calculate the expected number of infected children among 100 families who did receive a net.

(c) The χ^2 statistic for this 2 × 2 table is 1.66. Use this information to conduct a test of the null hypothesis of no effect of the use of a mosquito net on malaria infection in children.

(d) Compute and interpret the estimated relative risk of malaria infection, comparing the households without a net to those with a net.

---

[40]M. Lucas et al. "Coffee, caffeine, and risk of depression among women". In: *Archives of internal medicine* 171.17 (2011), p. 1571.

[41]A. O'Connor. "Coffee Drinking Linked to Less Depression in Women". In: *New York Times* (2011).

8.36 Health care fraud. Most errors in billing insurance providers for health care services involve honest mistakes by patients, physicians, or others involved in the health care system. However, fraud is a serious problem. The National Health Care Anti-Fraud Association estimates that approximately $68 billion is lost to health care fraud each year. Often when fraud is suspected, an audit of randomly selected billings is conducted. The selected claims are reviewed by experts and each claim is classified as allowed or not allowed. The claims not allowed are considered to be potentially fraudulent.

In general, the distribution of claims is highly skewed such that the majority of claims filed are small claims and only a few are large claims. Since simple random sampling would likely be overwhelmed by small claims, claims chosen for auditing are sampled in a stratified way: a set number of claims are sampled from each category of claim size: small, medium, and large. Here are data from an audit that used stratified sampling from three strata based on the claim size (i.e., monetary amount of the claim).

Stratum	Sampled Claims	Not Allowed
Small	100	10
Medium	50	17
Large	20	4

(a) Can these data be used to estimate the proportion of large claims for which fraud might be expected?

(b) Can these data be used to estimate the proportion of possibly fraudulent claims that are large claims?

(c) Construct a 2×3 contingency table of counts for these data and include the marginal totals, with the rows being the classification of claims and the columns being the size of the claim.

(d) Calculate the expected number of claims that would not be allowed among the large claims, under the hypothesis of no association of between size of claim and the claim not being allowed.

(e) Is the use of the chi-square statistic justified for these data?

(f) A chi-square test of no association between size of claim and whether it was allowed has value 12.93. How many degrees of freedom does the chi-square statistic have and what is the p-value for a test of no association?

(g) Compute the χ^2 residuals. Based on the residuals, interpret the findings in the context of the data.

8.37 Anxiety. Psychologists conducted an experiment to investigate the effect of anxiety on a person's desire to be alone or in the company of others (Schacter 1959; Lehmann 1975). A group of 30 individuals were randomly assigned into two groups; one group was designated the "high anxiety" group and the other the "low anxiety" group. Those in the high-anxiety group were told that in the "upcoming experiment", they would be subjected to painful electric shocks, while those in the low-anxiety group were told that the shocks would be mild and painless.[42] All individuals were informed that there would be a 10 minute wait before the experiment began, and that they could choose whether to wait alone or with other participants.

The following table summarizes the results:

	Wait Together	Wait Alone	Sum
High-Anxiety	12	5	17
Low-Anxiety	4	9	13
Sum	16	14	30

(a) Under the null hypothesis of no association, what are the expected cell counts?

(b) Under the assumption that the marginal totals are fixed and the null hypothesis is true, what is the probability of the observed set of results?

(c) Enumerate the tables that are more extreme than what was observed, in the same direction.

(d) Conduct a formal test of association for the results and summarize your findings. Let $\alpha = 0.05$.

---

[42]Individuals were not actually subjected to electric shocks of any kind

8.38 Salt intake and CVD. Suppose we are interested in investigating the relationship between high salt intake and death from cardiovascular disease (CVD). One possible study design is to identify a group of high- and low-salt users then follow them over time to compare the relative frequency of CVD death in the two groups. In contrast, a less expensive study design is to look at death records, identify CVD deaths from non-CVD deaths, collect information about the dietary habits of the deceased, then compare salt intake between individuals who died of CVD versus those who died of other causes. This design is called a retrospective design.

Suppose a retrospective study is done in a specific county of Massachusetts; data are collected on men ages 50-54 who died over a 1-month period. Of 35 men who died from CVD, 5 had a diet with high salt intake before they died, while of the 25 men who died from other causes, 2 had a diet with high salt intake. These data are summarized in the following table.

	CVD Death	Non-CVD Death	Total
High Salt Diet	5	2	7
Low Salt Diet	30	23	53
Total	35	25	60

(a) Under the null hypothesis of no association, what are the expected cell counts?

(b) Of the 35 CVD deaths, 5 were in the high salt diet group and 30 were in the low salt diet group. Under the assumption that the marginal totals are fixed, enumerate all possible sets of results (i.e., the table counts) that are more extreme than what was observed, in the same direction.

(c) Calculate the probability of observing each set of results from part (b).

(d) Evaluate the statistical significance of the observed data with a two-sided alternative. Let $\alpha = 0.05$. Summarize your results.

8.7.4 Chi-square tests for the fit of a distribution

8.39 Open source textbook. A professor using an open source introductory statistics book predicts that 60% of the students will purchase a hard copy of the book, 25% will print it out from the web, and 15% will read it online. At the end of the semester he asks his students to complete a survey where they indicate what format of the book they used. Of the 126 students, 71 said they bought a hard copy of the book, 30 said they printed it out from the web, and 25 said they read it online.

(a) State the hypotheses for testing if the professor's predictions were inaccurate.

(b) How many students did the professor expect to buy the book, print the book, and read the book exclusively online?

(c) This is an appropriate setting for a chi-square test. List the conditions required for a test and verify they are satisfied.

(d) Calculate the chi-squared statistic, the degrees of freedom associated with it, and the p-value.

(e) Based on the p-value calculated in part (d), what is the conclusion of the hypothesis test? Interpret your conclusion in this context.

8.40 Barking Deer. Microhabitat factors associated with foraging sites of barking deer in Hainan Island, China were examined. In this region, woods make up 4.8% of the land, cultivated grass plots make up 14.7%, and deciduous forests make up 39.6%. Of the 426 sites where the deer forage, 4 were categorized as woods, 16 as cultivated grass plots, and 61 as deciduous forests. The table below summarizes these data.[43]

woods	cultivated grassplot	deciduous forests	other	total
4	16	61	345	426

(a) Write the hypotheses for testing if barking deer prefer to forage in certain habitats over others.

(b) Check if the assumptions and conditions required for testing these hypotheses are reasonably met.

(c) Do these data provide convincing evidence that barking deer prefer to forage in certain habitats over others? Conduct an analysis and summarize your findings.

---

[43]Liwei Teng et al. "Forage and bed sites characteristics of Indian muntjac (Muntiacus muntjak) in Hainan Island, China". In: *Ecological Research* 19.6 (2004), pp. 675–681.

8.7.5   Outcome-based sampling: case-control studies

8.41   CVD and Diabetes. An investigator asked for the records of patients diagnosed with diabetes in his practice, then sampled 20 patients with cardiovascular disease (CVD) and 80 patients without CVD. For the sampled patients, he then recorded whether or not the age of onset of diabetes was at age 50 or younger. Of the 40 patients whose age of onset of diabetes was 50 years of age or earlier, 15 had cardiovascular disease. In the remaining 60 patients, 5 had cardiovascular disease.

(a) Write the contingency table that summarizes the result of this study.

(b) What is the relative odds of cardiovascular disease, comparing the older patients to those less than 50 years old at onset of diabetes?

(c) Interpret the relative odds of cardiovascular disease and comment on whether the relative odds cohere with what you might expect.

(d) In statistical terms, state the null hypothesis of no association between the presence of cardiovascular disease and age of onset of diabetes.

(e) What test can be used to test the null hypothesis? Are the assumptions for the test reasonably satisfied?

(f) The value of the chi-square test statistic for this table is 11. Identify the logical flaw in the following statement: "In this retrospective study of cardiovascular disease and diabetes, our study has demonstrated statistically significant evidence that diabetes increases the risk of cardiovascular disease."

8.42   Blood thinners. Cardiopulmonary resuscitation (CPR) is a procedure commonly used on individuals suffering a heart attack when other emergency resources are not available. This procedure is helpful in maintaining some blood circulation, but the chest compressions involved can also cause internal injuries. Internal bleeding and other injuries complicate additional treatment efforts following arrival at a hospital. For instance, while blood thinners may be used to help release a clot that is causing a heart attack, the blood thinner would have negative repercussions on any internal injuries.

This problem uses data from a study in which patients who underwent CPR for a heart attack and were subsequently admitted to a hospital. These patients were randomly divided into a treatment group where they received a blood thinner or the control group where they did not receive the blood thinner. The outcome variable of interest was whether the patients survived for at least 24 hours.

The study results are shown in the table below:

	Treatment	Control	Total
Survived	14	11	25
Died	26	39	65
Total	40	50	90

(a) For this table, calculate the odds ratio for survival, comparing treatment to control, and the relative risk of survival, comparing treatment to control.

(b) What is the interpretation of each of these two statistics?

(c) In this study, which of the two summary statistics in part (a) is the better description of the treatment effect? Why?

8.43 CNS disorder. Suppose an investigator has studied the possible association between the use of a weight loss drug and a rare central nervous system (CNS) disorder. He samples from a group of volunteers with and without the disorder, and records whether they have used the weight loss drug. The data are summarized in the following table:

	Drug Use	
CNS disorder	Yes	No
Yes	10	2000
No	7	4000

(a) Can these data be used to estimate the probability of a CNS disorder for someone taking the weight loss drug?

(b) For this study, what is an appropriate measure of association between the weight-loss drug and the presence of CNS disorder?

(c) Calculate the measure of association specified in part (b).

(d) Interpret the calculation from part (c).

(e) What test of significance is the best choice for analyzing the hypothesis of no association for these data?

8.44 Asthma risk. Asthma is a chronic lung disease characterized as hypersensitivity to a variety of stimuli, such as tobacco smoke, mold, and pollen. The prevalence of asthma has been increasing in recent decades, especially in children. Some studies suggest that children who either live in a farm environment or have pets become less likely to develop asthma later in life, due to early exposure to elevated amounts of microorganisms. A large study was conducted in Norway to investigate the association between early exposure to animals and subsequent risk for asthma.

Using data from national registers, researchers identified 11,585 children known to have asthma at age 6 years out of the 276,281 children born in Norway between January 1, 2006 and December 31, 2009. Children whose parents were registered as "animal producers and related workers" during the child's first year of life were defined as being exposed to farm animals. Of the 958 children exposed to farm animals, 19 had an asthma diagnosis at age 6.

(a) Do these data support previous findings that living in a farm environment is associated with lower risk of childhood asthma? Conduct a formal analysis and summarize your findings. Be sure to check any necessary assumptions.

(b) Is the relative risk an appropriate measure of association for these data? Briefly explain your answer.

(c) In language accessible to someone who has not taken a statistics course, explain whether these results represent evidence that exposure to farm animals reduces the risk of developing asthma. Limit your answer to no more than seven sentences.

8.45 Tea consumption and carcinoma. In a study examining the association between green tea consumption and esophageal carcinoma, researchers recruited 300 patients with carcinoma and 571 without carcinoma and administered a questionnaire about tea drinking habits. Out of the 47 individuals who reported that they regularly drink green tea, 17 had carcinoma. Out of the 824 individuals who reported they never drink green tea, 283 had carcinoma.

(a) Analyze the data to assess evidence for an association between green tea consumption and esophageal carcinoma from these data. Summarize your results.

(b) Report and interpret an appropriate measure of association.

Appendix A

End of chapter exercise solutions

1 Introduction to data

1.1 (a) Treatment: $10/43 = 0.23 \to 23\%$.

(b) Control: $2/46 = 0.04 \to 4\%$. (c) A higher percentage of patients in the treatment group were pain free 24 hours after receiving acupuncture. (d) It is possible that the observed difference between the two group percentages is due to chance.

1.3 (a) "Is there an association between air pollution exposure and preterm births?" (b) 143,196 births in Southern California between 1989 and 1993. (c) Measurements of carbon monoxide, nitrogen dioxide, ozone, and particulate matter less than $10\mu g/m^3$ (PM_{10}) collected at air-quality-monitoring stations as well as length of gestation. Continuous numerical variables.

1.5 (a) "Does explicitly telling children not to cheat affect their likelihood to cheat?". (b) 160 children between the ages of 5 and 15. (c) Four variables: (1) age (numerical, continuous), (2) sex (categorical), (3) whether they were an only child or not (categorical), (4) whether they cheated or not (categorical).

1.7 (a) Control: the group of 16 female birds that received no treatment. Treatment: the group of 16 female birds that were given supplementary diets.

(b) "Does egg coloration indicate the health of female collared flycatchers?"

(c) Darkness of blue color in female birds' eggs. Continuous numerical variable.

1.9 (a) Each row represents a participant.

(b) The response variable is colon cancer stage. The explanatory variables are the abundance levels of the five bacterial species.

(c) Colon cancer stage: ordinal categorical variable. Abundance levels of bacterial species: continuous numerical variable.

1.11 (a) The population of interest consists of babies born in Southern California. The sample consists of the 143,196 babies born between 1989 and 1993 in Southern California.

(b) Assuming that the sample is representative of the population of interest, the results of the study can be generalized to the population. The findings cannot be used to establish causal relationships because the study was an observational study, not an experiment.

1.13 (a) The population of interest consists of asthma patients who rely on medication for asthma treatment. The sample consists of the 600 asthma patients ages 18-69 who participated in the study.

(b) The sample may not be representative of the population because study participants were recruited, an example of a convenience sample. Thus, the results of the study may not be generalizable to the population. The findings can be used to establish causal relationships because the study is an experiment conducted with control, randomization, and a reasonably large sample size.

1.15 (a) Experiment.

(b) The experimental group consists of the chicks that received vitamin supplements. The control group consists of the chicks that did not receive vitamin supplements.

(c) Randomization ensures that there are not systematic differences between the control and treatment groups. Even if chicks may vary in ways that affect body mass and corticosterone levels, random allocation essentially evens out such differences, on average, between the two groups. This is essential for a causal interpretation of the results to be valid.

1.17 (a) Observational study.

(b) Answers may vary. One possible confounding variable is the wealth of a country. A wealthy country's citizens tend to have a higher life expectancy due to a higher quality of life, and the country tends to have a higher percentage of internet users because there is enough money for the required infrastructure and citizens can afford computers. Wealth of a country is associated with both estimated life expectancy and percentage of internet users. Omitting the confounder from the analysis distorts the relationship between the two variables, such that there may seem to be a direct relationship when there is not.

1.19 (a) Simple random sampling is reasonable if 500 students is a large enough sample size relative to the total student population of the university.

(b) Since student habits may vary by field of study, stratifying by field of study would be a reasonable decision.

(c) Students in the same class year may have more similar habits. Since clusters should be diverse with respect to the outcome of interest, this would not be a good approach.

1.21 (a) Non-responders may have a different response to this question, e.g. parents who returned the surveys likely don't have difficulty spending time with their children.

(b) It is unlikely that the women who were reached at the same address 3 years later are a random sample. These missing responders are probably renters (as opposed to homeowners) which means that they might be in a lower socio-economic class than the respondents.

(c) This is an observational study, not an experiment, so it is not advisable to draw conclusions about causal relationships. The relationship may be in the other direction; i.e., that these people go running precisely because they do not have joint problems. Additionally, the data are not even sufficient to provide evidence of an association between running and joint problems because data have only been collected from individuals who go running regularly. Instead, a sample of individuals should be collected that includes both people who do and do not regularly go running; the number of individuals in each group with joint problems can then be compared for evidence of an association.

1.23 The lead author's statements are not accurate because he or she drew conclusions about causation (that increased alcohol sales taxes lower rates of sexually transmitted infections) from an observational study. In addition, although the study observed that there was a decline in gonorrhea rate, the lead author generalized the observation to all sexually transmitted infections.

1.25 (a) Randomized controlled experiment. (b) Explanatory: treatment group (categorical, with 3 levels). Response variable: Psychological well-being. (c) No, because the participants were volunteers. (d) Yes, because it was an experiment. (e) The statement should say "evidence" instead of "proof".

1.27 (a) The two distributions have the same median since they have the same middle number when ordered from least to greatest. Distribution 2 has a higher IQR because its first and third quartiles are farther apart than in Distribution 2.

(b) Distribution 2 has a higher median since it has a higher middle number when ordered from least to greatest. Distribution 2 has a higher IQR because its first and third quartiles are farther apart than in Distribution 1.

(c) Distribution 2 has a higher median since all values in this distribution are higher than in Distribution 1. The two distributions have the same IQR since the distance between the first and third quartiles in each distribution is the same.

(d) Distribution 2 has a higher median since most values in this distribution are higher than those in Distribution 1. Distribution 2 has a higher IQR because its first and third quartiles are farther apart than those of Distribution 1.

1.29 (a) The distribution is bimodal, with peaks between 15-20 and 25-30. Values range from 0 to 65.

(b) The median AQI is about 30.

(c) I would expect the mean to be higher than the median, since there is some right skewing.

1.31 (a) The median is a much better measure of the typical amount earned by these 42 people. The mean is much higher than the income of 40 of the 42 people. This is because the mean is an arithmetic average and gets affected by the two extreme observations. The median does not get effected as much since it is robust to outliers. (b) The IQR is a much better measure of variability in the amounts earned by nearly all of the 42 people. The standard deviation gets affected greatly by the two high salaries, but the IQR is robust to these extreme observations.

1.33 (a) These data are categorical. They can be summarized numerically in either a frequency table or relative frequency table, and summarized graphically in a bar plot of either counts or proportions.

(b) The results of these studies cannot be generalized to the larger population. Individuals taking the survey represent a specific subset of the population that are conscious about dental health, since they are at the dentist's office for an appointment. Additionally, there may be response bias; even though the surveys are anonymous, it is likely that respondents will feel some pressure to give a "correct" answer in such a setting, and claim to floss more often than they actually do.

1.35 (a) Yes, there seems to be a positive association between lifespan and length of gestation. Generally, as gestation increases, so does life span.

(b) Positive association. Reversal of the plot axes does not change the nature of an association.

1.37 (a) 75% of the countries have an adolescent fertility rate less than or equal to 75.73 births per 1,000 adolescents.

(b) It is likely that the observations are missing due to the Iraq War and general instability in the region during this time period. It is unlikely that the five-number summary would have been affected very much, even if the values were extreme; the median and IQR are robust estimates, and the dataset is relatively large, with data from 188 other countries.

(c) The median and IQR decreases each year, with Q1 and Q3 also decreasing.

1.39 (a) $4{,}371/8{,}474 = 0.56 \rightarrow 56\%$

(b) $110/190 = 0.58 \rightarrow 58\%$

(c) $27/633 = 0.04 \rightarrow 4\%$

(d) $53/3{,}110 = 0.02 \rightarrow 2\%$

(e) Relative risk: $\frac{27/633}{53/3{,}110} = 2.50$. Yes, since the relative risk is greater than 1. A relative risk of 2.50 indicates that individuals with high trait anger are 2.5 times more likely to experience a CHD event than individuals with low trait anger.

(f) Side-by-side boxplots, since blood cholesterol level is a numerical variable and anger group is categorical.

2 Probability

2.1 (a) False. These are independent trials.

(b) False. There are red face cards.

(c) True. A card cannot be both a face card and an ace.

2.3 (a) $\frac{1}{4}$.

Solution 1: A colorblind male has genotype X^-Y. He must have inherited X^- from his mother (probability of $\frac{1}{2}$) and Y from his father (probability of $\frac{1}{2}$). Since these are two independent events, the probability of both occuring is $(\frac{1}{2})(\frac{1}{2}) = \frac{1}{4}$.

Solution 2: Determine the possibilities using a Punnett square. There are 4 equally likely possibilities, one of which is a colorblind male. Thus, the probability is $\frac{1}{4}$.

	X^+	Y
X^+	X^+X^+	X^+Y
X^-	X^+X^-	X^-Y

(b) True. An offspring of this couple cannot be both female and colorblind.

2.5 (a) 0.25. Let H represent the event of being a high school graduate and F represent the event of being a woman. $P(H) = P(H \text{ and } W) + P(H \text{ and } W^C) = P(H|W)P(W) + P(H|W^C)P(W^C) = (0.20)(0.50) + (0.30)(0.50) = 0.25$.

(b) 0.91. $(A^C) = P(A^C \text{ and } W) + P(A^C \text{ and } W^C) = (1 - 0.09) + (1 - 0.09) = 0.91$.

(c) 0.25. Let X represent the event of having at least a Bachelor's degree, where B represents the event of attaining at most a Bachelor's degree and G the event of attaining at most a graduate or professional degree. $P(X|W^C) = P(B|W^C) + P(G|W^C) = 0.16 + 0.09 = 0.25$.

(d) 0.26. $P(X|W) = P(B|W) + P(G|W) = 0.17 + 0.09 = 0.26$.

(e) 0.065. Let X_W be the event that a woman has at least a Bachelor's degree, and X_M be the event that a man has at least a Bachelor's degree. Assuming that the education levels of the husband and wife are independent, $P(X_W \text{ and } X_M) = P(X_W) \times P(X_M) = (0.25)(0.26) = 0.065$. This assumption is probably not reasonable, because people tend to marry someone with a comparable level of education.

2.7 (a) Let C represent the event that one urgent care center sees 300-449 patients in a week. Assuming that the number of patient visits are independent between urgent care centers in a given county for a given week, the probability that three random urgent care centers see 300-449 patients in a week is $[P(C)]^3 = (0.288)^3 = 0.024$. This assumption is not reasonable because a county is a small area with relatively few urgent care centers; if one urgent care center takes in more patients than usual during a given week, so might other urgent care centers in the same county (e.g., this could occur during flu season).

(b) 2.32×10^{-7}. Let D represent the event that one urgent care center sees 450 or more patients in a week. Assuming independence, the probability that 10 urgent care centers throughout a state all see 450 or more patients in a week is $[P(D)]^{10} = (0.217)^{10} = 2.32 \times 10^{-7}$. This assumption is reasonable because a state is a large area that contains many urgent care centers; the number of patients one urgent care center takes in is likely independent of the number of patients another urgent care center in the state takes in.

(c) No, it is not possible, because it is not reasonable to assume that the patient visits for a given week are independent of those for the following week.

2.9 (a) If the class is not graded on a curve, they are independent. If graded on a curve, then neither independent nor disjoint – unless the instructor will only give one A, which is a situation we will ignore in parts (b) and (c). (b) They are probably not independent: if you study together, your study habits would be related, which suggests your course performances are also related. (c) No. See the answer to part (a) when the course is not graded on a curve. More generally: if two things are unrelated (independent), then one occurring does not preclude the other from occurring.

2.11 (a) $0.60 + 0.20 - 0.18 = 0.62$

(b) $0.18/0.20 = 0.90$

(c) $0.11/0.33 = 0.33$

(d) No, because the answers to parts (c) and (d) are not equal. If global warming belief were independent of political party, then among liberal Democrats and conservative Republicans, there would be equal proportions of people who believe the earth is warming.

(e) $0.06/0.34 = 0.18$

2.13 (a) $375,264/436,968 = 0.859$

(b) $229,246/255,980 = 0.896$

(c) 0.896. This is equivalent to (b).

(d) $146,018/180,988 = 0.807$

(e) $4,719/7,394 = 0.638$

(f) No, because the answers to (c) and (d) are not equal. If gender and seat belt usage were independent, then among males and females, there would be the same proportion of people who always wear seat belts.

2.15 The PPV is 0.8248. The NPV is 0.9728.

$$P(D|T^+) = \frac{P(T^+|D)P(D)}{P(T^+|D)P(D)+P(T^+|D^C)P(D^C)} = \frac{(0.997)(0.0259)}{(0.997)(0.0259)+(1-0.926)(1-0.259)} = 0.8248.$$

$$P(D^C|T^-) = \frac{P(T^-|D^C)P(D^C)}{P(T^-|D^C)P(D^C)+P(T^-|D)P(D)} = \frac{(0.926)(1-0.259)}{(0.926)(1-0.259)+(1-0.997)(0.259)} = 0.9728.$$

2.17 0.0714. Even when a patient tests positive for lupus, there is only a 7.14% chance that he actually has

lupus. House may be right.

2.19 (a) Let E represent the event of agreeing with the idea of evolution and D be the event of being a Democrat. From the problem statement, $P(E|D) = 0.67$. $P(E^C|D) = 1 - P(E|D) = 1 - 0.67 = 0.33$.

(b) Let I represent the event of being an independent. $P(E|I) = 0.65$, as stated in the problem.

(c) Let R represent the event of being a Republican. $P(E|R) = 1 - P(E^C|R) = 1 - 0.48 = 0.52$.

(d) 0.35. $P(R|E) = \frac{P(E \text{ and } R)}{P(E)} = \frac{P(R)P(E|R)}{P(E)} = \frac{(0.40)(0.52)}{0.60} = 0.35$.

2.21 Mumps is the most likely disease state, since $P(B_3|A) = 0.563$, $P(B_1|A) = 0.023$, and $P(B_2|A) = .415$.
$P(B_i|A) = \frac{P(A|B_i)P(B_i)}{P(A)}$. $P(A) = P(A \text{ and } B_1) + P(A \text{ and } B_2) + P(A \text{ and } B_3) = P(A|B_1)P(B_1) + P(A|B_2)P(B_2) + P(A|B_3)P(B_3)$.

2.23 (a) Let A be the event of knowing the answer and B be the event of answering it correctly. Assume that if a participant knows the correct answer, they answer correctly with probability 1: $P(B|A) = 1$. If they guess randomly, they have 1 out of m chances to answer correctly, thus $P(B|A^C) = 1/m$. $P(A|B) = \frac{1 \cdot p}{(1 \cdot p) + (\frac{1}{m} \cdot (1-p))} = \frac{p}{p + \frac{1-p}{m}}$.

(b) 0.524. Let A be the event of having an IQ over 150 and B be the event of receiving a score indicating an IQ over 150. From the problem statement, $P(B|A) = 1$ and $P(B|A^C) = 0.001$. $P(A^C|B) = \frac{0.001 \cdot (1 - \frac{1}{1,100})}{(1 \cdot (\frac{1}{1,100})) + (0.001 \cdot (1 - \frac{1}{1,100}))} = 0.524$.

2.25 (a) In descending order on the table, the PPV for each age group is 0.003, 0.064, 0.175, 0.270; the NPV for each age group is 0.999, 0.983, 0.948, 0.914.

(b) As prevalence of prostate cancer increases by age group, PPV also increases. However, with rising prevalence, NPV decreases.

(c) The probability that a man has prostate cancer, given a positive test, necessarily increases as the overall probability of having prostate cancer increases. If more men have the disease, the chance of a positive test result being a true positive increases (and the chances of the result being a false positive decreases). The decreasing NPV values follow similar logic: if more men have the disease, the chance of a negative test being a true negative decreases (and the chances of the result being a false negative increases).

(d) Lowering the cutoff for a positive test would result in more men testing positive, since men with PSA values 2.5 ng/ml to 4.1 ng/ml were not previously classified as testing positive. Since the sensitivity of a test is the proportion who test positive among those who have disease, and the number with disease does not change, the proportion will increase, except in the rare and unlikely situation where the additional positive tests are among only men without the disease.

2.27 (a) Frequency of X^+X^+: 0.863. Frequency of X^+X^-: 0.132. Frequency of X^-X^-: 0.005. Frequency of X^-Y: 0.07. Frequency of X^+Y: 0.93. From frequency of X^-X^-, frequency of X^- allele is $\sqrt{0.005} = 0.071$; thus, frequency of X^+ allele is $1 - 0.071 = 0.929$. Frequency of X^+Y is $1 - 0.093 = 0.07$.

(b) 0.033. Let A be the event that two parents are not colorblind, and B represent the event of having a colorblind child. On the tree, $\times$ represents a mating between two genotypes. $P(B|A) = [P(X^+X^+ \times X^+Y|A) \cdot P(B|X^+X^+ \times X^+Y)] + [P(X^+X^- \times X^+Y|A) \cdot P(B|X^+X^- \times X^+Y)] = (0.867)(0) + (0.133)(1/4) = 0.033$.

$$
\begin{array}{c}
A \\
\diagup \quad \diagdown \\
X^+X^+ \times X^+Y \qquad X^+X^- \times X^+Y \\
\mid \qquad\qquad\qquad \mid \\
B \quad B^C \qquad\qquad B \quad B^C
\end{array}
$$

2.29 (a) Calculate $P(M \cap B)$, the probability a dog has a facial mask and a black coat. Note that the event M consists of having either a unilateral mask or a bilateral mask.

$$
\begin{aligned}
P(M \cap B) &= P(M_1 \cap B) + P(M_2 \cap B) \\
&= P(M_1|B)P(B) + P(M_2|B)P(B) \\
&= (0.25)(0.40) + (0.35)(0.40) \\
&= 0.24
\end{aligned}
$$

The probability an Australian cattle dog has a facial mask and a black coat is 0.31.

(b) Calculate $P(M_2)$, the prevalence of bilateral masks. The event of having a bilateral mask can be partitioned into either having a bilateral mask and a red coat or having a bilateral mask and a black coat.

$$
\begin{aligned}
P(M_2) &= P(R \cap M_2) + P(B \cap M_2) \\
&= P(M_2|R)P(R) + P(M_2|B)P(B) \\
&= (0.10)(0.60) + (0.35)(0.40) \\
&= 0.20
\end{aligned}
$$

The prevalence of bilateral masks in Australian cattle dogs is 0.20.

(c) Calculate $P(R|M_2)$, the probability of having a red coat given having a bilateral mask. Apply the definition of conditional probability.

$$P(R|M_2) = \frac{P(R \cap M_2)}{P(M_2)}$$
$$= \frac{P(M_2|R)P(R)}{P(M_2)}$$
$$= \frac{(0.10)(0.60)}{0.20}$$
$$= 0.30$$

The probability of being a Red Heeler among Australian cattle dogs with bilateral facial masks is 0.30.
(d) The following new information has been introduced:

- $P(D_1|R, M_0) = P(D_1|R, M_1) = 0.15$, $P(D^C|R, M_0) = P(D^C|R, M_1) = 0.60$.
- $P(M_2 \cap D_2 \cap R) = 0.012$, $P(M_2 \cap D_1 \cap R) = 0.045$
- $P(M_2 \cap D_2 \cap B) = 0.012$, $P(M_2 \cap D_1 \cap B) = 0.045$
- $P(D_1|M_0, B) = P(D_1|M_1, B) = 0.05$, $P(D_2|M_0, B) = P(D_2|M_1, B) = 0.01$

i. Calculate $P(M_2 \cap D^C \cap R)$.

$$P(M_2 \cap D^C \cap R) = P(D^C|M_2, R)P(M_2|R)P(R)$$
$$= P(D^C|M_2, R)(0.10)(0.60)$$

To calculate $P(D^C|M_2, R)$, first calculate $P(D_1|M_2, R)$ and $P(D_2|M_2, R)$ from the joint probabilities given in the problem, then apply the complement rule.

$$P(D_1|M_2, R) = \frac{P(M_2 \cap D_1 \cap R)}{P(M_2 \cap R)} = \frac{0.045}{(0.10)(0.60)} = 0.75$$

$$P(D_2|M_2, R) = \frac{P(M_2 \cap D_2 \cap R)}{P(M_2 \cap R)} = \frac{0.012}{(0.10)(0.60)} = 0.20$$

Back to the original question...

$$P(M_2 \cap D^C \cap R) = P(D^C|M_2, R)P(M_2|R)P(R)$$
$$= P(D^C|M_2, R)(0.10)(0.60)$$
$$= [1 - (0.75 + 0.20)](0.10)(0.60)$$
$$= (0.05)(0.10)(0.60)$$
$$= 0.003$$

The probability that an Australian cattle dog has a bilateral mask, no hearing deficits, and a red coat is 0.003.

ii. Calculate $P(D^C|M_2, B)$.

$$P(D^C|M_2, B) = 1 - [P(D_1|M_2, B) + P(D_2|M_2, B)]$$
$$= 1 - \left[\frac{P(D_1 \cap M_2 \cap B)}{P(M_2 \cap B)} + \frac{P(D_2 \cap M_2 \cap B)}{P(M_2 \cap B)} \right]$$
$$= 1 - \left[\frac{0.045}{P(M_2|B)P(B)} + \frac{0.012}{P(M_2|B)P(B)} \right]$$
$$= 1 - \left[\frac{0.045}{(0.35)(0.40)} + \frac{0.012}{(0.35)(0.40)} \right]$$
$$= 0.593$$

The proportion of bilaterally masked Blue Heelers without hearing deficits is 0.593.

iii. Calculate $P(D|R)$ and $P(D|B)$.

$$
\begin{aligned}
P(D|R) &= P(D \cap M_0|R) + P(D \cap M_1|R) + P(D \cap M_2|R) \\
&= [1 - P(D^C|R, M_0)](P(M_0|R) + [1 - P(D^C|R, M_1)](P(M_1|R) + [1 - P(D^C|M_2, R)](P(M_2|R) \\
&= (1 - 0.60)(0.50) + (1 - 0.60)(0.40) + (1 - 0.05)(0.10) \\
&= 0.455
\end{aligned}
$$

$$
\begin{aligned}
P(D|B) &= P(D \cap M_0|B) + P(D \cap M_1|B) + P(D \cap M_2|B) \\
&= [P(D_1|B, M_0) + P(D_2|B, M_0)](P(M_0|B) + [P(D_1|B, M_0) + P(D_2|B, M_0)](P(M_1|B) \\
&\quad + [1 - P(D^C|M_2, B)](P(M_2|B) \\
&= (0.05 + 0.01)(0.40) + (0.05 + 0.01)(0.25) + (1 - 0.593)(0.35) \\
&= 0.181
\end{aligned}
$$

The prevalence of deafness among Red Heelers is higher, at 0.455 versus 0.181 in Blue Heelers.
iv. Calculate $P(B|D^C)$.

$$
\begin{aligned}
P(B|D^C) &= \frac{P(B \cap D^C)}{P(D^C)} \\
&= \frac{P(D^C|B)P(B)}{P(D^C \cap B) + P(D^C \cap R)} \\
&= \frac{[1 - P(D|B)]P(B)}{[1 - P(D|B)]P(B) + [1 - P(D|R)]P(R)} \\
&= \frac{(1 - 0.181)(0.40)}{(1 - 0.181)(0.40) + (1 - 0.455)(0.60)} \\
&= 0.50
\end{aligned}
$$

The probability that a dog is a Blue Heeler given that it is known to have no hearing deficits is 0.50.

3 Distributions of random variables

3.1 (a) 13. (b) No, these 27 students are not a random sample from the university's student population. For example, it might be argued that the proportion of smokers among students who go to the gym at 9 am on a Saturday morning would be lower than the proportion of smokers in the university as a whole.

3.3 (a) The probability of drawing three hearts equals $(13/52)(12/51)(11/50) = 0.0129$, and the probability of drawing three black cards equals $(26/52)(25/51)(24/50) = 0.1176$; thus, the probability of any other draw is $1 - 0.0129 - 0.1176 = 0.8694$. $E(X) = 0.0129(50) + 0.1176(25) + 0.8694(0) = 3.589$. $Var(X) = 0.0129(50 - 3.589)^2 + 0.1176(25 - 3.589)^2 + 0.8694(0 - 3.589)^2 = 93.007$. $SD(X) = \sqrt{Var(X)} = 9.644$.
(b) Let Y represent the net profit/loss, where $Y = X - 5$. $E(Y) = E(X - 5) = E(X) - 5 = -1.412$. Standard deviation does not change from a shift of the distribution; $SD(Y) = SD(X) = 9.644$.
(c) It is not advantageous to play, since the expected winnings are lower than \$5.

3.5 (a) 215 eggs. Let X represent the number of eggs laid by one gull. $E(X) = 0.25(1) + 0.40(2) + 0.30(3) + 0.05(4) = 2.15$. $E(100X) = 100E(X) = 215$.
(b) 85.29 eggs. $Var(X) = 0.25(1 - 2.15)^2 + 0.40(2 - 2.15)^2 + 0.30(3 - 2.15)^2 + 0.05(4 - 2.15)^2 = 0.7275$. $Var(100X) = 100^2 Var(X) = 7275 \rightarrow \sqrt{7275} = 85.29$.

3.7 (a) Binomial conditions are met: (1) Independent trials: In a random sample across the US, it is reasonable to assume that whether or not one 18-20 year old has consumed alcohol does not depend on whether or not another one has. (2) Fixed number of trials: $n = 10$. (3) Only two outcomes at each trial: Consumed or did not consume alcohol. (4) Probability of a success is the same for each trial: $p = 0.697$.

(b) Let X be the number of 18-20 year olds who have consumed alcohol; $X \sim Bin(10, 0.697)$. $P(X = 6) = 0.203$.

(c) Let Y be the number of 18-20 year olds who have not consumed alcohol; $Y \sim Bin(10, 1 - 0.697)$. $P(Y = 4) = P(X = 6) = 0.203$.

(d) $X \sim Bin(5, 0.697)$. $P(X \leq 2) = 0.167$.

(e) $X \sim Bin(5, 0.697)$. $P(X \geq 1) = 1 - P(X = 0) = 0.997$.

3.9 (a) $\mu 34.85$, $\sigma = 3.25$ (b) $Z = \frac{45 - 34.85}{3.25} = 3.12$. 45 is more than 3 standard deviations away from the mean, we can assume that it is an unusual observation. Therefore yes, we would be surprised. (c) Using the normal approximation, 0.0009. With 0.5 correction, 0.0015.

3.11 (a) Both O+ and O- individuals can donate blood to a Type O+ patient; $n = 15$, $p = 0.45$. $\mu = np = 6.75$. $\sigma = \sqrt{np(1 - p)} = 1.93$.

(b) Only O- individuals can donate blood to a Type O- patient; $n = 15$, $p = 0.08$. $P(X \geq 3) = 0.113$.

3.13 0.132. Let X be the number of IV drug users who contract Hepatitis C within a month; $X \sim Bin(5, 0.30)$, $P(X = 3) = 0.132$.

3.15 (a) Let X represent the number of infected stocks in the sample; $X \sim Bin(250, 0.30)$. $P(X = 60) = 0.006$.

(b) $P(X \leq 60) = 0.021$.

(c) $P(X \geq 80) = 0.735$.

(D) 40% of 250 is 100. $P(X \leq 100) = 0.997$. Yes, this seems reasonable; it is essentially guaranteed that within a sample of 250, no more than 40% will be infected.

3.17 (a) $(200)(0.12) = 24$ cases of hyponatremia are expected during the marathon.

(b) Let X represent the number of cases of hyponatremia during the marathon. $P(X > 30) = 0.082$.

3.19 (a) 8.85%. (b) 6.94%. (c) 58.86%. (d) 4.56%.

3.21 (a) 0.005. (b) 0.911. (c) 0.954. (d) 1.036. (e) -0.842

3.23 (a) Verbal: $N(\mu = 151, \sigma = 7)$, Quant: $N(\mu = 153, \sigma = 7.67)$. $Z_{VR} = 1.29$, $Z_{QR} = 0.52$. She did better on the Verbal Reasoning section since her Z-score on that section was higher.

(b) $Perc_{VR} = 0.9007 \approx 90\%$, $Perc_{QR} = 0.6990 \approx 70\%$. $100\% - 90\% = 10\%$ did better than her on VR, and $100\% - 70\% = 30\%$ did better than her on QR.

(c) 159. (d) 147.

3.25 (a) 0.115. (b) The coldest 10% of days are colder than 70.59°F.

3.27 (a) 0.023. (b) 72.66 mg/dL.

3.29 (a) 82.4%. (b) About 38 years of age.

3.31 (a) $n = 50$, and $p = 0.70$. $\mu = np = 35$. $\sigma = \sqrt{np(1-p)} = 3.24$.
(b) Both np and $n(1-p)$ are greater than 10. Thus, it is valid to approximate the distribution as $X \sim N(35, 3.24)$, where X is the number of 18-20 year olds who have consumed alcohol. $P(X \geq 45) = 0.001$.

3.33 Let X represent the number of students who accept the offer; $X \sim \text{Bin}(2500, 0.70)$. This distribution can be approximated by a $N(1750, 22.91)$. The approximate probability that the school does not have enough dorm room spots equals $P(X \geq 1,786) = 0.06$.

3.35 The data appear to follow a normal distribution, since the points closely follow the line on the normal probability plot. There are some small deviations, but this is to be expected for such a small sample size.

3.37 (a) $P(X = 2) = \frac{\exp^{-2}(2^2)}{2!} = 0.271$. (b) $P(X \leq 2) = P(X = 0) + P(X = 1) + P(X = 2) = 0.677$. (c) $P(X \geq 3) = 1 - P(X \leq 2) = 0.323$.

3.39 (a) $\mu = \lambda = 75$, $\sigma = \sqrt{\lambda} = 8.66$. (b) $Z = -1.73$. Since 60 is within 2 standard deviations of the mean, it would not generally be considered unusual. Note that we often use this rule of thumb even when the normal model does not apply. (c) Using Poisson with $\lambda = 75$: 0.0402.

3.41 (a) The expected number of cases of osteosarcoma in NYC in a given year is 11.2. (b) Let X represent the number of osteosarcoma cases diagnosed. The probability that 15 or more cases will be diagnosed in a given year is the quantity $P(X \geq 15) = 1 - P(X < 15) = 1 - P(X \leq 14) = 0.161$. (c) First, calculate λ_B given that $n = 450,000$ for Brooklyn: 3.6. The probability of observing 10 or more cases in Brooklyn in a given year is the quantity $P(X_B \geq 10) = 1 - P(X_B < 10) = 1 - P(X_B \leq 9) = 0.004$. (d) No, he is not correct. The probability calculated in c) deals only with Brooklyn: the probability that there are 10 or more cases in Brooklyn for a single year. It does not say anything about cases in other boroughs. If we assume independence between boroughs, the probability that the official is referring to is:

$P(X = 0$ in other boroughs$) \times P(X \geq 10$ in Brooklyn$)$.

There is no reason to expect that $P(X = 0$ in other boroughs$)$ should equal 1, so this probability is different from the one in part c). (e) o, this probability is not equal to the probability calculated in part c). Over five years, there are five opportunities for the event of 10 or more cases in Brooklyn in a single year to occur. Let Y represent the event that in a single year, 10 or more cases of osteosarcoma are observed in Brooklyn. If we assume independence between years, then Y follows a binomial distribution with $n = 5$ and p of success as caculated in part c); $P(Y = 1) = 0.020$.

3.43 (a) λ for a population of 2,000,000 male births is 400. The probability of at most 380 newborn males with hemophilia is $P(X \leq 380)$, where $X \sim \text{Pois}(400)$: 0.165.
(b) $P(X \geq 450) = 0.0075$.
(c) The number of male births is $(1/2)(1,500,000) = 750,000$. The rate λ for one year is 150. Over 5 years, the rate λ is 750. The expected number of hemophilia births over 5 years is 750 and the standard deviation is $\sqrt{750} = 27.39$.

3.45 (a) On average, 2 women would need to be sampled in order to select a married woman ($\mu = 1/p = 2.123$), with standard deviation 1.544 ($\sigma = \sqrt{\frac{(1-p)}{p^2}}$).
(b) $\mu = 3.33$. $\sigma = 2.79$.
(c) Decreasing the probability increases both the mean and the standard deviation.

3.47 (a) Let X represent the number of stocks that must be sampled to find an infected stock; $X \sim \text{Geom}(0.30)$. $P(X \leq 5) = 0.832$.
(b) $P(X \leq 6) = 0.882$.
(c) $P(X \geq 3) = 1 - P(X \leq 2) = 0.49$.

3.49 (a) $0.875^2 \times 0.125 = 0.096$. (b) $\mu = 8$, $\sigma = 7.48$.

3.51 (a) 0.0804. (b) 0.0322. (c) 0.0193.

3.53 (a) 0.102, geometric with $p = 1994/14,604 = 0.137$.
(b) 0.854, binomial with $n = 10$, $p = 0.137$.
(c) 0.109, binomial with $n = 10$, $p = 0.137$.
(d) The mean and standard deviation of a negative binomial random variable with $r = 4$ and $p = 0.137$ are 29.30 and 13.61, respectively.

3.55 (a) $\mu = 2.05$; $\sigma^2 = 1.77$.
(b) Let X represent the number of soapy-taste detectors; $X \sim \text{HGeom}(1994, 14604 - 1994, 15)$. $P(X = 4) = 0.09435$.
(c) $P(X \le 2) = 0.663$.
(d) 0.09437, from the binomial distribution. With a large sample size, sampling with replacement is highly unlikely to result in any particular individual being sampled again. In this case, the hypergeometric and binomial distributions will produce equal probabilities.

3.57 (a) The marginal distributions for X is obtained by summing across the two rows, and for Y by summing the columns. The marginal probabilities for $X = 0$ and $X = 1$ are 0.60 and 0.40, and for $Y = -1$ and $Y = 1$ are both 0.50; i.e., $p_X(0) = 0.60$, $p_X(1) = 0.40$, $p_Y(-1) = p_Y(1) = 0.50$ (b) The mean and variance of X are calculated using the formulas in Section 3.1.2 and 3.1.3 and are

$$\mu_X = (0)(0.60) + (1)(0.40) = 0.40$$
$$\sigma_X^2 = (0 - 0.40)^2(0.60) + (1 - 0.40)^2(0.40) = 0.24$$

The standard deviation of X is $\sqrt{0.24} = 0.49$. (c) The two standardized values of X are obtained by subtracting the mean of X from each value and dividing by the standard deviation. The two standardized values are -0.82 and 1.23. (d) The correlation between X and Y adds the 4 products of the standardized values, weighted by the values in the joint distribution:

$$\rho_{X,Y} = (-0.82)(-1)(0.20) + (-0.83)(1)(0.40) + (1.23)(-1)(0.30) + (1.23)(1)(0.10) = -.41$$

(e) No. The correlation between X and Y is not zero.

3.59 (a) Sum over the margins to calculate the marginal distributions.

$$p_Y(-1) = 0.25 \quad p_Y(0) = 0.20 \quad p_Y(1) = 0.55$$

$$p_X(-1) = 0.45 \quad p_X(0) = 0.20 \quad p_X(1) = 0.35$$

(b) The expected value of X is calculated as follows:

$$E(X) = \sum_i x_i P(X = x_i) = (-1)(0.45) + (0)(0.20) + (1)(0.35) = -0.10$$

(c) The variance of Y is calculated by first calculating $E(Y)$, then using that in the formula for a variance of a random variable.
$$E(Y) = \sum_i y_i P(Y = y_i) = (-1)(0.25) + (0)(0.20) + (1)(0.55) = 0.30$$

$$\text{Var}(Y) = \sum_i (y_i - E(Y))^2 P(Y = y_i) = (-1 - 0.30)^2(0.25) + (0 - 0.30)^2(0.20) + (1 - 0.30)^2(0.55) = 0.71$$

(d) $P(X = -1 | Y = 0) = 0/0.20 = 0$; $P(X = 0 | Y = 0) = 0.10/0.20 = 0.50$; $P(X = 1 | Y = 0) = 0.10/0.20 = 0.5$.

3.61 (a) No. The new marginal distributions for the costs for the two members of the couple are shown in the following table. The values and the marginal distribution for the partner's cost do not change, so the expected value and standard deviation will not change. The previous values for the mean and standard deviation were $980 and $9.80.

	Partner Costs, Y		
Employee costs, X	$968	$988	Marg. Dist., X
$968	0.18	0.12	0.30
$1,008	0.15	0.25	0.40
$1,028	0.07	0.23	0.30
Marg. Dist., Y	0.40	0.60	1.00

(b) The expected value and standard deviation of the employee's costs are calculated as in Example 3.6, but using the new marginal distribution. The new values for the mean and standard deviation are $1,002 and $23.75. (c) The expected total cost is $1,002 + $980 = $1,982. (d) The calculation correlation depends on the standardized costs for each member of the couple and the joint probabilities. The new standardized values for the employee costs are -1.43, 0.25, and 1.09; the corresponding values for the partner are -1.22 and 0.82. The correlation is the weighted sum of the 6 products, weighted by the joint probabilities: $\rho_{X,Y} = 0.29$. (e) The new variance for the total cost will be $(23.80)^2 + (9.80)^2 + (2)(23.8)(9.80)(0.29) = 796.00$ The new standard deviation is $\sqrt{796.00} = \$28.21$.

4 Foundations for inference

4.1 (a) $\bar{x} = 0.6052$.

(b) $s = 0.0131$.

(c) $Z_{0.63} = \frac{0.63 - 0.6052}{0.0131} = 1.893$. No, this level of BGC is within 2 SD of the mean.

(d) The standard error of the sample mean is given by $\frac{s}{\sqrt{n}} = \frac{0.0131}{\sqrt{70}} = 0.00157$.

4.3 (a) This is the sampling distribution of the sample mean.

(b) The sampling distribution will be normal and symmetric, centered around the theoretical population mean μ of the number of eggs laid by this hen species during a breeding period.

(c) The variability of the distribution is the standard error of the sample mean: $\frac{s}{\sqrt{n}} = \frac{18.2}{\sqrt{45}} = 2.71$.

(d) The variability of the new distribution will be greater than the variability of the original distribution. Conceptually, a smaller sample is less informative, which leads to a more uncertain estimate. This can be shown concretely with a calculation: $\frac{18.2}{\sqrt{10}} = 5.76$ is larger than 2.71.

4.5 (a) We are 95% confident that the mean number of hours that U.S. residents have to relax or pursue activities that they enjoy is between 3.53 and 3.83 hours.

(b) A larger margin of error with the same sample occurs with a higher confidence level (i.e., larger critical value).

(c) The margin of error of the new 95% confidence interval will be smaller, since a larger sample size results in a smaller standard error. (d) A 90% confidence interval will be smaller than the original 95% interval, since the critical value is smaller and results in a smaller margin of error. The interval will provide a more precise estimate, but have an associated lower confidence of capturing μ.

4.7 (a) False. Provided the data distribution is not very strongly skewed ($n = 64$ in this sample, so we can be slightly lenient with the skew), the distribution of the sample mean will be nearly normal, allowing for the normal approximation.

(b) False. Inference is made on the population parameter, not the point estimate. The point estimate is always in the confidence interval.

(c) True.

(d) False. The confidence interval is not about a sample mean.

(e) False. A wider interval is required to be more confident about capturing the parameter.

(f) True. The margin of error is half the width of the interval, and the sample mean is the midpoint of the interval.

(g) False. To halve the margin of error requires sampling $2^2 = 4$ times the number of people in the initial sample.

4.9 (a) i. False. There is a 5% chance that any 95% confidence interval does not contain the true population mean days out of the past 30 days that U.S. adults experienced poor mental health. ii. False. The population parameter μ is either inside or outside the interval; there is no probability associated with whether the fixed value μ is in a certain calculated interval. The randomness is associated with the interval (and the method for calculating it), not the parameter μ. Thus, it would not be reasonable to say there is a 95% chance that the particular interval (3.40, 4.24) contains μ; this interpretation is coherent with the statement in part iii. of this question. iii. True. This is the definition of what it means to be 95% confident. iv. True. The interval corresponds to a two-sided test, with $H_0 : \mu = 4.5$ days and $H_A : \mu \neq 4.5$ days and $\alpha = 1 - 0.95 = 0.05$. Since μ_0 of 4.5 days is outside the interval, the sample provides sufficient evidence to reject the null hypothesis and accept the alternative hypothesis. v. False. We can only be confident that 95% of the time, the entire interval calculated contains μ. It is not possible to make this statement about $\bar{x}$ or any other point within the interval. vi. False. The confidence interval is a statement about the population parameter μ, the mean days out of the past 30 days that all US adults experienced poor mental health. The sample mean $\bar{x}$ is a known quantity.

(b) The 90% confidence interval will be smaller than the 95% confidence interval. If we are less confident that an interval contains μ, this implies that the interval is less wide; if we are more confident, the interval is wider. Think about a theoretical "100%" confidence interval—to be 100% confident of capturing μ, then the range must be all possible numbers that μ could be. (c) (3.47, 4.17) days

4.11 (a) The null hypothesis is that New Yorkers sleep an average of 8 hours of night ($H_0 : \mu = 8$ hours). The alternative hypothesis is that New Yorkers sleep less than 8 hours a night on average ($H_A : \mu < 8$ hours).

(b) The null hypothesis is employees spend on average 15 minutes on non-business activities in a day ($H_0 : \mu = 15$ minutes). The alternative hypothesis is that employees spend on average more than 15 minutes on non-business activities in a day ($H_A : \mu > 15$ minutes).

4.13 Hypotheses are always made about the population parameter μ, not the sample mean $\bar{x}$. The correct value of μ_0 is 10 hours, as based on the previous evidence; both hypotheses should include μ_0. The correct hypotheses are $H_0 : \mu = 10$ hours and $H_A : \mu > 10$ hours.

4.15 (a) This claim is not supported by the confidence interval. 3 hours corresponds to a time of 180 minutes; there is evidence that the average waiting time is lower than 3 hours.

(b) 2.2 hours corresponds to 132 minutes, which is within the interval. It is plausible that μ is 132 minutes, since we are 95% confident that the interval (128 minutes, 147 minutes) contains the average wait time.

(c) Yes, the claim would be supported based on a 99% interval, since the 99% interval is wider than the 95% interval.

4.17 $H_0 : \mu = 130$ grams, $H_A : \mu \neq 130$ grams. Test the hypothesis by calculating the test statistic: $t = \frac{\bar{x} - \mu_0}{s/\sqrt{n}} = \frac{130 - 134}{17/sqrt35} = 1.39$. This results in a p-value of 0.17. There is insufficient evidence to reject the null hypothesis. There is no evidence that the nutrition label does not provide an accurate measure of calories.

4.19 (a) The 95% confidence interval is $3,150 \pm (1.96 \times 250/\sqrt{50}) = (3080.7, 3219.3)$ grams.

(b) She will conduct a test of the null against the two-sided alternative $H_A : \mu \neq 3250$ grams. Calculate the test statistic: $t = \frac{\bar{x}-\mu_0}{s/\sqrt{n}} = \frac{3150-3250}{250/\sqrt{50}} = -2.83$. The p-value is 0.007. There is sufficient evidence to reject the null hypothesis and conclude that the mean birthweight of babies from inner-city teaching hospitals is lower than 3,260 grams.

4.21 (a) H_0: Anti-depressants do not help symptoms of fibromyalgia. H_A: Anti- depressants do treat symptoms of fibromyalgia. (b) Concluding that anti-depressants work for the treatment of fibromyalgia symptoms when they actually do not. (c) Concluding that anti-depressants do not work for the treatment of fibromyalgia symptoms when they actually do. (d) If she makes a Type 1 error, she will continue taking medication that does not actually treat her disorder. If she makes a Type 2 error, she will stop taking medication that could treat her disorder.

4.23 (a) The standard error is larger under scenario I; standard error is larger for smaller values of n.

(b) The margin of error is larger under scenario I; to be more confidence of capturing the population parameter requires a larger confidence interval.

(c) The p-value from a Z-statistic only depends on the value of the Z-statistic; the value is equal under the scenarios.

(d) The probability of making a Type II error and falsely rejecting the alternative is higher under scenario I; it is easier to reject the alternative with a high α.

5 Inference for numerical data

5.1 (a) $df = 6 - 1 = 5$, $t_5^\star = 2.02$ (column with two tails of 0.10, row with $df = 5$). (b) $df = 21 - 1 = 20$, $t_{20}^\star = 2.53$ (column with two tails of 0.02, row with $df = 20$). (c) $df = 28$, $t_{28}^\star = 2.05$. (d) $df = 11$, $t_{11}^\star = 3.11$.

5.3 On a z-distribution, the cutoff value for the upper 5% of values is 1.96. A t-distribution has wider tails than a normal distribution but approaches the shape of a standard normal as degrees of freedom increases. Thus, 1.98 corresponds to the cutoff for a t-distribution with 100 degrees of freedom, 2.01 the cutoff for 50 degrees of freedom, and 2.23 the cutoff for 10 degrees of freedom.

5.5 The mean is the midpoint: $\bar{x} = 20$. Identify the margin of error: $ME = 1.015$, then use $t_{35}^\star = 2.03$ and $SE = s/\sqrt{n}$ in the formula for margin of error to identify $s = 3$.

5.7 (a) H_0: $\mu = 8$ (New Yorkers sleep 8 hrs per night on average.) H_A: $\mu \neq 8$ (New Yorkers sleep less or more than 8 hrs per night on average.) (b) Independence: The sample is random. The min/max suggest there are no concerning outliers. $T = -1.75$. $df = 25 - 1 = 24$. (c) p-value = 0.093. If in fact the true population mean of the amount New Yorkers sleep per night was 8 hours, the probability of getting a random sample of 25 New Yorkers where the average amount of sleep is 7.73 hours per night or less (or 8.27 hours or more) is 0.093. (d) Since p-value > 0.05, do not reject H_0. The data do not provide strong evidence that New Yorkers sleep more or less than 8 hours per night on average. (e) No, since the p-value is smaller than $1 - 0.90 = 0.10$.

5.9 T is either -2.09 or 2.09. Then $\bar{x}$ is one of the following:

$$-2.09 = \frac{\bar{x}-60}{\frac{8}{\sqrt{20}}} \rightarrow \bar{x} = 56.26$$

$$2.09 = \frac{\bar{x}-60}{\frac{8}{\sqrt{20}}} \rightarrow \bar{x} = 63.74$$

5.11 (a) We will conduct a 1-sample t-test. H_0: $\mu = 5$. H_A: $\mu \neq 5$. We'll use $\alpha = 0.05$. This is a random sample, so the observations are independent. To proceed, we assume the distribution of years of piano lessons is approximately normal. $SE = 2.2/\sqrt{20} = 0.4919$. The test statistic is $T = (4.6 - 5)/SE = -0.81$. $df = 20 - 1 = 19$. The one-tail area is about 0.21, so the p-value is about 0.42, which is bigger than $\alpha = 0.05$ and we do not reject H_0. That is, we do not have sufficiently strong evidence to reject the notion that the average is 5 years. (b) Using $SE = 0.4919$ and $t^{\star}_{df=19} = 2.093$, the confidence interval is $(3.57, 5.63)$. We are 95% confident that the average number of years a child takes piano lessons in this city is 3.57 to 5.63 years. (c) They agree, since we did not reject the null hypothesis and the null value of 5 was in the t-interval.

5.13 If the sample is large, then the margin of error will be about $1.96 \times 100/\sqrt{n}$. We want this value to be less than 10, which leads to $n \geq 384.16$, meaning we need a sample size of at least 385 (round up for sample size calculations!).

5.15 (a) Since it's the same students at the beginning and the end of the semester, there is a pairing between the datasets; for a given student their beginning and end of semester grades are dependent. (b) Since the subjects were sampled randomly, each observation in the men's group does not have a special correspondence with exactly one observation in the other (women's) group. (c) Since it's the same subjects at the beginning and the end of the study, there is a pairing between the datasets; for a subject their beginning and end of semester artery thickness are dependent. (d) Since it's the same subjects at the beginning and the end of the study, there is a pairing between the datasets; for a subject their beginning and end of semester weights are dependent.

5.17 (a) For each observation in one data set, there is exactly one specially corresponding observation in the other data set for the same geographic location. The data are paired. (b) $H_0 : \mu_{\text{diff}} = 0$ (There is no difference in average number of days exceeding 90°F in 1948 and 2018 for NOAA stations.) $H_A : \mu_{\text{diff}} \neq 0$ (There is a difference.) (c) Locations were randomly sampled, so independence is reasonable. The sample size is at least 30, so we're just looking for particularly extreme outliers: none are present (the observation off left in the histogram would be considered a clear outlier, but not a particularly extreme one). Therefore, the conditions are satisfied. (d) $SE = 17.2/\sqrt{197} = 1.23$. $T = \frac{2.9-0}{1.23} = 2.36$ with degrees of freedom $df = 197 - 1 = 196$. This leads to a one-tail area of 0.0096 and a p-value of about 0.019. (e) Since the p-value is less than 0.05, we reject H_0. The data provide strong evidence that NOAA stations observed more 90°F days in 2018 than in 1948. (f) Type 1 Error, since we may have incorrectly rejected H_0. This error would mean that NOAA stations did not actually observe a decrease, but the sample we took just so happened to make it appear that this was the case. (g) No, since we rejected H_0, which had a null value of 0.

5.19 (a) $SE = 1.23$ and $t^{\star} = 1.65$. $2.9 \pm 1.65 \times 1.23 \rightarrow (0.87, 4.93)$.

(b) We are 90% confident that there was an increase of 0.87 to 4.93 in the average number of days that hit 90°F in 2018 relative to 1948 for NOAA stations.

(c) Yes, since the interval lies entirely above 0.

5.21 (a) Each of the 36 mothers is related to exactly one of the 36 fathers (and vice-versa), so there is a special correspondence between the mothers and fathers. (b) $H_0 : \mu_{diff} = 0$. $H_A : \mu_{diff} \neq 0$. Independence: random sample from less than 10% of population. Sample size of at least 30. The skew of the differences is, at worst, slight. $Z = 2.72 \rightarrow$ p-value $= 0.0066$. Since p-value < 0.05, reject H_0. The data provide strong evidence that the average IQ scores of mothers and fathers of gifted children are different, and the data indicate that mothers' scores are higher than fathers' scores for the parents of gifted children.

5.23 (a) Since $p < 0.05$, there is statistically significant evidence that the population difference in BGC is not 0. Since the observed mean BGC is higher in the food supplemented group, these data suggest that food supplemented birds have higher BGC on average than birds that are not food supplemented. (b) The 95% confidence interval is $\bar{d} \pm t^\star \frac{s_d}{\sqrt{n}}$. Since the mean of the differences is equal to the difference of the means, $\bar{d} = 1.70 - 0.586 = 1.114$. The test statistic is $t = \frac{\bar{d}}{s_d/\sqrt{n}}$, so the standard error ($s_d/\sqrt{n}$) can be solved for: $s_d/\sqrt{n} = \bar{d}/t = 1.114/2.64 = 0.422$. The critical t-value for a 95% confidence interval on a t-distribution with $16-1 = 15$ degrees of freedom is 2.13. Thus, the 95% confidence interval is $1.114 \pm (2.13 \times 0.422) \rightarrow (0.215, 2.01)$ grams. With 95% confidence, the interval (0.215, 2.01) grams contains the population mean difference in egg mass between food supplemented birds and non supplemented birds.

5.25 (a) These data are paired. For example, the Friday the 13th in say, September 1991, would probably be more similar to the Friday the 6th in September 1991 than to Friday the 6th in another month or year.
(b) Let $\mu_{diff} = \mu_{sixth} - \mu_{thirteenth}$. $H_0 : \mu_{diff} = 0$. $H_A : \mu_{diff} \neq 0$.
(c) Independence: The months selected are not random. However, if we think these dates are roughly equivalent to a simple random sample of all such Friday 6th/13th date pairs, then independence is reasonable. To proceed, we must make this strong assumption, though we should note this assumption in any reported results. Normality: With fewer than 10 observations, we would need to see clear outliers to be concerned. There is a borderline outlier on the right of the histogram of the differences, so we would want to report this in formal analysis results.
(d) $T = 4.93$ for $df = 10 - 1 = 9 \rightarrow$ p-value $= 0.001$.
(e) Since p-value < 0.05, reject H_0. The data provide strong evidence that the average number of cars at the intersection is higher on Friday the 6th than on Friday the 13th. (We should exercise caution about generalizing the interpretation to all intersections or roads.)
(f) If the average number of cars passing the intersection actually was the same on Friday the 6th and 13th, then the probability that we would observe a test statistic so far from zero is less than 0.01.
(g) We might have made a Type 1 Error, i.e. incorrectly rejected the null hypothesis.

5.27 (a) $H_0 : \mu_{diff} = 0$. $H_A : \mu_{diff} \neq 0$. $T = -2.71$. $df = 5$. p-value $= 0.042$. Since p-value < 0.05, reject H_0. The data provide strong evidence that the average number of traffic accident related emergency room admissions are different between Friday the 6th and Friday the 13th. Furthermore, the data indicate that the direction of that difference is that accidents are lower on Friday the 6th relative to Friday the 13th.
(b) (-6.49, -0.17).
(c) This is an observational study, not an experiment, so we cannot so easily infer a causal intervention implied by this statement. It is true that there is a difference. However, for example, this does not mean that a responsible adult going out on Friday the 13th has a higher chance of harm than on any other night.

5.29 (a) Chicken fed linseed weighed an average of 218.75 grams while those fed horsebean weighed an average of 160.20 grams. Both distributions are relatively symmetric with no apparent outliers. There is more variability in the weights of chicken fed linseed. (b) $H_0 : \mu_{ls} = \mu_{hb}$. $H_A : \mu_{ls} \neq \mu_{hb}$. We leave the conditions to you to consider. $T = 3.02$, $df = min(11, 9) = 9 \rightarrow 0.01 <$ p-value < 0.02. Since p-value < 0.05, reject H_0. The data provide strong evidence that there is a significant difference between the average weights of chickens that were fed linseed and horsebean. (c) Type 1 Error, since we rejected H_0. (d) Yes, since p-value > 0.01, we would have failed to reject H_0.

5.31 $H_0 : \mu_C = \mu_S$. $H_A : \mu_C \neq \mu_S$. $T = 3.27$, $df = 11 \rightarrow$ p-value < 0.01. Since p-value < 0.05, reject H_0. The data provide strong evidence that the average weight of chickens that were fed casein is different than the average weight of chickens that were fed soybean (with weights from casein being higher). Since this is a randomized experiment, the observed difference can be attributed to the diet.

5.33 $H_0 : \mu_T = \mu_C$. $H_A : \mu_T \neq \mu_C$. $T = 2.24$, $df = 21 \rightarrow 0.02 <$ p-value < 0.05. Since p-value < 0.05, reject H_0. The data provide strong evidence that the average food consumption by the patients in the treatment and control groups are different. Furthermore, the data indicate patients in the distracted eating (treatment) group consume more food than patients in the control group.

5.35 Let $\mu_{diff} = \mu_{pre} - \mu_{post}$. $H_0 : \mu_{diff} = 0$: Treatment has no effect. $H_A : \mu_{diff} \neq 0$: Treatment has an effect on P.D.T. scores, either positive or negative. Conditions: The subjects are randomly assigned to treatments, so independence within and between groups is satisfied. All three sample sizes are smaller than 30, so we look for clear outliers. There is a borderline outlier in the first treatment group. Since it is borderline, we will proceed, but we should report this caveat with any results. For all three groups: $df = 13$. $T_1 = 1.89 \rightarrow$ p-value $= 0.081$, $T_2 = 1.35 \rightarrow$ p-value $= 0.200$), $T_3 = -1.40 \rightarrow$ (p-value $= 0.185$). We do not reject the null hypothesis for any of these groups. As earlier noted, there is some uncertainty about if the method applied is reasonable for the first group.

5.37 Difference we care about: 40. Single tail of 90%: $1.28 \times SE$. Rejection region bounds: $\pm 1.96 \times SE$ (if 5% significance level). Setting $3.24 \times SE = 40$, subbing in $SE = \sqrt{\frac{94^2}{n} + \frac{94^2}{n}}$, and solving for the sample size n gives 116 plots of land for each fertilizer.

5.39 H_0: $\mu_1 = \mu_2 = \cdots = \mu_6$. H_A: The average weight varies across some (or all) groups. Independence: Chicks are randomly assigned to feed types (presumably kept separate from one another), therefore independence of observations is reasonable. Approx. normal: the distributions of weights within each feed type appear to be fairly symmetric. Constant variance: Based on the side-by-side box plots, the constant variance assumption appears to be reasonable. There are differences in the actual computed standard deviations, but these might be due to chance as these are quite small samples. $F_{5,65} = 15.36$ and the p-value is approximately 0. With such a small p-value, we reject H_0. The data provide convincing evidence that the average weight of chicks varies across some (or all) feed supplement groups.

5.41 (a) H_0: The population mean of MET for each group is equal to the others. H_A: At least one pair of means is different. (b) Independence: We don't have any information on how the data were collected, so we cannot assess independence. To proceed, we must assume the subjects in each group are independent. In practice, we would inquire for more details. Normality: The data are bound below by zero and the standard deviations are larger than the means, indicating very strong skew. However, since the sample sizes are extremely large, even extreme skew is acceptable. Constant variance: This condition is sufficiently met, as the standard deviations are reasonably consistent across groups. (c) See below, with the last column omitted:

	Df	Sum Sq	Mean Sq	F value
coffee	4	10508	2627	5.2
Residuals	50734	25564819	504	
Total	50738	25575327		

(d) Since p-value is very small, reject H_0. The data provide convincing evidence that the average MET differs between at least one pair of groups.

5.43 (a) H_0: Average GPA is the same for all majors. H_A: At least one pair of means are different. (b) Since p-value > 0.05, fail to reject H_0. The data do not provide convincing evidence of a difference between the average GPAs across three groups of majors. (c) The total degrees of freedom is $195 + 2 = 197$, so the sample size is $197 + 1 = 198$.

5.45 (a) False. As the number of groups increases, so does the number of comparisons and hence the modified significance level decreases. (b) True. (c) True. (d) False. We need observations to be independent regardless of sample size.

5.47 (a) H_0: Average score difference is the same for all treatments. H_A: At least one pair of means are different. (b) We should check conditions. If we look back to the earlier exercise, we will see that the patients were randomized, so independence is satisfied. There are some minor concerns about skew, especially with the third group, though this may be acceptable. The standard deviations across the groups are reasonably similar. Since the p-value is less than 0.05, reject H_0. The data provide convincing evidence of a difference between the average reduction in score among treatments. (c) We determined that at least two means are different in part (b), so we now conduct $K = 3 \times 2/2 = 3$ pairwise t-tests that each use $\alpha = 0.05/3 = 0.0167$ for a significance level. Use the following hypotheses for each pairwise test. H_0: The two means are equal. H_A: The two means are different. The sample sizes are equal and we use the pooled SD, so we can compute $SE = 3.7$ with the pooled $df = 39$. The p-value for Trmt 1 vs. Trmt 3 is the only one under 0.05: p-value = 0.035 (or 0.024 if using s_{pooled} in place of s_1 and s_3, though this won't affect the final conclusion). The p-value is larger than $0.05/3 = 1.67$, so we do not have strong evidence to conclude that it is this particular pair of groups that are different. That is, we cannot identify if which particular pair of groups are actually different, even though we've rejected the notion that they are all the same!

6 Simple linear regression

6.1 (a) Strong relationship, but a straight line would not fit the data. (b) Strong relationship, and a linear fit would be reasonable. (c) Weak relationship, and trying a linear fit would be reasonable. (d) Moderate relationship, but a straight line would not fit the data. (e) Strong relationship, and a linear fit would be reasonable. (f) Weak relationship, and trying a linear fit would be reasonable.

6.3 (a) There is a moderate, positive, and linear relationship between shoulder girth and height. (b) Changing the units, even if just for one of the variables, will not change the form, direction or strength of the relationship between the two variables.

6.5 Over-estimate. Since the residual is calculated as $observed - predicted$, a negative residual means that the predicted value is higher than the observed value.

6.7 (a) $\widehat{murder} = -29.901 + 2.559 \times poverty\%$. (b) Expected murder rate in metropolitan areas with no poverty is -29. 901 per million. This is obviously not a meaningful value, it just serves to adjust the height of the regression line. (c) For each additional percentage increase in poverty, we expect murders per million to be higher on average by 2.559. (e) $\sqrt{0.7052} = 0.8398$.

6.9 (a) The slope of -1.26 indicates that on average, an increase in age of 1 year is associated with a lower RFFT score by 1.26 points. The intercept of 137.55 represents the predicted mean RFFT score for an individual of age 0 years; this does not have interpretive meaning since the RFFT cannot be reasonably administered to a newborn. (b) RFFT score differs on average by $10(-1.26) = 12.6$ points between an individual who is 60 years old versus 50 years old, with the older individual having the lower score. (c) According to the model, average RFFT score for a 70-year-old is $137.55 - 1.26(70) = 49.3$ points. (d) No, it is not valid to use the linear model to estimate RFFT score for a 20-year-old. As indicated in the plot, data are only available for individuals as young as about 40 years old.

6.11 (a) The residual plot will show randomly distributed residuals around 0. The variance is also approximately constant. (b) The residuals will show a fan shape, with higher variability for smaller x. There will also be many points on the right above the line. There is trouble with the model being fit here.

6.13 (a) The points with the lowest and highest values for height have relatively high leverage. They do not seem particularly influential because they are not outliers; the one with a low x-value has a low y-value and the one with a high x-value has a high y-value, which follows the positive trend visible in the data. (b) Yes, since the data show a linear trend, it is appropriate to use R^2 as a metric for describing the strength of the model fit. (c) Height explains about 72% of the observed variability in length.

6.15 There is an upwards trend. However, the variability is higher for higher calorie counts, and it looks like there might be two clusters of observations above and below the line on the right, so we should be cautious about fitting a linear model to these data.

6.17 (a) There is an outlier in the bottom right. Since it is far from the center of the data, it is a point with high leverage. It is also an influential point since, without that observation, the regression line would have a very different slope.

(b) There is an outlier in the bottom right. Since it is far from the center of the data, it is a point with high leverage. However, it does not appear to be affecting the line much, so it is not an influential point.

(c) The observation is in the center of the data (in the x-axis direction), so this point does *not* have high leverage. This means the point won't have much effect on the slope of the line and so is not an influential point.

6.19 (a) Linearity is satisfied; the data scatter about the horizontal line with no apparent pattern. The variability seems constant across the predicted length values. (b) The fish were randomly sampled from a river, so without additional details about the life cycle of the fish, it seems reasonable to assume the height and length of any one fish does not provide information about the height and length of another fish. This could be violated, if, for example, the fish in a river tend to be closely related and height and length are highly heritable. (c) The residuals are approximately normally distributed, with some small deviations from normality in the tails. There are more outliers in both tails than expected under a normal distribution.

6.21 One possible equation is $\widehat{price} = 44.51 + 12.3(carat_{1.00})$, where the explanatory variable is a binary variable taking on value 1 if the diamond is 1 carat.

6.23 (a) The relationship is positive, moderate-to-strong, and linear. There are a few outliers but no points that appear to be influential.
(b) $\widehat{weight} = -105.0113 + 1.0176 \times height$.
Slope: For each additional centimeter in height, the model predicts the average weight to be 1.0176 additional kilograms (about 2.2 pounds).
Intercept: People who are 0 centimeters tall are expected to weigh - 105.0113 kilograms. This is obviously not possible. Here, the y- intercept serves only to adjust the height of the line and is meaningless by itself.
(c) H_0: The true slope coefficient of height is zero ($\beta_1 = 0$).
H_A: The true slope coefficient of height is different than zero ($\beta_1 \neq 0$).
The p-value for the two-sided alternative hypothesis ($\beta_1 \neq 0$) is incredibly small, so we reject H_0. The data provide convincing evidence that height and weight are positively correlated. The true slope parameter is indeed greater than 0.
(d) $R^2 = 0.72^2 = 0.52$. Approximately 52% of the variability in weight can be explained by the height of individuals.

6.25 (a) H_0: $\beta_1 = 0$. H_A: $\beta_1 \neq 0$. The p-value, as reported in the table, is incredibly small and is smaller than 0.05, so we reject H_0. The data provide convincing evidence that wives' and husbands' heights are positively correlated.

(b) $\widehat{height}_W = 43.5755 + 0.2863 \times height_H$.

(c) Slope: For each additional inch in husband's height, the average wife's height is expected to be an additional 0.2863 inches on average. Intercept: Men who are 0 inches tall are expected to have wives who are, on average, 43.5755 inches tall. The intercept here is meaningless, and it serves only to adjust the height of the line.

(d) The slope is positive, so r must also be positive. $r = \sqrt{0.09} = 0.30$.

(e) 63.33. Since R^2 is low, the prediction based on this regression model is not very reliable.

(f) No, we should avoid extrapolating.

(g) Yes, the p-value for the slope parameter is less than $\alpha = 0.05$. There is sufficient evidence to accept the alternative hypothesis, $H_A : \beta_1 \neq 0$. These data suggest that wife height and husband height are positively associated at the population level.

(h) No, a 95% confidence interval for β_1 would not be expected to contain the null value 0, since the p-value is less than 0.05.

6.27 (a) The point estimate and standard error are $b_1 = 0.9112$ and $SE = 0.0259$. We can compute a T-score: $T = (0.9112 - 1)/0.0259 = -3.43$. Using $df = 168$, the p-value is about 0.001, which is less than $\alpha = 0.05$. That is, the data provide strong evidence that the average difference between husbands' and wives' ages has actually changed over time. (b) $\widehat{age}_W = 1.5740 + 0.9112 \times age_H$. (c) Slope: For each additional year in husband's age, the model predicts an additional 0.9112 years in wife's age. This means that wives' ages tend to be lower for later ages, suggesting the average gap of husband and wife age is larger for older people. Intercept: Men who are 0 years old are expected to have wives who are on average 1.5740 years old. The intercept here is meaningless and serves only to adjust the height of the line. (d) $R = \sqrt{0.88} = 0.94$. The regression of wives' ages on husbands' ages has a positive slope, so the correlation coefficient will be positive. (e) $\widehat{age}_W = 1.5740 + 0.9112 \times 55 = 51.69$. Since R^2 is pretty high, the prediction based on this regression model is reliable. (f) No, we shouldn't use the same model to predict an 85 year old man's wife's age. This would require extrapolation. The scatterplot from an earlier exercise shows that husbands in this data set are approximately 20 to 65 years old. The regression model may not be reasonable outside of this range.

6.29 (a) Yes, since $p < 0.01$. $H_0 : \beta_1 = 0$, $H_A : \beta_1 \neq 0$, where β_1 represents the population average change in RFFT score associated with a change in 1 year of age. There is statistically significant evidence that age is negatively associated with RFFT score. (b) With 99% confidence, the interval (-1.49, -1.03) points contains the population average difference in RFFT score between individuals who differ in age by 1 year; the older individual is predicted to have a lower RFFT score.

6.31 (a) First, compute the standard error: s.e.$(E(age_{wife}|\widehat{age_{husband}} = 55)) = 3.95\sqrt{\frac{1}{170} + \frac{(55 - 42.92)^2}{(170-1)11.76^2}} = 0.435$. The critical value is $t^{\star}_{0.975, df=169} = 1.97$. Thus, the 95% confidence interval is $51.69 \pm (1.97)(0.435) = (50.83, 52.55)$ years. (b) First, compute the standard error: s.e.$(age_{wife}|\widehat{age_{husband}} = 55) = 3.95\sqrt{1 + \frac{1}{170} + \frac{(55 - 42.92)^2}{(170-1)11.76^2}} = 3.97$. The 95% prediction interval is $51.69 \pm (1.97)(3.97) = (43.85, 59.54)$ years. (c) For the approximate 95% confidence interval, use $s/\sqrt{n} = 3.95/\sqrt{170} = 0.303$ as the approximate standard error: $(51.09, 52.29)$ years. For the approximate 95% prediction interval, use $s\sqrt{1 + 1/n} = 3.95\sqrt{1 + 1/170} = 4.25$ as the approximate standard error: $(43.30, 60.09)$ years.

7 Multiple linear regression

7.1 Although the use of statins appeared to be associated with lower RFFT score when no adjustment was made for possible confounders, statin use is not significantly associated with RFFT score in a model that adjusts for age. After adjusting for age, the estimated difference in mean RFFT score between statin users and non-users is 0.85 points; there is a 74% chance of observing such a difference if there is no difference between mean RFFT score in the population of statin users and non-users.

7.3 (a) $\widehat{baby\_weight} = 123.57 - 8.96(smoke) - 1.98(parity)$ (b) A child born to a mother who smokes has a birth weight about 9 ounces less, on average, than one born to a mother who does not smoke, holding birth order constant. A child who is the first born has birth weight about 2 ounces less, on average, than one who is not first born, when comparing children whose mothers were either both smokers or both nonsmokers. The intercept represents the predicted mean birth weight for a child whose mother is not a smoker and who was not the first born. (c) The estimated difference in mean birth weight for two infants born to non-smoking mothers, where one is first born and the other is not, is -1.98. (d) This is the same value as in part (c). (e) $123.57 - 8.96(0) - 1.98(1) = 121.59$ ounces.

7.5 (a) $\widehat{baby\_weight} = -80.41 + 0.44 \times gestation - 3.33 \times parity - 0.01 \times age + 1.15 \times height + 0.05 \times weight - 8.40 \times smoke$. (b) $\beta_{gestation}$: The model predicts a 0.44 ounce increase in the birth weight of the baby for each additional day of pregnancy, all else held constant. β_{age}: The model predicts a 0.01 ounce decrease in the birth weight of the baby for each additional year in mother's age, all else held constant. (c) Parity might be correlated with one of the other variables in the model, which complicates model estimation. (d) $\widehat{baby\_weight} = 120.58$. $e = 120 - 120.58 = -0.58$. The model over-predicts this baby's birth weight. (e) $R^2 = 0.2504$. $R^2_{adj} = 0.2468$.

7.7 Nearly normal residuals: With so many observations in the data set, we look for particularly extreme outliers in the histogram and do not see any. Variability of residuals: The scatterplot of the residuals versus the fitted values does not show any overall structure. However, values that have very low or very high fitted values appear to also have somewhat larger outliers. In addition, the residuals do appear to have constant variability between the two parity and smoking status groups, though these items are relatively minor.
Independent residuals: The scatterplot of residuals versus the order of data collection shows a random scatter, suggesting that there is no apparent structures related to the order the data were collected.
Linear relationships between the response variable and numerical explanatory variables: The residuals vs. height and weight of mother are randomly distributed around 0. The residuals vs. length of gestation plot also does not show any clear or strong remaining structures, with the possible exception of very short or long gestations. The rest of the residuals do appear to be randomly distributed around 0.
All concerns raised here are relatively mild. There are some outliers, but there is so much data that the influence of such observations will be minor.

7.9 (b) True. (c) False. This would only be the case if the data was from an experiment and x_1 was one of the variables set by the researchers. (Multiple regression can be useful for forming hypotheses about causal relationships, but it offers zero guarantees.) (d) False. We should check normality like we would for inference for a single mean: we look for particularly extreme outliers if $n \geq 30$ or for clear outliers if $n < 30$.

7.11 (a) (-0.32, 0.16). We are 95% confident that male students on average have GPAs 0.32 points lower to 0.16 points higher than females when controlling for the other variables in the model. (b) Yes, since the p-value is larger than 0.05 in all cases (not including the intercept).

7.13 (a) $\widehat{eggs.laid} = -17.88 + 4.28(wolbachia) + 0.272(tibia)$ (b) An increase in *Wolbachia* density of one unit is associated with on average 4.28 more eggs laid over a lifetime, assuming body size is held constant. (c) In a multiple regression model adjusting for body size as a potential confounder, increase in *Wolbachia* density was significantly positively associated with realized fitness, measured as the number of eggs laid over a female's full lifetime ($p = 0.002$). These data are consistent with the scientific hypothesis that *Wolbachia* is beneficial for its host in nature. (d) $(1.85, 7.05)$ eggs (e) As a group, the predictors *Wolbachia* density and tibia length are useful for predicting the number of eggs laid over a lifetime.

7.15 (a) Since the difference is taken in the direction (pre - post), a positive value for trt.effect indicates that the post-intervention score is lower than the pre-intervention score, which represents efficacy of the intervention. A negative value would represent a patient's deviant T scores increasing after the intervention. (b) Let Y be the change in MMPI score for a participant in this study, $X_{neutral}$ a variable with value 1 for participants assigned to the neutral tape and 0 otherwise, and $X_{therapeutic}$ a variable with value 1 for participants in the emotional neutral group and 0 otherwise. The population-level equation is $E(Y) = \beta_0 + \beta_{neutral}X_{neutral} + \beta_{therapeutic}X_{therapeutic}$. For these data, the estimated model equation is $\widehat{y} = -3.21 + 6.07X_{neutral} + 9.43X_{therapeutic}$. (c) The predicted difference scores $\widehat{y}$ for a patient receiving the neutral tape will be $\widehat{y} = b_0 + b_{neutral}X_{neutral} + b_{therapeutic}X_{therapeutic} = -3.21 + 6.07 + 0 = 2.86$. (d) Yes. The intercept is the average of the score difference for the group that did not hear a taped message. (e) The two slopes represent the change in average MMPI score difference from the average for the group that did not receive a tape. The Absent category is the reference group. (f) The p–value for the intercept corresponds to a test of the null hypothesis that the average difference score was 0 in the group that did not hear a taped message. The slope p-values correspond to tests of the null hypotheses of (on average) no change in difference scores between the intervention with no tape and each of the other two interventions.

7.17 (a) Let *pre* and *post* denote the pre- and post-intervention scores, respectively. The estimated equation for the model is $\widehat{post} = 28.41 + 0.66(pre) - 5.73X_{neutral} - 9.75X_{therapeutic}$. (b) Since the coefficient of the pre-intervention score is positive, post-intervention scores tend to increase as the pre-intervention score increases. (c) Yes. The t-statistic for the coefficient of pre is 4.05 and is statistically significant. (d) In this model, treatment is a factor variable with three levels and the intervention with no tape is the baseline treatment that does not appear in the model. For a participant with pre = 70 and no tape, the predicted value of post is $28.41 + 0.66(73) - 5.73(1) = 70.86$ (e) For a given value of pre, the coefficient of treatmentNeutral is the predicted change in post between an participant without a tape and one with the emotionally neutral tape. The model implies that post will be 5.7 points lower with the emotionally neutral tape. The evidence for a treatment effect of the emotionally neutral tape is weak; the coefficient is not statistically significant at $\alpha = 0.05$.

7.19 (a) $\widehat{post} = -17.58 + 1.28(pre) + 67.75(neutral) + 64.42(therapeutic) - 0.99(pre \times neutral) - 1.01(pre \times therapeutic)$
(b) The coefficient for pre is the predicted increase in post score associated with a 1 unit increase in pre-score for individuals in the absent arm, while the coefficients of the interaction terms for neutral and therapeutic represent the difference in association between pre and post scores for individuals in those groups. For example, an individual in the neutral group is expected to have a $1.28 - 0.99 = 0.29$ point increase in post score, on average, per 1 point increase in pre-score. The coefficients of the slopes for neutral and therapeutic are differences in intercept values relative to the intercept for the model, which is for the baseline group (absent).
(c) Absent: $\widehat{post} = -17.58 + 1.28(pre)$ Neutral: $\widehat{post} = -17.58 + 67.75 + 1.28(pre) - 0.99(pre) = 50.17 + 0.29(pre)$ Therapeutic: $\widehat{post} = -17.58 + 64.42 + 1.28(pre) - 1.01(pre) = 46.84 + 0.27(pre)$ (d) These data suggest there is a statistically significant difference in association between pre- and post-intervention scores by treatment group relative to the group that did not receive any treatment. The coefficients of both interaction terms are statistically significant at $\alpha = 0.05$. Since the slopes are smaller than the slope for the treatment absent group, the data demonstrate that individuals in either treatment group show less increase in MMPI score than occurs when no treatment is applied.

7.21 (a) $\widehat{RFFT} = 140.20 - 13.97(Statin) - 1.31(Age) + 0.25(Statin \times Age)$ (b) The model intercept represents the predicted mean RFFT score for a statin non-user of age 0 years; the intercept does not have a meaningful interpretation. The slope coefficient for age represents the predicted change in RFFT score for a statin non-user; for non-users, a one year increase in age is associated with a 1.32 decrease in RFFT score. The slope coefficient for statin use represents the difference in intercept between the regression line for users and the regression line for non-users; the intercept for users is -13.97 points lower than that of non-users. The interaction term coefficient represents the difference in the magnitude of association between RFFT score and age between users and non-users; in users, the slope coefficient representing predicted change in RFFT score per 1 year change in age is higher by 0.25 points. (c) No, there is not evidence that the association between RFFT score and age differs by statin use. The p-value of the interaction coefficient is 0.32, which is higher than $\alpha = 0.05$.

7.23 Age should be the first variable removed from the model. It has the highest p-value, and its removal results in an adjusted R^2 of 0.255, which is higher than the current adjusted R^2.

7.25 (a) The strongest predictor of birth weight appears to be gestational age; these two variables show a strong positive association. Both parity and smoker status show a slight association with gestational age; the first born child tends to be a lower birth weight and children from mothers who smoke tend to have lower birth weight. While there does not appear to be an association between birth weight and age of the mother, there may be a slight positive association between both birth weight and height and birth weight and weight. All predictor variables with exception of age seem potentially useful for inclusion in an initial model. (b) Height and weight appear to be positively associated.

7.27 (a) The F-statistic for the model corresponds to a test of $H_0 : \beta_{neutral} = \beta_{therapeutic} = 0$. (b) The intercept coefficient is the estimated mean difference score for the no intervention group, and the estimated mean difference score for the other two groups can be calculated by adding each of the slope estimates to the intercept. (c) Under the null hypothesis that the two slope coefficients are 0, all three interventions would have the same mean difference in MMPI scores. This is the same as the null hypothesis for an ANOVA with three groups ($H_0 : \mu_1 = \mu_2 = \mu_3$), which states that all three population means are the same. (d) The assumptions for multiple regression and ANOVA are outlined in Sections 7.3.1 and 5.5, respectively. The assumptions for the two models are the same, though they may be phrased differently. The first assumption in multiple regression is linear change of the mean response variable when one predictor changes and the others do not change. Since each of the two predictor variables in this model can only change from 0 to 1, this assumption is simply that the means in the three groups are possibly different, which is true in ANOVA. The second assumption in regression is that the variance of the residuals is approximately constant. Since the predicted response for an intervention group is its mean, the constant variance assumption in regression is the equivalent assumption in ANOVA that the three groups have approximately constant variance. Both models assume that the observations are independent and that the residuals follow a normal distribution. This is a very long way of saying that the two models are identical!

8 Inference for categorical data

8.1 (a) False. Doesn't satisfy success-failure condition. (b) True. The success-failure condition is not satisfied. In most samples we would expect $\hat{p}$ to be close to 0.08, the true population proportion. While $\hat{p}$ can be much above 0.08, it is bound below by 0, suggesting it would take on a right skewed shape. Plotting the sampling distribution would confirm this suspicion. (c) False. $SE_{\hat{p}} = 0.0243$, and $\hat{p} = 0.12$ is only $\frac{0.12-0.08}{0.0243} = 1.65$ SEs away from the mean, which would not be considered unusual. (d) True. $\hat{p} = 0.12$ is 2.32 standard errors away from the mean, which is often considered unusual. (e) False. Decreases the SE by a factor of $1/\sqrt{2}$.

8.3 (a) False. A confidence interval is constructed to estimate the population proportion, not the sample proportion. (b) True. 95% CI: 82% ± 2%. (c) True. By the definition of the confidence level. (d) True. Quadrupling the sample size decreases the SE and ME by a factor of $1/\sqrt{4}$. (e) True. The 95% CI is entirely above 50%.

8.5 With a random sample, independence is satisfied. The success-failure condition is also satisfied. $ME = z^{\star}\sqrt{\frac{\hat{p}(1-\hat{p})}{n}} = 1.96\sqrt{\frac{0.56\times0.44}{600}} = 0.0397 \approx 4\%$

8.7 (a) No. The sample only represents students who took the SAT, and this was also an online survey. (b) $(0.5289, 0.5711)$. We are 90% confident that 53% to 57% of high school seniors who took the SAT are fairly certain that they will participate in a study abroad program in college. (c) 90% of such random samples would produce a 90% confidence interval that includes the true proportion. (d) Yes. The interval lies entirely above 50%.

8.9 (a) We want to check for a majority (or minority), so we use the following hypotheses:

$$H_0 : p = 0.5 \qquad\qquad\qquad H_A : p \neq 0.5$$

We have a sample proportion of $\hat{p} = 0.55$ and a sample size of $n = 617$ independents.

Since this is a random sample, independence is satisfied. The success-failure condition is also satisfied: 617×0.5 and $617\times(1-0.5)$ are both at least 10 (we use the null proportion $p_0 = 0.5$ for this check in a one-proportion hypothesis test).

Therefore, we can model $\hat{p}$ using a normal distribution with a standard error of

$$SE = \sqrt{\frac{p(1-p)}{n}} = 0.02$$

(We use the null proportion $p_0 = 0.5$ to compute the standard error for a one-proportion hypothesis test.) Next, we compute the test statistic:

$$Z = \frac{0.55 - 0.5}{0.02} = 2.5$$

This yields a one-tail area of 0.0062, and a p-value of $2 \times 0.0062 = 0.0124$.

Because the p-value is smaller than 0.05, we reject the null hypothesis. We have strong evidence that the support is different from 0.5, and since the data provide a point estimate above 0.5, we have strong evidence to support this claim by the TV pundit.

(b) No. Generally we expect a hypothesis test and a confidence interval to align, so we would expect the confidence interval to show a range of plausible values entirely above 0.5. However, if the confidence level is misaligned (e.g. a 99% confidence level and a $\alpha = 0.05$ significance level), then this is no longer generally true.

8.11 Since a sample proportion ($\hat{p} = 0.55$) is available, we use this for the sample size calculations. The margin of error for a 90% confidence interval is $1.65 \times SE = 1.65 \times \sqrt{\frac{p(1-p)}{n}}$. We want this to be less than 0.01, where we use $\hat{p}$ in place of p:

$$1.65 \times \sqrt{\frac{0.55(1-0.55)}{n}} \leq 0.01$$

$$1.65^2 \frac{0.55(1-0.55)}{0.01^2} \leq n$$

From this, we get that n must be at least 6739.

8.13 (a) $H_0 : p = 0.5$. $H_A : p \neq 0.5$. Independence (random sample) is satisfied, as is the success-failure conditions (using $p_0 = 0.5$, we expect 40 successes and 40 failures). $Z = 2.91 \rightarrow$ the one tail area is 0.0018, so the p-value is 0.0036. Since the p-value < 0.05, we reject the null hypothesis. Since we rejected H_0 and the point estimate suggests people are better than random guessing, we can conclude the rate of correctly identifying a soda for these people is significantly better than just by random guessing. (b) If in fact people cannot tell the difference between diet and regular soda and they were randomly guessing, the probability of getting a random sample of 80 people where 53 or more identify a soda correctly (or 53 or more identify a soda incorrectly) would be 0.0036.

8.15 (a) Yes, it is reasonable to use the normal approximation to the binomial distribution. The sample observations are independent and the expected numbers of successes and failures are greater than 10: $n\hat{p} = (100)(.15) = 15$ and $n(1 - \hat{p}) = (100)(0.85) = 85$. (b) An approximate 95% confidence interval is $\hat{p} \pm 1.96\sqrt{\frac{\hat{p}(1-\hat{p})}{n}} \rightarrow (0.08, 0.22)$. (c) The interval does not support the claim. Since the interval does not contain 0.05, there is statistically significant evidence at $\alpha = 0.05$ that the proportion of young women in the neighborhood who use birth control is different than 0.05. The interval is above 0.05, which is indicative of evidence that more than 5% of young women in the neighborhood use birth control.

8.17 This is not a randomized experiment, and it is unclear whether people would be affected by the behavior of their peers. That is, independence may not hold. Additionally, there are only 5 interventions under the provocative scenario, so the success-failure condition does not hold. Even if we consider a hypothesis test where we pool the proportions, the success-failure condition will not be satisfied. Since one condition is questionable and the other is not satisfied, the difference in sample proportions will not follow a nearly normal distribution.

8.19 (a) Standard error:

$$SE = \sqrt{\frac{0.79(1-0.79)}{347} + \frac{0.55(1-0.55)}{617}} = 0.03$$

Using $z^\star = 1.96$, we get:

$$0.79 - 0.55 \pm 1.96 \times 0.03 \rightarrow (0.181, 0.299)$$

We are 95% confident that the proportion of Democrats who support the plan is 18.1% to 29.9% higher than the proportion of Independents who support the plan. (b) True.

8.21 (a) Test $H_0 : p_1 = p_2$ against $H_A : p_1 \neq p_2$, where p_1 represents the population proportion of clinical improvement in COVID-19 patients treated with remdesivir and p_2 represents the population proportion of clinical improvement in COVID-19 patients treated with placebo. Let $\alpha = 0.05$. The p-value is 0.328, which is greater than α; there is insufficient evidence to reject the null hypothesis of no difference. Even though the proportion of patients who experienced clinical improvement about 7% higher in the remdesivir group, this difference is not extreme enough to represent sufficient evidence that remdesivir is more effective than placebo. (b) The 95% confidence interval is (-0.067, 0.217); with 95% confidence, this interval captures the difference in population proportion of clinical mortality between COVID-19 patients treated with remdesivir and those treated with placebo. The interval contains 0, which is consistent with no statistically significant evidence of a difference. The interval reflects the lack of precision around the effect estimate that is characteristic of an insufficiently large sample size.

8.23 (a) False. The entire confidence interval is above 0. (b) True. (c) True. (d) True. (e) False. It is simply the negated and reordered values: (-0.06,-0.02).

8.25 Subscript C means control group. Subscript T means truck drivers. $H_0 : p_C = p_T$. $H_A : p_C \neq p_T$. Independence is satisfied (random samples), as is the success-failure condition, which we would check using the pooled proportion ($\hat{p}_{pool} = 70/495 = 0.141$). $Z = -1.65 \rightarrow$ p-value = 0.0989. Since the p-value is high (default to alpha = 0.05), we fail to reject H_0. The data do not provide strong evidence that the rates of sleep deprivation are different for non-transportation workers and truck drivers.

8.27 (a) False. The chi-square distribution has one parameter called degrees of freedom. (b) True. (c) True. (d) False. As the degrees of freedom increases, the shape of the chi-square distribution becomes more symmetric.

8.29 (a) Two-way table:

Treatment	Quit Yes	No	Total
Patch + support group	40	110	150
Only patch	30	120	150
Total	70	230	300

(b-i) $E_{row_1,col_1} = \frac{(row\ 1\ total)\times(col\ 1\ total)}{table\ total} = 35$. This is lower than the observed value.
(b-ii) $E_{row_2,col_2} = \frac{(row\ 2\ total)\times(col\ 2\ total)}{table\ total} = 115$. This is lower than the observed value.

8.31 (a) H_0: There is no association between statin use and educational level. H_A: There is an association between statin use and educational level
(b) It is reasonable to assume the counts are independent. The smallest expected value in the table is 39.27, so the success-failure condition is reasonably met. (c) There is statistically significant evidence at $\alpha = 0.05$ of an association between educational level and statin use. Individuals with a higher educational level are less likely to be statin users.

8.33 (a)

	No Default	Default	Sum
Non-Diabetic	1053	127	1180
Diabetic	54	0	54
Sum	1107	127	1234

(b) $H_0 : p_1 = p_2$ versus $H_A : p_1 \neq p_2$, where p_1 represents the population proportion of treatment default in diabetics and p_2 represents the population proportion of treatment default in non-diabetics. (c) It is reasonable to assume the counts are independent. The smallest expected value is 5.56, which is not smaller than 5. (d) The χ^2 test statistic is 5.37, with 1 degree of freedom. The p-value of the test statistic is 0.02. There is sufficient evidence to conclude that the proportion of treatment default is higher in non-diabetics than in diabetics.

8.35 (a) One possible 2×2 contingency table:

	Mosquito Nets		
	No	**Yes**	**Total**
Malaria	30	22	52
No Malaria	70	78	148
Total	100	100	200

(b) Expected number of infected children among 100 families who did receive a net: $\frac{52 \times 100}{200} = 26$.

(c) The null hypothesis is H_0 : Using a mosquito net and being infected with malaria are not associated. The alternative is H_A : using a net and being infected with malaria are associated. The χ^2 statistic (1.66) has 1 degree of freedom and the table A3 can be used to show that $p > 0.10$. There is not statistically significant evidence of an association between malaria infection and use of a net in children.

(d) Because this is a prospective study, the relative risk can be calculated directly from the table. Let $p_{\text{No Nets}}$ be the probability that a child without a net will be infected with malaria: $\hat{p}_{\text{No Nets}} = \frac{30}{100} = 0.30$. Let p_{Nets} be the probability that a child with a net will be infected with malaria: $\hat{p}_{\text{Nets}} = \frac{22}{100} = 0.22$. The estimated relative risk: $\widehat{RR} = \frac{\hat{p}_{\text{No Nets}}}{\hat{p}_{\text{Nets}}} = \frac{0.30}{0.22} = 1.36$. The risk of malaria infection for children in the control group is 36% higher than risk for children in the treatment group.

8.37 (a) Under the null hypothesis of no association, the expected cell counts are 9.07 and 7.93 in the wait together and wait alone groups, respectively, for those considered "high anxiety" and 6.93 and 6.07 in the wait together and wait alone groups, respectively, for those considered "low anxiety". (b) Use the hypergeometric distribution with parameters $N = 30$, $m = 16$, and $n = 17$; calculate $P(X = 12)$. Consider the "successes" to be the individuals who wait together, and the "number sampled" to be the people randomized to the high-anxiety group. The probability of the observed set of results, assuming the marginal totals are fixed and the null hypothesis is true, is 0.0304. (c) More individuals than expected in the high-anxiety group were observed to wait together; thus, tables that are more extreme in the same direction also consist of those where more people in the high-anxiety group wait together than observed. These are tables in which 13, 14, 15, or 16 individuals in the high-anxiety group wait together.

	Wait Together	Wait Alone	Sum
High-Anxiety	13	4	17
Low-Anxiety	3	10	13
Sum	16	14	30

	Wait Together	Wait Alone	Sum
High-Anxiety	14	3	17
Low-Anxiety	2	11	13
Sum	16	14	30

	Wait Together	Wait Alone	Sum
High-Anxiety	15	2	17
Low-Anxiety	1	12	13
Sum	16	14	30

	Wait Together	Wait Alone	Sum
High-Anxiety	16	1	17
Low-Anxiety	0	13	13
Sum	16	14	30

(d) Let p_1 represent the population proportion of individuals waiting together in the high-anxiety group and p_2 represent the population proportion of individuals waiting together in the low-anxiety group. Test $H_0 : p_1 = p_2$ against $H_A : p_1 \neq p_2$. Let $\alpha = 0.05$. The two-sided p-value is 0.063. There is insufficient evidence to reject the null hypothesis; the data do not suggest there is an association between high anxiety and a person's desire to be in the company of others.

8.39 (a) H_0: The distribution of the format of the book used by the students follows the professor's predictions. H_A: The distribution of the format of the book used by the students does not follow the professor's predictions. (b) $E_{hard\ copy} = 126 \times 0.60 = 75.6$. $E_{print} = 126 \times 0.25 = 31.5$. $E_{online} = 126 \times 0.15 = 18.9$. (c) Independence: The sample is not random. However, if the professor has reason to believe that the proportions are stable from one term to the next and students are not affecting each other's study habits, independence is probably reasonable. Sample size: All expected counts are at least 5. (d) $\chi^2 = 2.32$, $df = 2$, p-value = 0.313. (e) Since the p-value is large, we fail to reject H_0. The data do not provide strong evidence indicating the professor's predictions were statistically inaccurate.

8.41 (a)

	CVD	No CVD
Age Onset ≤ 50 Years	15	25
Age Onset > 50 Years	5	55

(b) The odds of CVD for patients older than 50 years when diagnosed with diabetes is $5/55 = 0.09$. The odds of CVD for the patients younger than 50 years at diabetes onset is $15/25 = 0.60$. The relative odds (or odds ratio, OR) is $0.09/0.60 = 0.15$.

(c) The odds of CVD for someone with late onset diabetes is less than $1/5$ that of people with earlier onset diabetes. This can be explained by the fact that people with diabetes tend to build up plaque in their arteries; with early onset diabetes, plaque has longer time to accumulate, eventually causing CVD.

(d) $H_0 : OR = 1$.

(e) The chi-square test can be used to test H_0 as long as the conditions for the test have been met. The observations are likely independent; knowing one person's age of diabetes onset and CVD status is unlikely to provide information about another person's age of diabetes onset and CVD status. Under H_0, the expected cell count for the lower left cell is $(60)(20)/100 = 12$, which is bigger than 5; all other expected cell counts will be larger.

(f) Since the study is not a randomized experiment, it cannot demonstrate causality. It may be the case, for example, that CVD presence causes earlier onset of diabetes. The study only demonstrates an association between cardiovascular disease and diabetes.

8.43 (a) No. This is an example of outcome dependent sampling. Subjects were first identified according to presence or absence of the CNS disorder, then queried about use of the drug. It is only possible to estimate the probability that someone had used the drug, given they either did or did not have a CNS disorder.

(b) The appropriate measure of association is the odds ratio.

(c) The easiest way of calculating the OR for the table is the cross-product of the diagonal elements of the table: $[(10)(4000)]/[(2000)(7)] = 2.86$. Using the definition, it can be calculated as:

$$\hat{OR} = \frac{\dfrac{\hat{P}(\text{CNS}|\text{ Usage})}{1-\hat{P}(\text{CNS}|\text{ Usage})}}{\dfrac{\hat{P}(\text{CNS}|\text{ No Usage})}{1-\hat{P}(\text{CNS}|\text{ No Usage})}} = \frac{ad}{bc} = \frac{(10)(4000)}{(2000)(7)} = 2.86$$

(d) The odds ratio has the interpretation of the relative odds of presence of a CNS disorder, comparing people who have used the weight loss drug to those who have not. People who have used the weight loss drug have odds of CNS that are almost three times as large as those for people who have not used the drug.

(e) Fisher's exact test is better than the chi-square test. The independence assumption is met, but the expected cell count corresponding the presence of a CNS disorder and the use of the drug is 5.68, so not all the expected cell counts are less than 10.

8.45 (a) The p-value is 0.92; there is insufficient evidence to reject the null hypothesis of no association. These data are plausible with the null hypothesis that green tea consumption is independent of esophageal carcinoma. (b) Since the study uses outcome-dependent sampling, the odds ratio should be used as a measure of association rather than relative risk. The odds ratio of esophageal carcinoma, comparing green tea drinkers to non-drinkers, is 1.08; the odds of carcinoma for those who regularly drink green tea are 8% larger than the odds for those who never drink green tea.

Appendix B

Distribution tables

B.1 Normal Probability Table

The area to the left of Z represents the percentile of the observation. The normal probability table always lists percentiles.

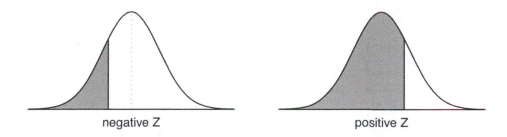

negative Z   positive Z

To find the area to the right, calculate 1 minus the area to the left.

1.0000   —   0.6664   =   0.3336

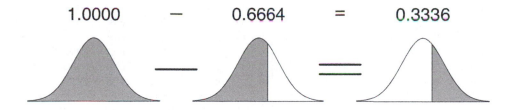

For additional details about working with the normal distribution and the normal probability table, see Section 3.3, which starts on page 152.

negative Z

Second decimal place of Z										
0.09	0.08	0.07	0.06	0.05	0.04	0.03	0.02	0.01	0.00	Z
0.0002	0.0003	0.0003	0.0003	0.0003	0.0003	0.0003	0.0003	0.0003	0.0003	−3.4
0.0003	0.0004	0.0004	0.0004	0.0004	0.0004	0.0004	0.0005	0.0005	0.0005	−3.3
0.0005	0.0005	0.0005	0.0006	0.0006	0.0006	0.0006	0.0006	0.0007	0.0007	−3.2
0.0007	0.0007	0.0008	0.0008	0.0008	0.0008	0.0009	0.0009	0.0009	0.0010	−3.1
0.0010	0.0010	0.0011	0.0011	0.0011	0.0012	0.0012	0.0013	0.0013	0.0013	−3.0
0.0014	0.0014	0.0015	0.0015	0.0016	0.0016	0.0017	0.0018	0.0018	0.0019	−2.9
0.0019	0.0020	0.0021	0.0021	0.0022	0.0023	0.0023	0.0024	0.0025	0.0026	−2.8
0.0026	0.0027	0.0028	0.0029	0.0030	0.0031	0.0032	0.0033	0.0034	0.0035	−2.7
0.0036	0.0037	0.0038	0.0039	0.0040	0.0041	0.0043	0.0044	0.0045	0.0047	−2.6
0.0048	0.0049	0.0051	0.0052	0.0054	0.0055	0.0057	0.0059	0.0060	0.0062	−2.5
0.0064	0.0066	0.0068	0.0069	0.0071	0.0073	0.0075	0.0078	0.0080	0.0082	−2.4
0.0084	0.0087	0.0089	0.0091	0.0094	0.0096	0.0099	0.0102	0.0104	0.0107	−2.3
0.0110	0.0113	0.0116	0.0119	0.0122	0.0125	0.0129	0.0132	0.0136	0.0139	−2.2
0.0143	0.0146	0.0150	0.0154	0.0158	0.0162	0.0166	0.0170	0.0174	0.0179	−2.1
0.0183	0.0188	0.0192	0.0197	0.0202	0.0207	0.0212	0.0217	0.0222	0.0228	−2.0
0.0233	0.0239	0.0244	0.0250	0.0256	0.0262	0.0268	0.0274	0.0281	0.0287	−1.9
0.0294	0.0301	0.0307	0.0314	0.0322	0.0329	0.0336	0.0344	0.0351	0.0359	−1.8
0.0367	0.0375	0.0384	0.0392	0.0401	0.0409	0.0418	0.0427	0.0436	0.0446	−1.7
0.0455	0.0465	0.0475	0.0485	0.0495	0.0505	0.0516	0.0526	0.0537	0.0548	−1.6
0.0559	0.0571	0.0582	0.0594	0.0606	0.0618	0.0630	0.0643	0.0655	0.0668	−1.5
0.0681	0.0694	0.0708	0.0721	0.0735	0.0749	0.0764	0.0778	0.0793	0.0808	−1.4
0.0823	0.0838	0.0853	0.0869	0.0885	0.0901	0.0918	0.0934	0.0951	0.0968	−1.3
0.0985	0.1003	0.1020	0.1038	0.1056	0.1075	0.1093	0.1112	0.1131	0.1151	−1.2
0.1170	0.1190	0.1210	0.1230	0.1251	0.1271	0.1292	0.1314	0.1335	0.1357	−1.1
0.1379	0.1401	0.1423	0.1446	0.1469	0.1492	0.1515	0.1539	0.1562	0.1587	−1.0
0.1611	0.1635	0.1660	0.1685	0.1711	0.1736	0.1762	0.1788	0.1814	0.1841	−0.9
0.1867	0.1894	0.1922	0.1949	0.1977	0.2005	0.2033	0.2061	0.2090	0.2119	−0.8
0.2148	0.2177	0.2206	0.2236	0.2266	0.2296	0.2327	0.2358	0.2389	0.2420	−0.7
0.2451	0.2483	0.2514	0.2546	0.2578	0.2611	0.2643	0.2676	0.2709	0.2743	−0.6
0.2776	0.2810	0.2843	0.2877	0.2912	0.2946	0.2981	0.3015	0.3050	0.3085	−0.5
0.3121	0.3156	0.3192	0.3228	0.3264	0.3300	0.3336	0.3372	0.3409	0.3446	−0.4
0.3483	0.3520	0.3557	0.3594	0.3632	0.3669	0.3707	0.3745	0.3783	0.3821	−0.3
0.3859	0.3897	0.3936	0.3974	0.4013	0.4052	0.4090	0.4129	0.4168	0.4207	−0.2
0.4247	0.4286	0.4325	0.4364	0.4404	0.4443	0.4483	0.4522	0.4562	0.4602	−0.1
0.4641	0.4681	0.4721	0.4761	0.4801	0.4840	0.4880	0.4920	0.4960	0.5000	−0.0

*For $Z \leq -3.50$, the probability is less than or equal to 0.0002.

positive Z

Z	Second decimal place of Z									
	0.00	0.01	0.02	0.03	0.04	0.05	0.06	0.07	0.08	0.09
0.0	0.5000	0.5040	0.5080	0.5120	0.5160	0.5199	0.5239	0.5279	0.5319	0.5359
0.1	0.5398	0.5438	0.5478	0.5517	0.5557	0.5596	0.5636	0.5675	0.5714	0.5753
0.2	0.5793	0.5832	0.5871	0.5910	0.5948	0.5987	0.6026	0.6064	0.6103	0.6141
0.3	0.6179	0.6217	0.6255	0.6293	0.6331	0.6368	0.6406	0.6443	0.6480	0.6517
0.4	0.6554	0.6591	0.6628	0.6664	0.6700	0.6736	0.6772	0.6808	0.6844	0.6879
0.5	0.6915	0.6950	0.6985	0.7019	0.7054	0.7088	0.7123	0.7157	0.7190	0.7224
0.6	0.7257	0.7291	0.7324	0.7357	0.7389	0.7422	0.7454	0.7486	0.7517	0.7549
0.7	0.7580	0.7611	0.7642	0.7673	0.7704	0.7734	0.7764	0.7794	0.7823	0.7852
0.8	0.7881	0.7910	0.7939	0.7967	0.7995	0.8023	0.8051	0.8078	0.8106	0.8133
0.9	0.8159	0.8186	0.8212	0.8238	0.8264	0.8289	0.8315	0.8340	0.8365	0.8389
1.0	0.8413	0.8438	0.8461	0.8485	0.8508	0.8531	0.8554	0.8577	0.8599	0.8621
1.1	0.8643	0.8665	0.8686	0.8708	0.8729	0.8749	0.8770	0.8790	0.8810	0.8830
1.2	0.8849	0.8869	0.8888	0.8907	0.8925	0.8944	0.8962	0.8980	0.8997	0.9015
1.3	0.9032	0.9049	0.9066	0.9082	0.9099	0.9115	0.9131	0.9147	0.9162	0.9177
1.4	0.9192	0.9207	0.9222	0.9236	0.9251	0.9265	0.9279	0.9292	0.9306	0.9319
1.5	0.9332	0.9345	0.9357	0.9370	0.9382	0.9394	0.9406	0.9418	0.9429	0.9441
1.6	0.9452	0.9463	0.9474	0.9484	0.9495	0.9505	0.9515	0.9525	0.9535	0.9545
1.7	0.9554	0.9564	0.9573	0.9582	0.9591	0.9599	0.9608	0.9616	0.9625	0.9633
1.8	0.9641	0.9649	0.9656	0.9664	0.9671	0.9678	0.9686	0.9693	0.9699	0.9706
1.9	0.9713	0.9719	0.9726	0.9732	0.9738	0.9744	0.9750	0.9756	0.9761	0.9767
2.0	0.9772	0.9778	0.9783	0.9788	0.9793	0.9798	0.9803	0.9808	0.9812	0.9817
2.1	0.9821	0.9826	0.9830	0.9834	0.9838	0.9842	0.9846	0.9850	0.9854	0.9857
2.2	0.9861	0.9864	0.9868	0.9871	0.9875	0.9878	0.9881	0.9884	0.9887	0.9890
2.3	0.9893	0.9896	0.9898	0.9901	0.9904	0.9906	0.9909	0.9911	0.9913	0.9916
2.4	0.9918	0.9920	0.9922	0.9925	0.9927	0.9929	0.9931	0.9932	0.9934	0.9936
2.5	0.9938	0.9940	0.9941	0.9943	0.9945	0.9946	0.9948	0.9949	0.9951	0.9952
2.6	0.9953	0.9955	0.9956	0.9957	0.9959	0.9960	0.9961	0.9962	0.9963	0.9964
2.7	0.9965	0.9966	0.9967	0.9968	0.9969	0.9970	0.9971	0.9972	0.9973	0.9974
2.8	0.9974	0.9975	0.9976	0.9977	0.9977	0.9978	0.9979	0.9979	0.9980	0.9981
2.9	0.9981	0.9982	0.9982	0.9983	0.9984	0.9984	0.9985	0.9985	0.9986	0.9986
3.0	0.9987	0.9987	0.9987	0.9988	0.9988	0.9989	0.9989	0.9989	0.9990	0.9990
3.1	0.9990	0.9991	0.9991	0.9991	0.9992	0.9992	0.9992	0.9992	0.9993	0.9993
3.2	0.9993	0.9993	0.9994	0.9994	0.9994	0.9994	0.9994	0.9995	0.9995	0.9995
3.3	0.9995	0.9995	0.9995	0.9996	0.9996	0.9996	0.9996	0.9996	0.9996	0.9997
3.4	0.9997	0.9997	0.9997	0.9997	0.9997	0.9997	0.9997	0.9997	0.9997	0.9998

*For $Z \geq 3.50$, the probability is greater than or equal to 0.9998.

B.2   t-Probability Table

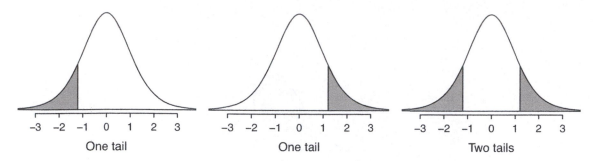

Figure B.1: Tails for the *t*-distribution.

one tail	0.100	0.050	0.025	0.010	0.005
two tails	0.200	0.100	0.050	0.020	0.010
df 1	3.08	6.31	12.71	31.82	63.66
2	1.89	2.92	4.30	6.96	9.92
3	1.64	2.35	3.18	4.54	5.84
4	1.53	2.13	2.78	3.75	4.60
5	1.48	2.02	2.57	3.36	4.03
6	1.44	1.94	2.45	3.14	3.71
7	1.41	1.89	2.36	3.00	3.50
8	1.40	1.86	2.31	2.90	3.36
9	1.38	1.83	2.26	2.82	3.25
10	1.37	1.81	2.23	2.76	3.17
11	1.36	1.80	2.20	2.72	3.11
12	1.36	1.78	2.18	2.68	3.05
13	1.35	1.77	2.16	2.65	3.01
14	1.35	1.76	2.14	2.62	2.98
15	1.34	1.75	2.13	2.60	2.95
16	1.34	1.75	2.12	2.58	2.92
17	1.33	1.74	2.11	2.57	2.90
18	1.33	1.73	2.10	2.55	2.88
19	1.33	1.73	2.09	2.54	2.86
20	1.33	1.72	2.09	2.53	2.85
21	1.32	1.72	2.08	2.52	2.83
22	1.32	1.72	2.07	2.51	2.82
23	1.32	1.71	2.07	2.50	2.81
24	1.32	1.71	2.06	2.49	2.80
25	1.32	1.71	2.06	2.49	2.79
26	1.31	1.71	2.06	2.48	2.78
27	1.31	1.70	2.05	2.47	2.77
28	1.31	1.70	2.05	2.47	2.76
29	1.31	1.70	2.05	2.46	2.76
30	1.31	1.70	2.04	2.46	2.75

one tail	0.100	0.050	0.025	0.010	0.005
two tails	0.200	0.100	0.050	0.020	0.010
df 31	1.31	1.70	2.04	2.45	2.74
32	1.31	1.69	2.04	2.45	2.74
33	1.31	1.69	2.03	2.44	2.73
34	1.31	1.69	2.03	2.44	2.73
35	1.31	1.69	2.03	2.44	2.72
36	1.31	1.69	2.03	2.43	2.72
37	1.30	1.69	2.03	2.43	2.72
38	1.30	1.69	2.02	2.43	2.71
39	1.30	1.68	2.02	2.43	2.71
40	1.30	1.68	2.02	2.42	2.70
41	1.30	1.68	2.02	2.42	2.70
42	1.30	1.68	2.02	2.42	2.70
43	1.30	1.68	2.02	2.42	2.70
44	1.30	1.68	2.02	2.41	2.69
45	1.30	1.68	2.01	2.41	2.69
46	1.30	1.68	2.01	2.41	2.69
47	1.30	1.68	2.01	2.41	2.68
48	1.30	1.68	2.01	2.41	2.68
49	1.30	1.68	2.01	2.40	2.68
50	1.30	1.68	2.01	2.40	2.68
60	1.30	1.67	2.00	2.39	2.66
70	1.29	1.67	1.99	2.38	2.65
80	1.29	1.66	1.99	2.37	2.64
90	1.29	1.66	1.99	2.37	2.63
100	1.29	1.66	1.98	2.36	2.63
150	1.29	1.66	1.98	2.35	2.61
200	1.29	1.65	1.97	2.35	2.60
300	1.28	1.65	1.97	2.34	2.59
400	1.28	1.65	1.97	2.34	2.59
500	1.28	1.65	1.96	2.33	2.59
∞	1.28	1.65	1.96	2.33	2.58

B.3   Chi-Square Probability Table

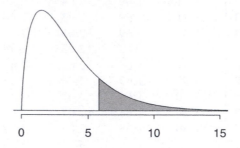

Figure B.2: Areas in the chi-square table always refer to the right tail.

Upper tail		0.3	0.2	0.1	0.05	0.02	0.01	0.005	0.001
df	1	1.07	1.64	2.71	3.84	5.41	6.63	7.88	10.83
	2	2.41	3.22	4.61	5.99	7.82	9.21	10.60	13.82
	3	3.66	4.64	6.25	7.81	9.84	11.34	12.84	16.27
	4	4.88	5.99	7.78	9.49	11.67	13.28	14.86	18.47
	5	6.06	7.29	9.24	11.07	13.39	15.09	16.75	20.52
	6	7.23	8.56	10.64	12.59	15.03	16.81	18.55	22.46
	7	8.38	9.80	12.02	14.07	16.62	18.48	20.28	24.32
	8	9.52	11.03	13.36	15.51	18.17	20.09	21.95	26.12
	9	10.66	12.24	14.68	16.92	19.68	21.67	23.59	27.88
	10	11.78	13.44	15.99	18.31	21.16	23.21	25.19	29.59
	11	12.90	14.63	17.28	19.68	22.62	24.72	26.76	31.26
	12	14.01	15.81	18.55	21.03	24.05	26.22	28.30	32.91
	13	15.12	16.98	19.81	22.36	25.47	27.69	29.82	34.53
	14	16.22	18.15	21.06	23.68	26.87	29.14	31.32	36.12
	15	17.32	19.31	22.31	25.00	28.26	30.58	32.80	37.70
	16	18.42	20.47	23.54	26.30	29.63	32.00	34.27	39.25
	17	19.51	21.61	24.77	27.59	31.00	33.41	35.72	40.79
	18	20.60	22.76	25.99	28.87	32.35	34.81	37.16	42.31
	19	21.69	23.90	27.20	30.14	33.69	36.19	38.58	43.82
	20	22.77	25.04	28.41	31.41	35.02	37.57	40.00	45.31
	25	28.17	30.68	34.38	37.65	41.57	44.31	46.93	52.62
	30	33.53	36.25	40.26	43.77	47.96	50.89	53.67	59.70
	40	44.16	47.27	51.81	55.76	60.44	63.69	66.77	73.40
	50	54.72	58.16	63.17	67.50	72.61	76.15	79.49	86.66

Index

Made in United States
North Haven, CT
09 June 2023

37569888R00259